THE CRIME OF CLAUDIUS PTOLEMY

Other books by Robert R. Newton

*Ancient Astronomical Observations and the Accelerations of
the Earth and Moon*
Medieval Chronicles and the Rotation of the Earth

Robert R. Newton

THE CRIME OF CLAUDIUS PTOLEMY

THE JOHNS HOPKINS UNIVERSITY PRESS
BALTIMORE AND LONDON

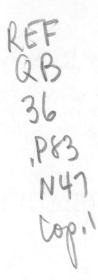
The Johns Hopkins University Press, Baltimore, Maryland 21218
The Johns Hopkins Press Ltd., London

Library of Congress Catalog Card Number 77-4211
ISBN 0-8018-1990-3

Library of Congress Cataloging in Publication data
will be found on the last printed page of this book.

CONTENTS

CONTENTS (continued)

CONTENTS (continued)

FIGURES

TABLES

PREFACE

This is the story of a scientific crime. By this, I do not mean a crime planned with the care and thoroughness that scientists like to think of as a characteristic of their profession, nor do I mean a crime carried out with the aid of technological gadgetry like hidden microphones and coded messages on microdots. I mean a crime committed by a scientist against his fellow scientists and scholars, a betrayal of the ethics and integrity of his profession that has forever deprived mankind of fundamental information about an important area of astronomy and history.

I have demonstrated the existence of this crime in four earlier publications scattered through the literature over a period of about three years. My purpose when I began this work was to pull these scattered publications into a single work that would be self-contained and that would treat the subject from a uniform instead of an evolving point of view. However, when I had written about a third of this work, I discovered new evidence which showed that the crime was far more extensive than I had suspected. Thus this work actually presents both the old and the new evidence of the crime in a single cohesive source.

In order to make the study of the crime intelligible, it is necessary to understand the background of ancient Greek astronomy against which it was committed. Hence the first part of this work is a presentation of some of the major achievements of that astronomy. This is not intended to be a history of Greek astronomy. It is merely a description of the heights to which Greek astronomy rose, along with enough background information to make the subject, and the story of the crime, intelligible.

It might be better to use "Hellenistic" rather than "Greek". However, much of the development that will be described took place in the culture that is properly called Greek, and so I shall use the latter term. Besides, "Greek" has five letters and one syllable while "Hellenistic" has eleven letters and four syllables.

The story of the crime begins in Chapter V. The first chapter is devoted to the elements of astronomy, described from the modern viewpoint, that the reader needs to understand this work. The reader who has studied elementary astronomy at the college level can probably omit this chapter, although he should look at Section I.7 to acquaint himself with the terminology and notation that will be used. The next three chapters are devoted to the necessary background in Greek astronomy and mathematics, and the reader who has studied these topics can pass by these chapters. In passing, however, he should pause at Section II.3, which brings out a point about Greek trigonometry that has not been mentioned in earlier literature, I believe.

If a source appears in the list of references, it means that I have directly consulted and used the source in question. In some cases, a topic has not seemed important enough to warrant the labor of locating a hard-to-get reference and I have relied on secondary sources for these minor topics. These cases are explicitly identified in the text, and the relevant primary sources do not appear in the list of references.

I identify a source by giving two items. The first is the underlined name of the author. The second is the date of writing, in square brackets after the author's name. If it is necessary to identify a specific place such as a section or a page of a source, that information appears in the brackets after the date. When the exact year of writing is not known, I indicate this fact by putting "ca." in front of the year used. Thus, for example, "Menelaos [ca. 100, p. 194]" refers to page 194 in a work written by the mathematician Menelaos sometime near the year 100; this happens to be a work on spherical trigonometry.

When a source is cited frequently within a short space, I often omit the year and use the author's name without underlining, followed by an identification of an exact place in brackets, if necessary. Some works, such as the American Ephemeris and Nautical Almanac, do not have authors. I designate these works by using part or all of the title, followed by the year in square brackets.

Five sources are cited often enough to justify the use of abbreviated citations for them. Four of them are the earlier publications that were mentioned above. The three journal articles that are cited in full as Newton [1973], Newton [1974a], and Newton [1974b] will be identified as Part I, Part II, and Part III, respectively. An earlier book [Newton, 1976] that deals in part with the subject of this work will be designated as APO, after the initials of the first three words in the title. Finally, the main astronomical work of Ptolemy [ca. 142] will usually be called the Syntaxis, one of the words in the title.

I thank Mr. Philip G. Couture of Santee, California for correspondence which led me to understand some of the relations between chronology and the work of Ptolemy. I also thank my colleagues B. B. Holland and R. E. Jenkins of the Applied Physics Laboratory for many fruitful discussions. I particularly thank V. L. Pisacane of the Applied Physics Laboratory for his critical reading of the typescript and for many valuable suggestions. Finally I thank J. W. Howe and Mrs. Mary Jane O'Neill, also of the Applied Physics Laboratory, for their dedicated work in preparing this book for the press.

This work, which is intimately connected with research into the secular accelerations of the earth and moon, and with the precise measurement of time, was supported by the Department of the Navy under contract with The Johns Hopkins University.

THE CRIME OF CLAUDIUS PTOLEMY

CHAPTER I

SAVE THE PHENOMENA

I don't believe what makes you red in the face
Is after explosion going away so fast.

Robert Frost[†]

1. Saving the Phenomena

According to several Greek writers, the purpose of
theory in astronomy is to "save the phenomena". That is,
astronomical theory is intended to produce a representation
or description that can be put into quantitative terms and
that agrees with astronomical observation.

This is essentially the same as the modern viewpoint,
and it may seem that the matter is too obvious to need mentioning. However, the matter is not obvious, and the fact
that it now seems to be is actually a considerable achieve-
ment. The Greeks did not apply the same principle to other
areas in the physical sciences. Many Greek philosophers
made a sharp distinction between physics and astronomy. They
taught [Pannekoek, 1961, p. 130, for example] that the astronomer is required only to "save the phenomena" while the
physicist must "deduce the truth". He must explain the phenomena on the basis of the first causes and working forces
of the universe.

As I understand it, this means that the goal of astronomy is description while the goal of physics is the truth.
The difficulty is, of course, that "physical truth" still
lacks a definition. Freedom from the need to pursue this
undefined and unobtainable goal may explain why Greek astronomy seems to be more successful than Greek physics.

If we cannot say that one theory is true and that another is false, how do we decide which theories to accept
and use and which to reject? This is a problem that has
exercised many modern philosophers of science. One aspect
of a theory that seems to influence its acceptability is
its simplicity. It is probably the case that almost any
theory can be made to work if it is made complicated enough.
Thus, perhaps on esthetic grounds and perhaps on lethargic
ones, we prefer, and accept as "true", that theory which is
simpler than its competitors while being at least as successful as them in saving the phenomena.

This criterion of acceptability is found in fact in
Greek astronomy. Ptolemy, in Chapter III.1 of the Syntaxis,
says: "We think that it is right to explain the phenomena
by the simplest possible hypotheses, provided that they do

[†]From "Skeptic" in the Complete Poems of Robert Frost, page
549, Henry Holt and Company, New York, 1949.

not conflict with the observations in any important way."
I believe that any modern scientist could accept this doctrine. He might perhaps object to the word "important",
and he might prefer to say "in any known way" or "in any
established way".

We may now turn to the main observations or phenomena
that the Greek astronomers had to save.

2.   The Stars

There are several thousand stars that are bright enough
to be seen with the naked eye by a person with ordinary
vision. About half of these can be seen at any one time on
a perfectly clear night. We sometimes refer to the main
collection of stars as the fixed stars. We know now that no
star is exactly fixed with respect to the others. In fact,
careful studies of stars made over a considerable period of
time show that each star known to us has two kinds of motion.

From measurements of the Doppler shift in the light received from a star, we find that most stars are travelling
away from us at high speed, and that, the farther away a
star is, the faster is its speed of recession. The current
picture, sometimes called the big bang theory, is that all
of the matter and energy in the universe was contained in a
small volume that began to expand about 11 billion years
ago.† The pieces of the universe that happened to get the
highest speeds at that time are now the farthest away. Those
that are receding from us have their light reddened slightly
by the Doppler shift, which, measurement shows, is approximately proportional to the distance of the star from us.
The poet is able to describe this theory in two lines, with
enough room left over to tell us that he doesn't believe it.‡

The eye is not able to tell us the distance to a star,
and all stars seem to lie on the surface of a sphere called
the celestial sphere. To us, a star seems to have only an
angular position. Measurements of stellar position (angular
position) made with high precision over an extended period of
time show that the angular positions, as well as the distances,
are changing. Each star is in fact moving with high speed,
but each star is so far away that the apparent position
changes very slowly. For the stars that can be seen with
the naked eye, a typical angular velocity is perhaps 20 seconds of arc per century. In the 20 centuries since the height
of Greek astronomy, such a star has moved about 400 seconds,

---

†Some writers use "big bang" only for that part of the
theory which says that most atomic nuclei were formed
almost immediately after the expansion began, literally
within the first few seconds.

‡The fact that the stars, with a few accidental exceptions,
are moving away from us does not mean that we are the
center of the universe. An observer anywhere in the universe would find the same thing. We are the center of the
universe that we can see, however.

about one tenth of a degree.† This is just about the limit
with which one can measure the position of a star with the
eye.

In the time of the greatest Greek astronomers, there
were no accurate measurements of stellar position (naked eye
measurements, of course) that were more than a few centuries
old. No measurable change in position had occurred within
that span, so that, to the Greeks, the stars were fixed.
Hence it was possible to make meaningful tables of stellar
positions that were good for all time, so far as they knew.

Of course the Greek astronomers had to establish the
apparent fixity of the stars by observation; they could not
make deductions from 20th century tables. Only a little of
the work in which they did this has survived. The great
astronomer Hipparchus wrote a work of commentary on the
stars [Hipparchus, ca. -135]‡; this is the only work of Hipparchus that has survived. Almost three centuries later, in
Chapter VII.1 of the Syntaxis,‡ Ptolemy gives Hipparchus's
descriptions of several stellar configurations and finds
that the same descriptions are valid in his own time. From
this, he infers that the stars form a fixed background.

From what some writers of antiquity have said, we believe that Hipparchus prepared a star table or catalogue,
meaning a work in which stars are identified and their positions stated. It is sometimes said that Hipparchus was in-
spired to prepare a star table because a nova had recently
appeared, but there seems to be no firm authority for this
[Pannekoek, 1961, p. 129, for example]. Ptolemy has also
left us a table of slightly more than 1000 stars in Books
VII and VIII of the Syntaxis. The centuries have seen a
great controversy about Ptolemy's table. One side of the
controversy holds that Ptolemy merely appropriated Hipparchus's table, while the other side holds that Ptolemy based
his table upon new observations made by himself. I shall
return to the controversy in Chapter IX, where I shall give
extensive reasons for concluding that the table in the Syntaxis is really that of Hipparchus, with some obvious modifications made by means of theory.

---

†However, the bright star called Rigil Kentaurus moves more
  than 7 seconds of arc per year. It has changed position
  by more than 4° in the past 2000 years.

‡I write years in astronomical rather than historical style.
  Both styles are the same for the years of the common era.
  In historical style, the year before the year 1 is called
  1 B. C. E., the year before that is called 2 B. C. E., and
  so on. In astronomical style, the year preceding 1 is
  called 0, the year before that is called -1, and so on.

‡I often use this term to denote the reference Ptolemy [ca.
  142]. Many writers use the term Almagest. However,
  Almagest has a specific meaning which is definitely inap-
  propriate for this reference. As a convenient designation,
  I therefore adopt Syntaxis, a word that occurs in Greek
  usage and that is appropriate. See Section XIII.10 for the
  reasons that lead me to reject the term Almagest.

We shall need to designate particular stars in later parts of this work, and this is a good place to describe the method of designation. The method used by the Greeks, which is still a method commonly used today, is based upon constellations. Originally a constellation was a set of stars that formed some recognizable pattern. The constellations best known to most people are probably the ones that we often call the Big Dipper and the Little Dipper. These names actually suggest the shapes of the corresponding constellations, but this situation is rare. In most cases, imagination has grafted an animal figure (Taurus, the bull, for example) or a mythological figure (the ship Argo, for example) onto a collection of stars, with little resemblance in shape between the figure named and the collection itself.†

Most systems of names, such as the names of the streets in many large cities, are arbitrary, but being arbitrary does not keep them from being useful. The constellations form unique patterns that can be learned, and, using them, anyone can learn to find his way around the celestial sphere of the stars just as he would learn to find his way around a city. To identify a star, then, we give the constellation it is in, and we must then locate it within its constellation. This is like using the street number to identify a particular house, once we have found the street itself.

This is the system used in Ptolemy's table, and, we may be sure, in the table of Hipparchus before him. It was used by Babylonian astronomers even earlier, and many Greek designations are direct translations of Babylonian ones. Ptolemy's table is arranged by constellations. Each star that belongs to a particular constellation is given a descriptive phrase. For example, in the constellation Scorpius (the Scorpion), one of the stars is called the "middle star on the brow of the Scorpion". A person who has learned to identify Scorpius, and who has learned how the figure of a scorpion is fitted over the collection of stars, would recognize this star immediately, and he would have no trouble in pointing to it.

After Ptolemy names all the stars in a constellation, and gives their positions, he describes and locates some stars that are nearby but that do not form part of the constellation proper. In this way, he identifies all the stars in his table.

We have changed Ptolemy's procedures slightly. Instead of having stars in a constellation plus those nearby, we have divided the celestial sphere into regions that surround the classical constellations with no space left over. Every star in the appropriate region belongs to the modern constellation. Instead of using a descriptive phrase like the middle star on the brow of the Scorpion, we assign a letter or number. The German astronomer Johann Bayer, around 1600, used α to

---

†Of course, as the reader probably knows, the standard astronomical names for the Big Dipper and the Little Dipper are Ursa Major and Ursa Minor, the Big Bear and the Small Bear.

denote the brightest star in a constellation, β to denote the next brightest, and so on. In his system, the middle star on the brow of the Scorpion becomes δ Scorpii.†

Casual observation does not tell us that δ Scorpii is the fourth brightest star in the Scorpion. In order to use Bayer's scheme, we must simply memorize the position of δ Scorpii and the other lettered stars, without the aid of the classical pictures. This disadvantage seems to be outweighed by the brevity of Bayer's scheme; for the stars that Bayer identified, his designations are still the ones most commonly used.

John Flamsteed, the first English astronomer royal, introduced a different scheme almost a century later. He used the number 1 to designate the star that is farthest west in a particular constellation, 2 to designate the one that is next to being farthest west, and so on. His criterion is more objective than Bayer's, but it is not really easier for a casual observer to apply. Since Bayer's system had a head start of about a century, it has survived for the stars that he identified. Flamsteed's numbers are generally used only for stars that he identified which Bayer did not. The middle star on the brow of the Scorpion is 7 Scorpii in Flamsteed's system. For the sake of definiteness, some writers combine the two systems in referring to stars that appear in both. Thus this star is sometimes called 7 δ Scorpii.

Other tabulators have introduced other systems, but it is not necessary to describe them. So far as I know, most of them use Latin letters rather than Greek letters or numerals. Becvar [1964] gives the designations of a large number of stars in the principal systems.

Of course, there are many complexities in designation that we do not need to go into here. The only reason for discussing this matter is to prepare the reader for the identifications of stars that will come up in later parts of this work.

3. The Sun

"Star" is sometimes used to mean any body in the heavens. The sun is sometimes called the star of the day and the moon is called the star of the night, although these usages are now uncommon, at least in English. We also have shooting stars, and even the comet was once called the "hairy star" (stella cometes).

If we ignore shooting stars and comets, we can divide the remaining stars into two classes. These are the fixed

---

†In the usual terminology, the name of the constellation follows the letter or number designation, using the genitive case of the Latin name. Thus δ Scorpii means δ of the Scorpion. The star that I called Rigil Kentaurus a moment ago is α Centauri in Bayer's system. Rigil Kentaurus is its Islamic name.

-5-

stars and the wandering stars, the planets. To the naked
eye, there are seven planets in this early meaning of the
word: the sun, the moon, Mercury, Venus, Mars, Jupiter, and
Saturn.

In the rest of this work, I shall follow modern usage.
That is, the sun and moon will be omitted from the list of
the planets, but the earth will be added. There will be no
need to talk about the planets beyond Saturn, since they
cannot be seen with the naked eye.

The daytime sky by its brilliance pales the stars to
invisibility, and we cannot directly see the position of the
sun among the stars.† However, we can notice which stars
are first visible in the western sky at sunset and which are
last visible in the morning at sunrise; the sun must be between them. We find that the position of the sun among the
stars moves eastward around a circle on the celestial sphere
called the ecliptic, and it takes slightly more than $365\frac{1}{4}$
days‡ for the sun to make a complete circuit of the ecliptic.

The point on the earth that is directly under the sun
is sometimes north of the equator and sometimes south of it,
and it therefore crosses the equator on occasion. When it
crosses the equator on its way to the north side, we have
the vernal equinox, which is the beginning of spring. When
it crosses the equator on its way south, we have the autumnal
equinox, which is the beginning of autumn. In between, we
have the summer solstice when the sub-solar point is farthest
north; this is the beginning of summer. We also have the
winter solstice when the point is farthest south; this begins
winter.

Suppose we measure the time interval from, say, one
vernal equinox to the next, or from one winter solstice to
the next. This time is slightly less than $365\frac{1}{4}$ days,♯ and
it is less than the time to make a circuit of the ecliptic.

The difference between the two times means that the sun
does not make a complete circuit of the ecliptic between
successive vernal equinoxes. The position of the sun at the
time of an equinox (or of a solstice) is moving westward a-
long the ecliptic at the rate of about 50 seconds of arc per
year (50 "/yr), and it will make a complete circuit in about
26 000 years. It has moved about 30° since Greek times.

---

†As I understand the matter, it is the sky and not the sun
 directly that keeps us from seeing most stars in the daytime. Outside the atmosphere, I believe, we can see stars
 even when the sun is visible.

‡Modern ephemeris publications give 365 days, 6 hours, 9
 minutes, and 9.5 seconds.

♯Modern publications give 365 days, 5 hours, 48 minutes,
 46.0 seconds for this interval in the year 1900. Because
 of the effect of the other planets on the motion of the
 earth, the interval is decreasing by about $\frac{1}{2}$ second per
 century, so that the number of seconds was about 36 in the
 time of the Greek astronomers.

This motion is called the precession of the equinoxes.

We have just used equinox and solstice to mean certain times. The words are also used to mean the positions of the sun at those times, and the context usually makes it clear whether a time or a position is meant.

Since the sun travels in the plane defined by the ecliptic, and since it is sometimes north and sometimes south of the equator, it is clear that the ecliptic is not in the same plane as the equator. The angle between the two is called the obliquity of the ecliptic, or the obliquity for short. The obliquity is often denoted by the symbol ε. In the year 1900, the value of ε was $23°.452$, and it is currently decreasing by about $0°.013$ (about $47''$) per century, because of the effects of the other planets on the earth's orbit. When the Greek astronomers flourished, ε was about $23°.71$.

The lengths of the seasons are not equal. In Greek times, the lengths were, beginning with summer, approximately 92.3 days, 88.7 days, 90.2 days, and 94.1 days.[†] The fact that these lengths are unequal means that the apparent motion of the sun is not uniform circular motion. To high accuracy, the sun moves around the earth on an ellipse.[‡] The eccentricity of the ellipse in Greek times was about 0.017 558, and the sun and earth were closest together (the sun was at perigee) when the sun was about $246°$ east of the vernal equinox, measured along the circle of the ecliptic. In 1900, the values were 0.016 751 and $281°$, respectively. If the other objects were not present, the values would be constant.[‡] The changes, like the changes in other parameters that have been mentioned, come from the small gravitational effects of the moon or the other planets. All the changes are presumably oscillatory, but with long periods of oscillation. During the past 2000 years, the changes have taken place at nearly constant rates.

On the average, the angle subtended by the diameter of the sun is close to $32'$, and the distance from the earth to the sun is 23 500 times the radius of the earth. From these data, we calculate that the radius of the sun is about 109 times that of the earth, and that its volume is about 1.3 million times that of the earth. However, the sun's mass is only about 330 000 times that of the earth, which means that the sun's density is only about one fourth that of the earth.

The Greek astronomers had no way to estimate the mass of the sun, but they could and did study the other properties of the sun that have been described.

[†] Because of rounding, these do not add up exactly to the length of the year.

[‡] Or the earth moves around the sun, if the reader prefers.

[‡] Except for some small relativistic effects.

4. The Moon

According to the first chapter of <u>Genesis</u>, day and
night were the first results of creation, and the first
alternation of day and night constituted the first day.†
It is interesting that the sun and moon were not created on
the first day. They were created on the fourth day, the sun
to rule over the day and the moon to rule over the night.
When this was written, it had not been realized that the
sun was the cause of daylight and not a mere symbol of it.

We may also question whether the moon can be said to
rule over the night. Only when the moon is full is it visible for the entire night, and when it is new it cannot be
seen at all. On the average, the moon is visible for only
half of the hours of the night.

In the first stages of Greek astronomy, the nature of
daylight and moonlight was still a matter of fierce debate.
By the time that concerns us, say by the time of Hipparchus,
these questions were pretty well settled in favor of the
modern viewpoint. The question that caused the most debate
was probably about the nature of moonlight: Does the moon
have independent light or does it shine by reflected sunlight?

Two considerations probably lead most strongly to the
conclusion that the moon shines by reflected light, and both
were known before the time of Hipparchus, although they may
not have received universal acceptance. When the moon is a
crescent, its convex side always points toward the sun.
When the moon is full, its light is interrupted when and only
when it moves into the shadow of the earth. The simplest way
to save both of these phenomena is to assume that the moon is
a sphere that shines only by reflected sunlight.

The moon may be considered to move in a plane, but the
plane is one that itself moves rapidly. The plane of the
moon's motion makes an angle of slightly more than 5° with
the ecliptic plane, and the two planes intersect in a line
that passes through the earth. The angle is called the inclination and the line of intersection is called the line of
nodes. The line of nodes is analogous to the line that joins
the equinox positions. When the moon is on the line of nodes,
it is passing through the plane of the ecliptic. One end of
the line, where the moon passes from the south to the north
side of the ecliptic, is called the ascending node. The
other end is called the descending node.

The line of nodes moves westward along the ecliptic,
making a complete revolution in about 18.61 years, far shorter
than the time that the equinox takes. While the line of nodes
is moving, the angle between the ecliptic and the plane of

---

†Day obviously has two meanings. One is the period of
light; the other is a full sequence of light and dark.
This duality of meaning is very old and still persists.

the lunar motion remains fixed.† Thus the plane of the
motion performs an odd kind of motion which is called precession, and the motion of the line of nodes is called the
precession of the nodes.‡

Within the plane, the motion of the moon may be approxi-
mated by an ellipse, but the approximation is not nearly as
good for the moon as it is for the sun. The greatest deviation from elliptical motion is less than 1 minute of arc for
the sun but it is more than 1° for the moon. In the standard ephemeris publications, the theory that is used to calculate the solar motion has slightly more than 100 terms,
while that used for the moon has more than 1500.

The ellipse that best fits the lunar motion has an eccentricity of 0.054 900. The axes of this ellipse rotate
rapidly in the plane of the motion, making a complete revolution in slightly less than 9 years.

Suppose we note the time of a new moon, when the moon
passes closest to the sun in its orbital motion,‡ and let us
also note the position of the moon among the stars at this
time. The moon next returns to this position among the
stars 27.321 582 days later, on the average; this interval
is called a sidereal month. However, the sun is no longer
at this position because it is also moving in the same direction as the moon. Thus we have not yet reached another new
moon. The average interval between new moons is 29.530 589
days. This is the interval concerning the moon that interests
us most in our ordinary life. When we refer to a month in
ordinary conversation, this is the interval that we mean.*
When it is necessary to be explicit, we call this a synodic
month.

On the average, the angle subtended by the diameter of
the moon is about 31' 5", slightly less than the corresponding quantity for the sun. The average distance to the moon
is about 60.3 times the radius of the earth. This tells us
that the radius of the moon is about 0.2725 times the radius
of the earth, and that the ratio of the volumes is about
0.0202. However, the ratio of the masses is about 0.0123,
so that the density of the moon, like that of the sun, is
less than that of the earth.

---

†More accurately, it remains fixed on the average. Perturbations produced by the gravitation of the sun and planets
cause it to oscillate about its average. The biggest deviation from the average is about 10 minutes of arc.

‡I do not know whether the term was first applied to the
motion of the plane or to the motion of the nodes.

‡We cannot actually see the moon at this time, but we can
see it at times shortly before and after this time. Thus
we can find the position and time of this event by interpolation.

*Unless, of course, we mean a month on the calendar.

TABLE I.1

SOME PROPERTIES OF THE PLANETS

Planet	Relative distance[a]	Relative diameter	Relative mass	Relative period	Orbital eccentricity
Mercury	0.387 099	0.381	0.055	0.240 84	0.205 614
Venus	0.723 332	0.958	0.807	0.615 19	0.006 821
Earth	1	1	1	1	0.016 751
Mars	1.523 691	0.533	0.106	1.880 81	0.093 313
Jupiter	5.202 803	10.846	314.497	11.861 76	0.048 338
Saturn	9.538 843	8.989	94.068	29.456 54	0.055 890

[a]From the sun.

[b]In 1900.

5.   The Planets

    Some important properties of the planets are listed in
Table I.1; the planets that are not visible to the naked
eye are omitted. The distances, diameters, masses and periods are given relative to those of the earth. The eccentricities of the heliocentric orbits are given for the year
1900; the eccentricities change slowly with time because of
the perturbations that the planets produce on each other.

    The information in Table I.1 is derived from Chapter 4
of the Explanatory Supplement [1961]. Seidelmann, Doggett,
and DeLuccia [1974] give more accurate values for some of
the tabulated quantities.

    The heliocentric orbits of the planets (other than the
earth's) lie in planes that do not quite coincide with the
ecliptic, which is the plane of the earth's orbit. Just as
we did with the moon, we can locate these planes by giving
an inclination angle and the location of the line of nodes.
We have little need for the planetary inclinations and the
nodes in our discussion of Greek astronomy, and I have
omitted them from the table. The interested reader can find
them in either of the sources just cited.

    Within the orbital planes, the orbits are nearly ellip-
tical. The maximum deviation of any planetary orbit from an
ellipse is† little more than 1′. In order to specify the
ellipses, we need the eccentricities, which are given in
Table I.1, and the directions of the major axes. The directions of the axes are not needed in discussing most of Greek
astronomy, and they are not listed in the table. They can
be found in the cited references.

----

†When I place the symbol ′ after a number, I always mean the
minute of arc, not the minute of time. Similarly, ″ will
always denote a second of arc.

At its closest approach to the earth, the diameter of Venus subtends an angle slightly larger than 1'. I do not know whether a person with extremely keen vision could see the diameter of Venus or not.† I am sure that he could not see the diameters of the other planets. Thus, with the possible exception of Venus, naked eye observations tell us nothing about the sizes of the planets.

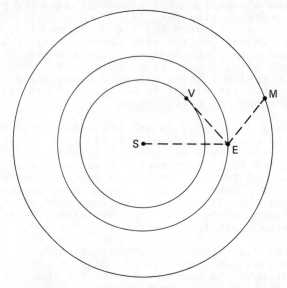

Figure I.1   The orbits of Venus and Mars, as seen from the earth.
E, S, V, and M denote the earth, the sun, Venus, and Mars, respectively.
The angle between Venus and the sun can never exceed the angle SEV,
but the angle between Mars and the sun can have any value.

Mercury and Venus are closer to the sun than the earth is, and thus they are called the inner planets. Correspondingly, Mars, Jupiter, and Saturn are called the outer planets. As seen from the earth, the motions of the inner and outer planets are quite different; we can see this with the aid of Figure I.1.

In the figure, E is the earth and S is the sun. The heliocentric orbits of Venus (V), earth (E), and Mars (M) are drawn approximately to scale, but with no attempt to show the eccentricities of the orbits. Venus can never appear farther from the sun than the angle SEV, which is about 46°. However, the angle SEM can have any value.

---

†Because of phenomena that occur within the eye, Venus, perhaps other planets on occasion, and the brightest stars appear to have a finite size. This apparent size, however, is a function of the brightness and not of the angle subtended by the object. We can tell this by viewing the object through a slightly darkened piece of film, for example. The size appears to decrease as the light reaching the eye is made to decrease.

The angle between a planet and the sun is called the planetary elongation, and the maximum value of this angle is called the maximum elongation. The maximum elongation of Mercury is about 28° and that of Venus is about 47°.[†] An outer planet also has an elongation, but its elongation can increase without limit.

When the elongation of a planet is zero, we have the particular configuration that is called a conjunction of the planet and the sun.[‡] With the aid of Figure I.1, we see that Mars is in conjunction with the sun when it is on the opposite side of the sun from us. The distance from the earth to Mars clearly has its greatest value when this occurs, so the conjunction of an outer planet with the sun takes place when the planet is at apogee. The same remarks can be made about any outer planet.

When an inner planet is in conjunction with the sun, it can be on the opposite side of the sun from us, or it can be on the same side. Thus we must distinguish two kinds of conjunction for an inner planet. When the planet is on the opposite side of the sun, we have a superior conjunction. As we see from Figure I.1, the planet is at apogee when this occurs. When the planet is on the same side of the sun, we have an inferior conjunction, and the planet is at perigee, its closest approach to the earth.

When the elongation of a planet is 180°, we have the configuration called opposition.[‡] Only an outer planet can be in opposition. We see from Figure I.1 that a planet is at perigee when it is in opposition.

An interesting thing happens when a planet is near perigee, whether it is an inner planet or an outer one. First, take Venus as an example of an inner planet in Figure I.1. Suppose that the planets revolve counterclockwise in the figure, so that the counterclockwise direction is the direction from west to east. During most of its orbit around the sun, Venus also appears to be moving counterclockwise (eastward) as it is seen from the earth. Consider what happens when Venus is on the line SE between the sun and the earth, however; this is inferior conjunction and perigee. In order to see what happens, we must know that Venus travels faster in its (heliocentric) orbit than the earth does. Thus, when Venus is near perigee, the line from the earth to Venus is actually revolving in the clockwise direction.

_____

[†]This is greater than the value in Figure I.1, because of the eccentricities of the orbits.

[‡]It is necessary to specify the sun as well as the planet. A planet can also have a conjunction or meeting with the moon, with another planet, or with a particular star.

[‡]A planet can be in opposition to any other body, so that, in principle, it is necessary to specify two bodies when one speaks of an opposition. However, the oppositions of interest almost always involve the sun, and the sun is assumed to be one of the bodies unless the contrary is explicitly stated.

The same thing happens with an outer planet, of which Mars is an example in Figure I.1. The earth travels faster in its orbit than Mars does. Thus, when Mars is near perigee (opposition), the line from the earth to Mars revolves in a clockwise direction, although it moves counterclockwise most of the time.

Thus, as seen from the earth, any planet travels eastward among the stars most of the time. As it approaches perigee, however, it stops its eastward motion and begins to travel westward. The point at which this happens is called the first turning point or first stationary point. The planet continues westward through perigee and for some time thereafter, until its westward motion stops and it resumes its normal eastward motion. The point at which this happens is called the second turning point or second stationary point. The motion between the two turning points is called retrograde.

Mercury and Venus have their turning points when they are fairly close to inferior conjunction. During much of the time when their motion is retrograde, they are invisible because they are too close to the sun, and the retrograde behavior is not particularly noticeable. The outer planets, on the other hand, have their retrograde motions when they are near opposition and hence when they are most readily seen. Their retrograde behavior is quite noticeable.

6. The Earth

To high accuracy, the earth is a sphere with an average radius of about 6365 kilometers. The maximum deviation of any point from a sphere is about 18 kilometers, about 1 part in 350.

The earth performs one rotation with respect to the sun in 1 day or 24 hours of solar time, rotating from west to east. During this time, the sun itself has moved eastward by a small amount with respect to the stars, so that the earth has made more than one full rotation with respect to them. Its period of rotation with respect to the stars is about 23 hours and 56 minutes.

7. Technical Terms and Notation

In an attempt to make this work be reasonably self-contained, I shall try to explain the technical terms in astronomy that it is necessary to use in discussing Greek astronomy. I have already introduced some technical terms such as the ecliptic, and it will be necessary to introduce others as the work progresses. Handling this introduction poses a problem of some difficulty.

It is sometimes possible to explain a term at the time it is introduced without interrupting the flow of the writing. At other times, however, breaking the development in order to explain a term makes a serious interruption. Further, even if topics are explained where they are first used, the reader is unlikely to remember all of them and hence he is likely to have a need to refer back to some of the explanations. Since the explanations are scattered throughout the work, he will probably have difficulty in locating a particular explanation that he needs.

For this reason, I have explained the technical topics needed in Appendix A, even those that are explained as they are introduced. This gives the reader a single source to which he can refer for the explanation of any topic that he has forgotten or which he has not yet met. If a technical topic is not found in Appendix A, it is an oversight.

I assume that the reader has a moderate knowledge of mathematics. He must be able to read expressions in ordinary algebra and to solve elementary equations. He must also be able to read elementary expressions involving trigonometric functions. It is desirable to emphasize here a general proposition that dominates the process of constructing astronomical theories from astronomical observations.

Consider the equation

$$7x + 3 = 17.$$

This particular equation is called a linear equation, and it can be solved by using the elementary operations of addition, subtraction, multiplication, and division, without the need for more complex operations such as finding a square root. In other equations, the "variable" or "unknown" quantity x occurs in a more complicated manner. For example, we might have

$$\sin^2 x + \tan x = 1.4,$$

an equation that cannot be solved by ordinary algebra.

Regardless of the complexity of the equation, there is one important principle: One equation determines one unknown. That is, for each of the equations just given, we can find a specific value of x that satisfies the equation.†
For the first equation, this value is x = 2; for the second it is approximately x = 43°.0472.

Now consider the equation

$$7x + 5y + 3 = 37,$$

which has two unknown quantities x and y. It is not possible to solve this equation, because we can give y (or x)

---

†In some cases, there may be two or more values of x that satisfy the equation. For example, $x^2 = 1$ is satisfied by x = +1 or -1. We can ignore this possibility here.

any value that we wish and still find a value of x (or y) that satisfies the equation. For example, if we let y = 1.2, we find x = 4 and if we let y = 4 we find x = 2. However, if we specify that x and y must simultaneously satisfy this equation and another equation, say

$$x + y = 6,$$

there is a unique solution. The solution is obviously x = 2 and y = 4.

Thus we have the general principle: In order to solve a set of equations for a set of unknowns, we must have exactly the same number of each.[†] This holds good whether the equations are linear or whether they are more complicated. If we have more unknowns than equations, there is an infinite set of solutions. If we have more equations than unknowns, there are no solutions. The latter situation often arises when we are trying to develop a theory from observations, and we must give it further study.

In developing a theory to agree with, say, astronomical observation, we usually reach a stage where we decide upon the mathematical form that a theory is to take. To take a simple example, the theory of an isolated object is that it moves along a straight line with constant speed. Let x denote its distance along this line from some identifiable point and let t denote the time. Then our "straight line – constant speed" theory takes the form

$$x = a + bt. \tag{I.1}$$

In this, a and b are numbers that we do not know a _priori_ and that we must determine from observations. Such numbers are called the parameters of the theory.

In this example, there are two parameters and hence we must make at least two observations. Suppose we find that x = 5 meters when t = 1 second and that x = 7 meters when t = 2 seconds. Then we obtain the pair of equations

$$5 = a + b,$$
$$7 = a + 2b, \tag{I.2}$$

by substituting the measured sets of values in turn into the formal Equation I.1. The parameters are the unknowns a and b, and the solution is obviously

$$a = 3 \text{ meters}, \quad b = 2 \text{ meters per second}. \tag{I.3}$$

However, all measurements contain some error, and we know that these values are not exactly correct. We can never

---

[†]To be rigorous, we must also specify some minor restrictions that do not need to detain us. For example, we must find a way to exclude situations like the pair of equations x + 2 = 4 and 2x + 4 = 8, which are satisfied by x = 2 even though there are two equations and only one unknown.

find exactly the correct values, but we should always be able to improve the accuracy by using more measurements. Suppose we measure again at t = 3 seconds and find x = 10 meters, giving the equation

$$10 = a + 3b. \tag{I.4}$$

Equation I.4 must now be solved simultaneously with Equations I.2, but there is obviously no solution. The values from Equation I.3 do not satisfy Equation I.4.

In this case, we find the "best-fitting" parameters. By these, we mean values of the unknowns that give the best overall agreement to the set of equations while, usually, satisfying none of them exactly.

There are various criteria of what constitutes the best fit. A criterion that is commonly used is called the "least squares" criterion. In it, suppose that A and B are the "best-fitting" values of a and b. Now substitute the values of the time at which we measured x successively into Equation I.1, using these values of A and B, which we do not yet know. This gives us the three values A + B, A + 2B, and A + 3B. The three errors in the measured values of x are then A + B - 5, A + 2B - 7, and A + 3B - 10. The sum Σ of the squares of the errors is

$$\Sigma = (A + B - 5)^2 + (A + 2B - 7)^2 + (A + 3B - 10)^2.$$

We say that the best fitting values of A and B are those that make Σ a minimum. These values are A = 7/3 meters and B = 5/2 meters per second. We can extend this method to any number of observations.

Two other matters will be mentioned briefly. A moment ago I wrote 43°.0472, meaning 43.0472 degrees. When an astronomer wants to write a quantity in decimal form, he frequently places the symbol that denotes the unit before the decimal point. Similarly, if he wants to write 14 hours plus 12 minutes, an astronomer might write $14^h 12^m$ (omitting the plus sign) or he might write $14^h.2$.

In a footnote near the beginning of this chapter, I mentioned the astronomical style of designating years. All dates in this work will be written in astronomical style, which has two aspects that need mentioning. The first is that the years will be written in the astronomical system as opposed to the historical system. The second is that the year will be written as the first component of a date, followed by the name of the month, which in turn is followed by the day of the month. The parts of a date will be written without the use of separating commas. Thus, for example, an historian would probably write that Julius Caesar was assassinated on March 15, 44 B. C. E. An astronomer would write this date as -43 March 15.

When the names of the months appear in tables, I shall use only the first three letters of the English name. This gives an orderly appearance to the columns in which dates appear.

-16-

CHAPTER II

GREEK MATHEMATICS

> . . . and then the different branches of
> Arithmetic - Ambition, Distraction, Ugli-
> fication, and Derision.
>
> Lewis Carroll[†]

1. Greek Arithmetic

Human nature has not changed in the past 2000 years,
and the Greeks, like us, had ambition, distraction, and de-
rision, but I am not sure about uglification. More impor-
tantly, they also had a firm command of the basic operations
of arithmetic.

Modern translations or discussions of Greek works on
astronomy often present the numbers in a modern form. However, knowing and understanding the exact form in which a
Greek astronomer wrote a number is sometimes important in
understanding what he did. For this reason, I shall devote
a little space to explaining Greek numerals. As it turns
out, this is the only topic in Greek arithmetic that we need
to study explicitly.

The Greek alphabet that was in use 2000 years ago contains 24 letters. Let us add 3 special symbols to these 24,
so that we now have 27 symbols.[‡] The first of the extra
symbols is added between ε and ζ, the next one is added between π and ρ, and the last one is added after ω. The Greeks
used the first nine members of this expanded set of symbols
for the integers 1 to 9, they used the next nine for the
multiples of 10, that is, for 10 to 90, and they used the
final nine for multiples of 100.[‡]

With these symbols, the Greeks could write integers
through 999, in a notation that is something like a place-

[†]From Chapter IX of Alice in Wonderland.

[‡]The three added symbols are letters in an early Greek
alphabet that were dropped in later times. From this fact,
we can locate the development of the "classical" Greek numerals to about -800 in time and, probably, to Asia Minor
in place [Neugebauer, 1957, p. 11].

[‡]The first extra symbol equals 6, the next one equals 90,
and the last one equals 900. I do not know whether the
Greeks gave them these positions arbitrarily, or whether
these positions are ancient ones in the alphabet. In discussing Greek numerals, I have used the "little" letters
like γ, because they are probably better known to most
readers. However, in classical times, they had probably
not been invented; classical writing probably used only
the "big" letters like Γ.

value notation. For example, the number 237 was written σλζ. The symbols were written in order from left to right, with the largest symbol on the left. However, we really know the value of a symbol by the set from which it was taken rather than by its position, as we can see by writing a number with a zero in it. Thus 207 was written σζ, while 230 was written σλ.

In order to write numbers in the thousands, the Greeks used the first nine symbols over again to represent the first nine multiples of 1000. For example, in a context that will be discussed in Section VIII.7, Ptolemy had occasion to write the number 1210. He wrote it as ασι. This allows writing numbers up to 9999. Since we shall not have to deal with larger numbers, I shall not describe them except to say that they follow a more complicated system.

When numerals occur as part of a sentence, as they do in the preceding paragraphs, the Greeks might draw bars over them, or put accent marks beside them, in order to indicate that numerals and not letters are meant.

The Greeks had two ways of writing fractions. One way was to write simple fractions, but with one peculiarity. Except for the fractions 2/3 and 2/5, the only fractions that I have noticed have unity for the numerator. The fraction 2/3 had a special symbol while 2/5 was written out in words. All fractions other than 2/3 and 2/5 were stated by writing the numeral for the denominator, with a symbol to indicate that a fraction and not an integer was meant. For example, 1/4 might be written δ″, with δ indicating 4 and ″ indicating the fraction.†

The other way was to use a place-value notation exactly analogous to our decimal fractions, but with the base 60 rather than 10. We shall frequently need to write numbers in this sexagesimal notation, using modern symbols for the digits, and I shall start by explaining how this will be done. The easiest explanation is by example. The number written as 359;45,24,45,21,8,35 means‡

$$359 \times 60^0 + 45 \times 60^{-1} + 24 \times 60^{-2} + 45 \times 60^{-3}$$
$$+ 21 \times 60^{-4} + 8 \times 60^{-5} + 35 \times 60^{-6}.$$

A semicolon separates the integral part of the number from the coefficients of the negative powers of 60, while commas separate the coefficients of the negative powers from each other.

A Greek astronomer would write this sexagesimal number by using his own symbols for the digits instead of ours. He might simply space them, with some special way of indicating

---

† Many Greek texts use a special symbol for ½, instead of writing β″.

‡ This is the number of degrees that the mean sun moves in 365 days, in Hipparchus's theory of the sun.

which part is the integral (coefficient of 60^0) part. He might also draw bars over the individual coefficients, perhaps combining this with spacing to separate the coefficients from each other. Or he might draw a bar over the integral part while putting marks beside the other coefficients. In the latter way, he would write the number as $\overline{\tau\vartheta}$ με′ κδ″ με‴ κα⁗ η′‴ λε‴‴.

This is obviously the origin of our way of writing angles and the time of day. The number of minutes is the coefficient that appears in the first fractional position (the coefficient of 60^{-1}) and the number of seconds is the coefficient that appears in the next or second position. We can continue this by referring to thirds, fourths, and so on. Thus the number could be read as 359 degrees plus 45 primes or minutes, 24 seconds, 45 thirds, 21 fourths, 8 fifths, and 35 sixths.

The Greeks took this notation, as they took much else in astronomy, from the Babylonians. Aside from the fact that they used different symbols for individual digits, the Greeks modified the Babylonian system in one important way, the way in which they wrote the integer part of the number. The Greeks usually wrote this in their decimal way that has already been explained. The Babylonians, however, used the sexagesimal notation for the integral part of the number also. Thus, up to the semicolon in the number illustrated above, they would write the equivalent of "5,59" rather than the equivalent of 359.†

The sexagesimal system, in both late Babylonian and Greek writing, was a true place-value system, because it used the zero. The zero was used only in a sexagesimal way and not in a decimal way. To explain what I mean, let me use the number 3;0,20. For the moment, let me write out zero instead of using the Greek symbol for it. A Greek astronomer might write this as $\overline{\gamma}$ zero′ κ″. That is, he would insert zero in the vacant sexagesimal position (the prime position), but he would not insert it in the vacant decimal position following κ (20) in the seconds position.

We can put the matter in another way. In any position (primes, seconds, and so on) in the sexagesimal system, we may have numbers from 0 to 59 in our notation. The Greek astronomer would write a special symbol for zero when it occurred. For the numbers from 1 to 59, he would use the alphabetic notation already described.

It has been speculated that the Greek symbol for zero is the initial letter of ονδεν, the neuter form of the word for no-one or none. This would make the symbol be an open circle, just as our zero is. However, it is doubtful that any Greek letter would have been introduced for the zero,

---

†These remarks apply only to technical writing, such as astronomical texts. In other writing, the Babylonians used number systems that are quite different from the sexagesimal one. I believe that the Greeks also used the sexagesimal system only in technical writing.

because the letters were already preempted for other numerals; o already represented 70. Further, the open circle was used for zero only in the Byzantine period of Greek writing [Neugebauer, 1957, p. 14]. In the period that interests us, the zero had various forms. One form was $\bar{o}$. This could be interpreted as o (omicron) with a line above it to distinguish it from the numeral for 70. However, other forms that are probably older look like o——o , or variants thereof.

We may never know the origin of the open circle for zero, but it is plausible that it was simply a way of drawing "nothing".

Our numerals are usually called Arabic numerals, but they are really Hindu in origin. It is interesting that the Hindu form of zero is also an open circle [Smith and Ginsburg, 1956]. Zero in genuine Arabic numerals, either now or in the medieval period, is a solid dot. In Arabic numerals, the open circle represents five.

Neugebauer [1957, Chapter I] gives a more extensive discussion of Greek numerals, including variant methods that are not involved in this work, and including further citations to the literature. Smith and Ginsburg [1956] also give an extensive discussion, including an interesting discourse on the applications for which Roman numerals are more useful than modern numerals.

When I am quoting from a Greek text, I shall write fractions in the way they are written in the original, except that I shall use modern digits rather than Greek ones. That is, if the fractions are written in words in the original, I shall use words; if the fractions are written in sexagesimal form, I shall use that form; and if the fractions are written as simple fractions, I shall use the same form, with one exception: Since the Greek astronomers rarely used a fraction whose numerator is anything other than unity, they often had to write a sum of fractions in place of a single fraction. For example, they might write $\frac{1}{2} + \frac{1}{4}$. In these cases, I shall replace the sum by the single fraction that is equivalent. Thus, in the example, I shall write $\frac{3}{4}$ instead of $\frac{1}{2} + \frac{1}{4}$.†

---

†Fractions whose numerator is unity are called unit fractions. Unit fractions were a development of Egyptian arithmetic, and their use in Greek arithmetic was presumably borrowed from the Egyptians. I do not know whether Greek mathematicians generally used unit fractions or not. Ptolemy used them extensively, but this is not surprising since he lived in Egypt and undoubtedly absorbed much of its culture, even though he is classed as a Greek astronomer. This is a good place to mention that Ptolemy had no known relation to the Egyptian rulers named Ptolemy, in spite of frequent statements that he was a member of the royal family.

2. Greek Geometry

The Greeks were preeminent in geometry, both plane geometry and solid geometry. Euclid's famous work on geometry, written around -300, has never really been superseded, although some educators in recent times have tried to find more effective ways of teaching the subject than by using his book.

Babylonians and Egyptians, and perhaps other peoples older than the Greeks, had a large body of knowledge about geometry. The Greeks attained to greater knowledge, but what seems to have distinguished them especially was the role that they assigned to logical proof. Earlier knowledge of geometry seems to have been largely or entirely pragmatic.

As we shall see, the theoretical structure of Greek astronomy was basically geometrical. That is, the Greek astronomers postulated that the heavenly bodies move in certain geometric patterns (Chapter IV). The problem that they set themselves was to discover these patterns by analyzing astronomical observations. The process of doing this was, to a large extent, a process of solving problems in geometry.

Specific examples will come up later. Here, it is useful to call attention to one class of problem that the Greeks had mastered. A triangle is fully described by six quantities, namely the lengths of the three sides and the sizes of the three angles. We know, and the Greek astronomers knew, that we may specify three quantities and then calculate the other three, with two exceptions. That is, with the exceptions to be noted, a triangle is really specified by only three quantities. These three may be assigned arbitrarily, provided that the fundamental laws of geometry are not violated.

The first exception arises if the three angles are given. Suppose we draw a line 10 centimeters long, and construct two of the angles at opposite ends of the line. The lines that form the other sides of the angles meet at some point, and we now have a triangle possessing the given angles. However, we could have made the first side 20 centimeters; had we done so, we would have obtained a triangle with the same angles, but with all sides twice as large. In other words, one of the three specified quantities must be a side.

The other exception arises if two sides and one angle are given, and if the angle is not the angle between the two sides. The reader should be able to establish, by trying to construct a triangle from this amount of information, that there are two and only two triangles possessing the given quantities.

A problem that is central to understanding some of Ptolemy's results in astronomy is that of finding the angles when the three sides are given. In his Chapter VI.7, Ptolemy solves this problem by the method shown in Figure II.1; this method was probably a standard one in Greek trigonometry.

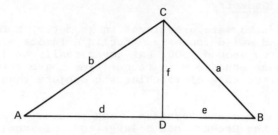

Figure II.1 Determining the angles of a triangle whose sides are given. The perpendicular from C to the opposite side creates two right triangles ACD and BCD. When we apply Pythagoras' theorem to the right triangles, we easily find the distances d, e, and f, and from these distances we easily find the angles.

In the figure, A, B, and C denote the three angles, which are to be found, while a, b, and c denote the sides, which are known. a is the side opposite A, b is the side opposite B, and c is the side opposite C. For clarity, the letter c is not shown in the drawing. Let c denote the longest side,† and let c be divided into the parts d and e by the perpendicular f drawn from C. Further, let a denote the shortest side. If we apply Pythagoras' theorem to the triangles ACD and BCD, we find $f^2 = b^2 - d^2 = a^2 - e^2$, whence

$$d^2 - e^2 = b^2 - a^2. \qquad (II.1)$$

Since a and b are given, $d^2 - e^2$ is known. Further,

$$d + e = c, \qquad (II.2)$$

and the sum $d + e$ is known since c was given. If we divide Equation II.1 by Equation II.2, we find

$$d - e = (b^2 - a^2)/c. \qquad (II.3)$$

It is now trivial to solve Equations II.2 and II.3 for d and e; knowing d and e also tells us f. Then $A = \sin^{-1}(f/b)$, $B = \sin^{-1}(f/a)$, and C is $180° - A - B$.

The Greek mathematicians did outstanding work in solid geometry as well as in plane geometry. Except for some elementary results in spherical trigonometry, however, we shall have no need for solid geometry.

3. Greek Plane Trigonometry

The usual trigonometric functions are shown in Figure II.2. The radius of the circle is unity. OB and OE are two perpendicular radii, BD and EF are tangents that are perpendicular to OB and OE, respectively, and CA is perpendicular

---

†The problem is trivial if the triangle is isosceles or equilateral.

to the radius OB. If we now use a pair of letters with a bar over them to denote the length of a line segment, for example, if we let $\overline{CA}$ denote the length of the corresponding segment, we have

$$\sin\theta = \overline{CA}, \qquad\qquad \csc\theta = \overline{OF},$$
$$\cos\theta = \overline{OA}, \qquad\qquad \sec\theta = \overline{OD}, \qquad\qquad (II.4)$$
$$\tan\theta = \overline{BD}, \qquad\qquad \cot\theta = \overline{EF}.$$

If we had let the radius $\overline{OB}$ be something other than unity, the functions would become ratios. That is, we would have had $\sin\theta = \overline{CA}/\overline{OB}$, and so on.

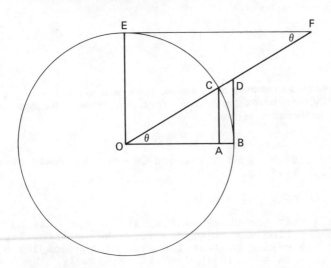

Figure II.2   The usual trigonometric functions of the angle θ.
If the radius $\overline{OB}$ of the circle is unity, the distance $\overline{CA}$ is the sine, the distance $\overline{OA}$ is the cosine, the distance $\overline{BD}$ is the tangent, the distance $\overline{EF}$ is the cotangent, the distance $\overline{OD}$ is the secant, and the distance $\overline{OF}$ is the cosecant.

The Greeks could easily solve problems involving triangles, as we have just seen. Interestingly, they did so by using only one trigonometric function, and that one is not among the six functions just defined.

The function they used is called the chord, which is illustrated in Figure II.3. If the radius of the circle is unity, then $\overline{CG}$ is the chord of the angle α. Since there will not be much occasion to deal with the chord function, I shall not introduce an abbreviation for it and shall simply write

$$\text{chord } \alpha = \overline{CG}. \qquad\qquad (II.5)$$

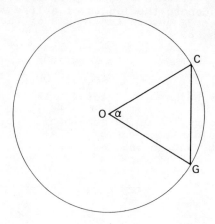

Figure II.3 The chord of the angle α. If the radius of the circle is unity, the distance $\overline{CG}$ is the chord. For brevity, "chord of the angle α" will be written as chord α.

If we take half of the chord $\overline{CG}$, we obviously have the sine of $\frac{1}{2}\alpha$. That is,

$$\text{chord } \alpha = 2 \sin \tfrac{1}{2}\alpha. \tag{II.6}$$

Saying that the Greek mathematicians had only one trig-onometric function may be misleading. We see from Figure II.2 that the cosine of θ is the sine of the complement of θ. That is, cos θ = sin (90° - θ). Similarly, the Greeks were aware that the chord of the supplement of α is a use-ful quantity. However, they did not give it a separate name. To put this more explicitly, let β = 180° - α. The Greeks used chord α and they used chord β, but they did not, as we would do, call chord β something like the "co-chord" of α. Nonetheless, they used chord β, which is analogous to cos $\frac{1}{2}\alpha$.

The biggest lack was that of an analogue to the tangent function. However, that was merely an inconvenience, and it did not limit the Greeks' ability to handle triangles. Basically, the sine or cosine (or the sine of the complement) involves one leg and the hypotenuse of a right triangle, while the tangent involves both legs. However, if we have the two legs, we can always find the hypotenuse from Pytha-gora's theorem, without using any information about the angles. Thence we can find the angles by using the sine or chord function only, even if we originally know only the legs.

Ptolemy, in his Chapter I.11, gives a table of chords for angles from 0° to 180°, with an increment of $\frac{1}{2}$ degree. This is equivalent to a table of sines from 0° to 90° in steps of $\frac{1}{4}$ degree. Nowadays, we would probably evaluate the functions by using infinite series, but this method was not available to the Greek mathematicians. They started from the chords that could be evaluated by elementary constructions. We can get chord 60° and chord 90° from the properties of equilateral triangles and squares. There was a known 'compass and straight edge' construction for the regular pentagon, which gave chord 72°. Successive bisections then give the chords for all angles of the form $60°/2^n$, $72°/2^n$, and $90°/2^n$, in which n is an integer. To go beyond these, we must be able to find chord $(\alpha \pm \beta)$ when chord α and chord β are both known.

To do this, Ptolemy uses the theorem shown in Figure II.4. In the figure, the segments $\overline{AB}$, $\overline{BG}$, $\overline{GD}$, and $\overline{DA}$ are the sides of a quadrilateral inscribed in a circle. If we let $\overset{\frown}{AB}$ denote the arc between points A and B, and so on, we clearly have $\overline{AB}$ = chord $\overset{\frown}{AB}$, and so on. Ptolemy proves the theorem

$$\overline{AG} \times \overline{BD} = \overline{AB} \times \overline{GD} + \overline{BG} \times \overline{DA}.$$

This is still called Ptolemy's theorem in some texts on geometry.

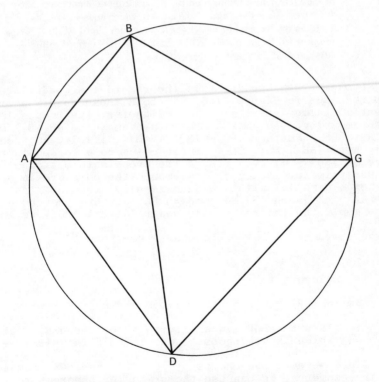

Figure II.4 Ptolemy's theorem for a quadrilateral inscribed in a circle. Form the product $\overline{AG} \times \overline{BD}$ of the diagonals, and the products $\overline{AB} \times \overline{GD}$ and $\overline{BG} \times \overline{DA}$ of pairs of opposite sides. Then $\overline{AG} \times \overline{BD} = \overline{AB} \times \overline{GD} + \overline{BG} \times \overline{DA}$.

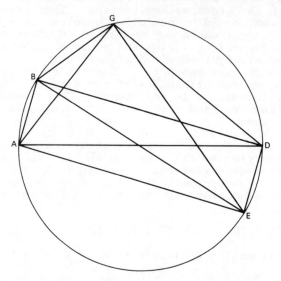

Figure II.5   The chord of the sum of two angles. AD is a diameter of the circle, whose radius is unity. BE is another diameter, so that $\overline{AB} = \overline{DE}$. $\overline{AG}$ is the chord of the arc $\overset{\frown}{AG}$, which equals $\overset{\frown}{AB} + \overset{\frown}{BG}$. By using the theorem of Fig. II.4, Ptolemy finds chord $\overset{\frown}{AG}$ in terms of $\overline{AB}$ (= chord $\overset{\frown}{AB}$), $\overline{BG}$ (= chord $\overset{\frown}{BG}$), $\overline{BD}$ (= chord $[180° - \overset{\frown}{AB}]$), and $\overline{GE}$ (= chord $[180° - \overset{\frown}{BG}]$).

Ptolemy then applies this theorem to the quadrilaterals ABGD and ABGE in Figure II.5. In both these quadrilaterals, either a diagonal or a side is a diameter; the radius is taken as unity. $\overline{AD}$ and $\overline{BE}$ are both diameters, so $\overline{AB} = \overline{DE}$ and $\overset{\frown}{AB} = \overset{\frown}{DE}$. Further, $\overset{\frown}{BD} = 180° - \overset{\frown}{AB}$. Clearly, $\overline{AG} =$ chord $\overset{\frown}{AG} =$ chord $(\overset{\frown}{AB} + \overset{\frown}{BG})$. Let me use co-chord α to denote the chord of $180° - \alpha$, even though the Greek mathematicians did not use this terminology; I introduce the co-chord in order to show a parallel with a modern result. Thus $\overline{AB} =$ chord $\overset{\frown}{AB}$, $\overline{BD} = \overline{AE} =$ co-chord $\overset{\frown}{AB}$, $\overline{BG} =$ chord $\overset{\frown}{BG}$, and $\overline{GE} =$ co-chord $\overset{\frown}{BG}$. If we apply the theorem to the quadrilateral ABGE, we get

2 chord $(\overset{\frown}{AB} + \overset{\frown}{BG})$ = chord $\overset{\frown}{AB}$ × co-chord $\overset{\frown}{BG}$

+ co-chord $\overset{\frown}{AB}$ × chord $\overset{\frown}{BG}$.

This is analogous to

$\sin (\alpha + \beta) = \sin \alpha \cos \beta + \cos \alpha \sin \beta$.

The quadrilateral ABGD gives a formula for chord $\overset{\frown}{BG}$ = chord $(\overset{\frown}{AG} - \overset{\frown}{AB})$ which is analogous to our formula for $\sin (\alpha - \beta)$.

Since we have the chords of 60°, 72° and 90° from elementary geometry, we can use the preceding theorems to find the chord of any angle that is an integral multiple or an integral fraction of $1\frac{1}{2}$ degrees, but we cannot find the chord of $\frac{1}{2}$ degree, which is what Ptolemy needs to complete

his table. Ptolemy starts the last phase of his job by
using†

$$\text{chord } 1\tfrac{1}{2} \text{ degrees} = 0;1,34,15,$$

$$\text{chord } \tfrac{3}{4} \text{ degrees} = 0;0,47,8. \tag{II.7}$$

He also uses the result that

$$\frac{\text{chord } \alpha}{\text{chord } \beta} < \frac{\alpha}{\beta}, \tag{II.8}$$

provided that $\alpha/\beta > 1$. This is equivalent to saying that
the chord increases less rapidly than the angle or the arc,
and there are many ways of proving it.‡

Ptolemy first uses inequality II.8, taking $\alpha = 1°$ and
$\beta = 3/4$ degrees, and he uses chord $3/4$ from Equations II.7.
This gives chord $1° < (4/3) \times 0;0,47,8$, which Ptolemy writes
as

$$\text{chord } 1° < 0;1,2,50. \tag{II.9}$$

The correct value of $(4/3) \times 0;0,47,8$, however, is $0;1,2,50,$
40, so that Ptolemy has not proved inequality II.9.

Ptolemy next uses inequality II.8 with $\alpha = 1\tfrac{1}{2}$ degrees
and $\beta = 1°$, and he takes chord $1\tfrac{1}{2}$ from Equations II.7. This
gives chord $1° > (2/3) \times 0;1,34,15$, which he writes correctly
as

$$\text{chord } 1° > 0;1,2,50. \tag{II.10}$$

The combination of the inequalities II.9 and II.10, in which
the right members have been evaluated closely but not exactly, leads to taking chord $1° = 0;1,2,50$. Ptolemy can now
complete his table.

Ptolemy was somewhat lucky, because his methods are
actually not accurate enough to establish his conclusion.
Let us go to the fourths position in Equations II.7. The
right members are then $0;1,34,14,42$ and $0;0,47,7,25$. Rounded
to the thirds position, the first value in II.7 is correct,
but the second value should be $0;0,47,7$. Now, if we carry
out the operations leading to inequalities II.9 and II.10
and round to the thirds position, we have simultaneously

---

†Actually Ptolemy takes the radius of his standard circle
to be 60 rather than 1. This shifts all values by one
sexagesimal position, so that Ptolemy would write the
equivalent of 1;34,15, for example, rather than 0;1,34,15.
I shall follow modern practice and take the radius to be
unity.

‡Some writers credit Ptolemy with the inequality II.8.
However, Heath [1913, p. 333] says that Aristarchus of
Samos knew this result four centuries earlier.

chord 1° < 0;1,2,49,

chord 1° > 0;1,2,50,

which are impossible. If Ptolemy had worked carefully, he would have realized that the value can be established in the thirds position only by keeping accuracy to the next position. If we do this, we find

chord 1° < 0;1,2,49,53,

chord 1° > 0;1,2,49,48.

These tell us that chord 1°, when rounded to the thirds position, is indeed 0;1,2,50 as Ptolemy took it, but his proof of the matter is not correct.

Ptolemy says, in his Chapter I.10, that we cannot find the chord of $\frac{1}{2}$ degree directly from the chord of $1\frac{1}{2}$ degrees, presumably because that would be equivalent to the geometrical trisection of an angle. It is true that there is no compass and straight edge construction that will solve the problem, but we can nonetheless find chord $\frac{1}{2}$ degrees directly from chord $1\frac{1}{2}$ degrees. Since Ptolemy could handle chord ($\alpha \pm \beta$), he could find chord 2α and thence chord 3α. That is, he should have been able to find the result that

$$\text{chord } 3\alpha = 3 \text{ chord } \alpha - (\text{chord } \alpha)^3,$$

which is analogous to the formula $\sin 3\theta = 3 \sin \theta - 4 \sin^3\theta$. By taking $3\alpha = 1\frac{1}{2}$ degrees, he would have found chord $\frac{1}{2}$ immediately by solving this relation, a task that was well within his powers. However, it generally takes a long time to find the easiest way to solve a problem, and we should not criticize Ptolemy for failing to find this way. We should only regret that the method he did use was not correct.

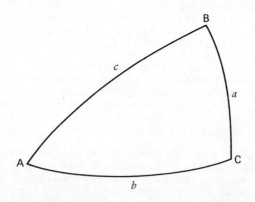

Figure II.6 The trigonometry of a spherical right triangle. The arcs a, b, and c are portions of great circles drawn on a sphere, and the arcs a and b intersect at a right angle. That is, the angle C equals 90°. We can find the remaining three quantities if any two of the quantities a, b, c, A, and B are given.

4. Greek Spherical Trigonometry

The only remaining part of Greek mathematics that will concern us is the trigonometry of a spherical right triangle. The situation is shown in Figure II.6. The reader must imagine that the arcs a, b, and c are portions of great circles that have been drawn on a sphere. These arcs will be called the sides of the spherical triangle. The arcs meet in the angles A, B, and C; C is taken to be a right angle. The situation is somewhat like that in a plane right triangle: All the quantities can be found provided that any two are given, besides the fact that C = 90°. However, it is not necessary for one of the given quantities to be a side, but the accuracy of finding the sides may be poor when only the angles are given.†

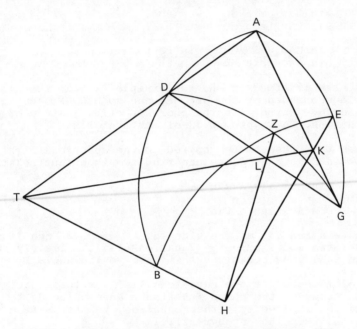

Figure II.7 An important theorem in spherical trigonometry. AEG and ADB are arcs of great circles on a sphere whose center is H. When D and G are joined by a great circle, and when E and B are also joined, the arcs DG and BE meet at Z. The theorem relates the sines of the arcs AB, AG, BE, and DG, and the various arcs into which they are divided.

---

†The relation here is that cos c = cot A cot B. If the sides are small, the triangle approaches a plane triangle and A approaches the complement of B. Thus cos c approaches unity and c is hard to find accurately.

The most important Greek work on spherical trigonometry that has survived is probably the treatise by Menelaos [ca. 100]. This work is lost in the original language but it was preserved through the medieval period in Arabic translation, and it became available in German translation in 1936. The main theorem that we need in our study of Greek astronomy is the theorem numbered III.1 in the translation, on pages 194-197. The accompanying figure, with the Latin equivalents of the Arabic letters used in it, appears in Figure II.7.

We start with a sphere whose center is H. On it, we draw arcs AB and AG of two great circles, and we mark arbitrary points D and E on the two arcs, as shown. We then draw the arcs DG and BE, which meet in the point Z. The straight lines that appear in the figure are used in the proof, but they are not needed in stating the theorem. The theorem is stated in two parts:

$$\frac{\sin\ GE}{\sin\ EA} = \frac{\sin\ GZ}{\sin\ ZD} \times \frac{\sin\ BD}{\sin\ BA} \ ,$$

$$\frac{\sin\ GA}{\sin\ AE} = \frac{\sin\ GD}{\sin\ DZ} \times \frac{\sin\ BZ}{\sin\ BE} \ .$$

Where I have written sin GE, for example, I mean the sine of the angle subtended by the arc GE, and so on. Ptolemy gives this theorem, stated with the chord function instead of the sine function, in Chapter I.13 of the Syntaxis.

When this theorem is applied to the spherical right triangle in Figure II.6, we can find two important relations:

$$\sin\ a = \sin\ A\ \sin\ c,$$

$$\cos\ A = \tan\ b/\tan\ c.$$

(II.11)

If we have A and one of the sides a, b, or c, we can find the other two sides by using Equations II.11. Clearly, we can find analogous equations that involve the angle B.

Ptolemy does not use Equations II.11 or their equivalents in terms of the chord function. When he needs to use the relation implied by either equation, he starts from the theorem that Menelaos numbers III.1.

5. Two General Comments

Two general comments need to be made. The first concerns the antiquity of the mathematical methods that were used in Greek astronomy. The sexagesimal notation was borrowed from the Babylonians, as I have already said. Also, the zero was "in full use" in Babylonian mathematical texts from -300 on [Neugebauer, 1957, p. 27]. Since this precedes most Greek work in astronomy, it is likely that the system of place-value notation, including the use of the zero, has a Babylonian origin.

The developments in geometry and plane trigonometry that have been described were probably all known to Hipparchus.† Much of this mathematics is even older; Euclid, for example, lived around -300. Hipparchus is nearly three centuries before Ptolemy, whose Syntaxis was written in or near the year +142. Thus a large part of the mathematics that Ptolemy used was already as old to him as the Elizabethan age is to us.

The second comment concerns the use of the term "Greek mathematics", quite aside from the question of using "Greek" rather than "Hellenistic". Some writers object to the use of such a term. Neugebauer [1957, p. 190], for example, calls it "more misleading than helpful", and his position is well taken. There are only a few Greek mathematicians whose work is well known to us. The work of others, even though it is sometimes of high quality, is known only from fragments that have happened to survive, or from quotations from other writers. This has happened to Hipparchus, for example. By any standard, he is a major figure and perhaps the most important figure in Greek astronomy. Unfortunately, only one work of his has survived [Hipparchus, ca. -135], and it is not one of his most important works. Otherwise, we know what he did only from what others said about him.

Thus the totality of what we call Greek or Hellenistic mathematics consists of the work of perhaps twenty persons, usually preserved only in part, spread over several centuries. The danger consists of assuming that this small sample is typical of mathematics among the Greeks. However, barring unlikely discoveries which show that it originated elsewhere, we may say safely that what we have discussed is a part of Greek mathematics. That is, we may fairly call it Greek mathematics, provided that we do not intend to be exclusive. We must always remember the possibility that there was other Greek mathematics which is now lost to us, probably forever.

---

†Neugebauer [1957, p. 161], says that spherical trigonometry as such was not available to Hipparchus, and that he solved spherical triangles by other means, such as projections.

CHAPTER III

THE EARTH

> If earth, industrious of herself, fetch day
> Travelling east, and with her part averse
> From the sun's beam meet night, her other part
> Still luminous by his ray.

John Milton[†]

1. Knowledge of the Earth Needed for Astronomy

Astronomers do not need nearly as much information about the earth as geographers or politicians do. Astronomers need to know the general shape and size of the earth, they need to locate the places on the earth's surface where they make their observations,[‡] and they need to know the motions that the earth has. In this chapter, I shall summarize the methods that the Greeks had for studying these matters, and the consequent beliefs that they had about them.

2. The Shape of the Earth

It is commonplace in much writing to say that people thought that the world was flat until the voyage of Columbus. Then Columbus found America by sailing west from Europe, and thus he proved that the world is round. These statements call for two comments.

First, it is not clear how the conclusion follows from the argument. Since Columbus found America by sailing west, this proves nothing about the shape of the earth. If he had found Asia, that would have been different.[‡]

Second, the starting premise is not correct. Informed people knew about 2000 years before Columbus that the world is round, and they have never lost this knowledge.

In early times, various peoples had various fanciful ideas about the shape of the earth. Since the earth is "obviously" flat, an important problem in early cosmology was how to connect the earth to the rest of the visible

---

[†]Paradise Lost, Book VIII, lines 137-140.

[‡]For some purposes, they also need to know the direction of the line defined by a suspended plumb bob.

[‡]In the rest of this work, except perhaps in special circumstances, I am going to ignore the problem that came up in Section I.1: What do we mean by physical truth? When I say that a physical hypothesis or theory is true, I shall mean that it agrees with observations, so far as we know. If two or more hypotheses or theories agree with the observations, I shall say that the one is true or correct which most modern scientists accept as true or correct.

universe, which is "obviously" not flat. In particular, it was a problem to explain how the sun, moon, and stars got from their settings in the west back to their risings in the east. Among the early cosmologies, my personal favorite is one that is Egyptian, I believe. In it, the earth is a flat object resting on the backs of four elephants, who in turn stand on the back of a tortoise that swims in the sea. This cosmology has the merit of giving a simple explanation of earthquakes.† Some other early ideas about the earth, particularly Greek ones, are summarized by Pannekoek [1961, Chapter 9] and Dreyer [1905, Chapter 1].

By around −500 or perhaps a bit later, the Greek thinkers were definitely working toward a spherical earth. By the time of Aristotle (ca. −383 to ca. −321), the idea was probably accepted as well established, at least by people who concerned themselves with such matters. Several considerations lead to the idea of a spherical earth:

a. On a flat earth, the heavenly bodies would appear in the same directions at the same times to all observers, except for the small effects of parallax.‡ This idea has several consequences that are denied by experience. Many stars that are visible in Egypt cannot be seen in Athens. Conversely, many stars that never set in Athens do rise and set in Egypt. In midsummer, the sun comes within about 7° of the zenith in Alexandria but only within about 14° in Athens. The longest day is measurably longer in Athens than in Alexandria, as is the longest night. The preceding all have to do with curvature in the north-south direction. Lunar eclipses show that there is also curvature in the east-west direction; a particular eclipse is always later in the night in eastern regions than in western regions. It is hard to save these phenomena if we assume that the earth is flat.

b. In looking at a distant object, we see only its upper parts. The usual example is a ship going away from us. We see its masts last, not its hull. However, if it vanished simply because of distance, we should see the big hull last, not the small masts. Similarly, if we look at a wooded shore line across a mile or so of water, we see the tops of the trees, but not their bases.

c. The shadow that the earth casts across the moon during a lunar eclipse is circular. This argument has cogency only if we already accept that the moon shines by reflected sunlight, and that it is eclipsed when the

---

†Obviously, a flea biting one of the elephants could cause an earthquake. I have seen this cosmology (but not this explanation of earthquakes) described in several places, but only one comes to hand at the moment of writing. This is the introductory article on the earth in the Encyclopaedia Britannica, Eleventh Edition, v. VIII, pp. 799–801, Encyclopaedia Britannica, Inc., New York, 1910.

‡I shall discuss the meaning of parallax in Section VI.1.

earth gets in the way. Thus it is a more sophisticated argument than the earlier ones.

  d. The various masses that make up the earth are always being urged downward by their natural tendency to fall. Thus the individual masses must end up by giving the earth a spherical shape. This argument suggests an awareness that every mass attracts every other mass. Otherwise, there is no obvious tendency toward sphericity.

  e. The sphere is an ideal or perfect shape. The modern mind probably rejects this as a valid argument, but it seems to have been highly persuasive to many Greek philosophers, such as Plato [ca. -380, 33B-34A].†

There are a few educated people even in this century who insist that we live on a flat earth, or even on the inside of a sphere, rather than on the outside of a sphere. Just so, there may have been a few Greek astronomers, mathematicians, or philosophers who did not accept the idea of a spherical earth,‡ but in general the idea was accepted by educated Greeks from, say, the year -350 on. This knowledge or acceptance passed from the Greeks to other peoples including the Romans, from whom western Europe had certainly received it by the early Middle Ages.

A few early Christian writers wrote works denying the sphericity of the earth, presumably because the doctrine seems to conflict with some passages in the Christian scriptures. Not all Christian writers denied it, however. For example, Isidorus Hispalensis, Bishop of Seville, wrote an encyclopedic work around 600 in which he refers to the spherical shape of the earth.‡ In contrast, a person called the "anonymous geographer of Ravenna", writing around 675, describes the earth as a flat surface bounded on the west by the ocean, on the east by a boundless desert, and on the north by great mountains which screened the heavenly bodies as they went from the west back to the east in order to rise again.

However, as Dreyer says, this is "the last writer of note who refuses obstinately to listen to common sense." About the same time, the Venerable Bede wrote reasonably detailed descriptions, mostly taken from the works of Pliny, of the main astronomical phenomena, including the shape of the earth and the fact that the sun is much larger than the earth. From about this time on, as Dreyer says, "the rotundity of the earth and the geocentric system of planetary

---

†See also the discussion by Cornford [1937]. Plato also says [44D] that our heads are globe-shaped because the head is the most divine part of us, being the part that rules over the rest of the body.

‡With us on the outside of it.

‡Quoted by Dreyer [1905, Chapter X]. In this chapter, called "Medieval Cosmology", Dreyer gives an extensive discussion of early medieval writing on astronomy.

motions may be considered to have been reinstated in the places they had held among the philosophers of Greece from the time of Plato." Thus, when Columbus set out on his famous voyage, he based his plans upon the accepted astronomical knowledge of his time. All educated people recognized that India lay both west and east of Europe, and the main question was how far it would be to India by the western route.

3. The Measurement of Latitude and the Obliquity

The latitude of a place can be measured with rather simple equipment, by either of two methods, and both methods were known to the Greek astronomers by around -400 or earlier.

In the first method, suppose for the moment that the star Polaris is exactly above the North Pole. It would then always appear in the same position to a particular observer. If an observer were at the Equator, Polaris would be on the horizon and its elevation angle would be zero. When he moves north, Polaris appears to rise above the horizon, and its elevation angle is just equal to the observer's latitude. The fact that Polaris is not exactly over the pole complicates the matter only slightly. The observer measures its greatest and smallest elevation angles, and his latitude is the average of these.†

The other method, which seems to have been more common, is based upon the sun. Suppose for the moment that the time is the vernal equinox, so that the sun is directly above the Equator. For an observer at the Equator (latitude = 0°), the sun is directly overhead at noon. That is, its distance from the zenith at noon is zero. As the observer moves north from the Equator, the zenith distance at noon increases and it is always equal to the observer's latitude. When he reaches the North Pole, the zenith distance of the sun at noon at the time of the equinox is 90°; that is, the sun just reaches the horizon but does not rise above it.

At the time of the equinox, the sun has its greatest rate of motion in the north-south direction. With this method an error in calculating the time of the equinox, or in measuring the time, makes an appreciable error in the latitude. For this reason, it was more customary to use the solstices, when the sun is not moving in the north-south direction.

The method of using the solstices is shown in Figure III.1. Suppose for example that the observer is at latitude 30° north, and that the obliquity of the ecliptic is 24°. This means that the point beneath the sun is at latitude 24° north at the summer solstice. Hence the sun, at noon on the day of the solstice, is 6° south of the zenith for our observer. At the winter solstice, the zenith distance is 54°.

---

†In almost all discussions of Greek astronomical measurements, I shall ignore refraction and other small effects. The errors introduced by these effects can be as much as 30', but they are usually much smaller. A typical error is around 1' or so.

Thus the latitude is the average of the two figures.

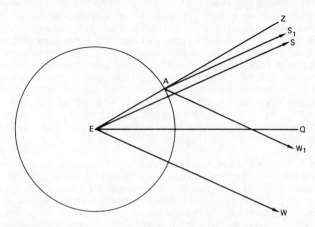

Figure III.1   Measuring the obliquity and the latitude. The line
EQ represents the equator. The observer is at A, and his latitude
ϕ equals the angle AEQ. His zenith is in the direction AZ. At the
summer solstice, the sun is in the direction ES, and at the winter
solstice, it is in the direction EW. The obliquity ϵ equals the angles
SEQ and WEQ. At the summer solstice, the observer sees the sun in
the direction AS_1, and the angle ZAS_1 equals $\phi - \epsilon$. At the
winter solstice, the observer sees the sun in the direction AW_1,
and the angle ZAW_1 equals $\phi + \epsilon$.

If the obliquity were considered to be a well-known
quantity, the observer could use a measurement at a single
solstice only. However, during much of Greek astronomy,
the obliquity was not taken to be well known, and some observers thought it necessary to measure it themselves, at
the same time that they measured their latitude. As we
have seen, the latitude is the average (half of the sum) of
the winter and summer zenith distances, and we also see that
the obliquity is half of the difference.

Newcomb's theoretical expression [Newcomb, 1895] for
the obliquity ϵ is

$$\epsilon = 23°.452\ 294 - 0°.013\ 012\ 5T$$
$$- 0°.000\ 001\ 64T^2 + 0°.000\ 000\ 503T^3,$$

(III.1)

in which T denotes time measured in centuries from noon on
1900 January 0, Greenwich time. In the year −225, when T =
−21.25, ϵ was $23°.723\ 242$ (= $23°\ 43'\ 23''.7$), while in 1900
it was only $23°.452\ 294$ ($23°\ 27'\ 8''.3$). It has decreased
by slightly more than a quarter of a degree since −225.

I use the year −225 for an example because that is about
the time when the most famous Greek measurement of the obliquity was made. It was made by Eratosthenes, an astronomer

and geographer, whose dates are approximately -275 to -195.
He was a native of Cyrene [Dreyer, 1905, p. 174], who studied in Alexandria and Athens before he settled in Alexandria.
His work has been lost except for a few fragments, and it is
known to us mostly through quotations and comments made by
later writers. We do not know what kind of instrument he
used for making his measurement. His result is stated by
saying that he found the difference between the elevations
of the sun at the solstices to be 11/83 of a full circle;
thus he found the obliquity to be 11/166 of a circle, or
23°.855 (= 23° 51′ 20″). By comparison with the value given
above, we see that his error is almost 8′, which is about
half of the radius of the apparent solar disk.

We need to ask how the measurement could have been made.
A common device in later Greek times was the mural quadrant.
This is a quarter of a circle, clamped in such a way that it
lies in the meridian plane. The user first hangs a plumb
bob from its center and marks where the plumb line crosses
the circle; this gives him the vertical. He then graduates
the circle in degrees or whatever units suit his fancy.
Finally, he puts a small marker at the center, so that its
shadow at noon falls across the graduations, and he gets the
elevation angle of the sun by reading his graduated circle.

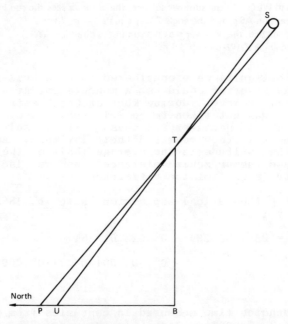

Figure III.2   The principle of the gnomon. The gnomon is a
vertical pole represented by the line TB, along with a graduated
scale BUP running north from B along a level surface. It is used
to find the elevation angle of the sun above the horizontal. The
sun is not a point source. The ray from the upper edge strikes the
scale at point U and that from the lower edge strikes at point
P. The region between U and P is called the penumbra.

A person who uses such a device with care should be able to find the obliquity with an error around 2′ or 3′. The fact that Eratosthenes' error is so large suggests that he used an older type of instrument called a gnomon, which is illustrated in Figure III.2. The gnomon is simply a vertical pole, mounted on a level surface, with a scale running north from its base. When the shadow lies on the scale, the time is noon. The user measures the length of the shadow and calculates the elevation angle of the sun by solving the right triangle.

The quadrant is superior to the gnomon in two important ways. First, if the user makes an error in finding the vertical direction with a quadrant, this does not make any error in the obliquity, although it will make an error in the latitude that he finds. An error in the vertical cancels when the readings at the winter and summer solstices are subtracted, although it does not cancel from the average, which gives the latitude. However, if he makes an error in putting the gnomon vertical, this affects the scale readings in a different fashion in summer and winter, and makes an error in both the obliquity and the latitude.

Second, the shadow cast by the central marker in the quadrant is symmetrical, and the user does not make a systematic error in reading the angle. However, the shadow cast by the gnomon is not symmetrical, as the figure shows, because the sun is not a point source. The ray from the upper edge of the sun strikes the level surface at U while the ray from the lower edge strikes it at P. The part of the scale between B and U is completely shadowed, the part beyond P is not shadowed at all, and the shadow falls from full to nothing between U and P. The transition zone, called the penumbra, is clearly different when the sun is high in the sky and when it is low.

The point in the shadow that a person should read is the point midway between P and U. However, an early Greek astronomer may not have realized this and, even if he did, it is not easy to pick out this point. Thus, when using a gnomon, a person is likely to make an error comparable with the apparent radius of the sun's disk, which is about 16′. That is, Eratosthenes' error of 8′ in the obliquity is easy to understand if he used a gnomon, but not if he used a quadrant or a similar instrument that casts a symmetrical shadow.

Eratosthenes also measured the latitude of Alexandria, as we shall see in the next section. His value is wrong by about 10′, again the size we expect from a gnomon but not from an instrument with a graduated circle.

Of course, the size of Eratosthenes' errors does not prove that he used a gnomon. It is possible that he used, for example, a quadrant of a circle with a small radius, so small that he could not read the position of the shadow accurately.

Eratosthenes' value is larger than that from Newcomb's theory, and Greek astronomers before him had used 24°.† It also happens that Chinese measurements of the obliquity made before the year zero give values as large as that of Eratosthenes [Needham, 1959, pp. 287-291]. This led Needham to suggest that Newcomb's theory may need revision for times 2000 years ago.

Luckily, we have better information about the obliquity 2000 years ago than we can get from measurements of the obliquity made by astronomers at that time. We have about 80 measurements of the declinations of various stars‡ made by Greek astronomers between the years -290 and +140, approximately. From these measurements, we can calculate how the plane of the Equator was positioned at the average year -114, and hence we can calculate the obliquity [Newton, 1974] at that time. The result is 23° 43' ± 3', which is much lower than the measurements made at the time and which agrees closely with Newcomb's theory. Thus there is no reason to suspect a large error in the theory.

It is interesting that the Greek astronomers never improved on Eratosthenes' value, so far as we know.‡ His value was the one which Greek astronomy used during the more than seven centuries remaining to it, and the one which Greek astronomy bequeathed to its successors.

Although the Greek astronomers could measure latitude fairly easily, they did not have an extensive knowledge of the latitudes of specific places. The reason is that the necessary measurements had not been made in many places by the end of Greek astronomy. The latitudes of several places in Egypt, of Rhodes, of Athens, and of a few other places were probably all that had been measured carefully by the year 500, say.

4. The Size of the Earth

The Greek astronomers could measure the latitude of a place without knowing the size of the earth. However, with the methods that were available to them, they could not find the longitude without knowing the size. Hence I take up the size of the earth next.

The earliest figure for the size of the earth that we know of is given by Aristotle [ca. -350, Chapter II.14], who

---

†This is probably a rounded number rather than an attempt at an accurate measurement; it is a fifteenth of a circle.

‡The declination of a star is the amount by which it is north or south of the equator. See Appendix A for more details.

‡Needham quotes a measurement of about 23° 50' made by an astronomer named Pytheas about a century before Eratosthenes. However, subsequent astronomers used the value of Eratosthenes, even though it is not as good. I have not tried to locate the source of the value attributed to Pytheas

says: "Mathematicians who try to calculate the circumference put it at 400,000 stades."† Some writers have attributed this value to Aristotle, but it seems more likely that he merely took it from some unstated source that has since been lost.‡

There are other early measurements,‡ but I shall pass to the one that has received the most publicity in modern writing. This was made by Eratosthenes, who also measured the obliquity (Section III.3). The basic method used by Eratosthenes is simple; it was used by other Greek astronomers and, in its essential features, it is also used in modern geodesy.

In this method, we need two points, one of which is north of the other. We need the latitude of each point, or at least the difference in latitude between them, and we need the distance between them. Suppose that the distance between them is D and that the difference between their latitudes is β degrees. Then the circumference C of the earth is

$$C = D(360/\beta). \qquad\qquad (III.2)$$

Differences between astronomers arise from differences in measuring D and β.

We know how Eratosthenes measured β. He said that the latitude of Syene (the modern Aswan) in Egypt was equal to the obliquity, namely 23° 51′ 20″, while the latitude of Alexandria was greater than the obliquity by 1/50 of a circle, that is, by 7° 12′. He found the latitudes from the position of the sun at the summer solstice. These figures imply 31° 3′ 20″ for the latitude of Alexandria.* The correct latitude is about 31° 13′ [Times Atlas, 1955], so

---

†The stade (σταδιον) was a unit of length used in many and perhaps all places in the Greek world, but it did not have the same size everywhere. One stade used in Egypt was about 157.5 meters, while another, sometimes called the Royal Egyptian stade, was about 210 meters. The Olympic stade was about 185 meters [Dreyer, 1905, p. 176]. There are still other possible sizes, all in this general range.

‡Just before he gives the size, Aristotle refers to those who suggest that the Atlantic joins on to the regions of India, and he remarks that this is not incredible. Thus, in suggesting that India could be reached by sailing west, Columbus was not advancing a new idea. His great contribution was to take action on an old idea.

‡The interested reader can consult Dreyer [1905, Chapter VIII] or Fischer [1975]. A comparison of these sources gives some idea of the controversy that surrounds the Greek measurements of the size of the earth.

*In Section V.6, I shall give some grounds for believing that Eratosthenes' latitude for Alexandria was actually 30° 58′. If so, his difference in latitude was 7° 6′ 40″ [Newton, 1973], which is about 1/50.6 of a circle. Eratosthenes would naturally have rounded this to 1/50.

-41-

that Eratosthenes' error is about 10'.

We do not know how he measured the distance D, because ancient sources disagree about the matter.† According to one version, he used the distance that had been determined by the surveyors employed by the crown.‡ According to the other version, he used the distance that had been estimated from the experience of camel caravans. Whatever the method of estimating was, the value that Eratosthenes used was 5000 stades.‡ Thus the circumference of the earth was 250 000 stades.

If the caravan method is mentioned by ancient writers, it seems to me that there is only one explanation of the fact that ancient writers describe two methods. Eratosthenes is quoted uniformly on the latitudes of Syene and Alexandria and on the distance between them; this must mean that he stated the values in his now-lost work. Since he is not quoted uniformly about the method used to find the distance, it seems highly unlikely that he stated the basis for the distance; later commentators must have supplied the method that they thought probable, on the basis of what they would have done. It is probable that the distance of 5000 stades was a conventional value.

Much ink has been shed* about the accuracy of Eratosthenes' result. If we use 157.5 meters for the length of the stade, we get 39 375 kilometers for the circumference; the correct value is about 40 000 kilometers. However, if we use the Olympic stade of 185 meters, we get 46 250 kilometers for the circumference, and if we use the Royal Egyptian stade, we get 52 500 kilometers. Most writers who have tried to assess the accuracy have chosen one of these

---

†Eratosthenes' writing has been lost, so we are dependent upon what other ancient writers said about him. All sources agree that he used Syene and Alexandria, and that he used 1/50 of a circle for the difference in latitude, but they do not agree about how he measured the distance involved.

‡The Hellenistic rulers of Egypt had many distances measured by professional pacers who estimated distance by counting paces. So far as I know, we have no direct knowledge that the distance between Syene and Alexandria was measured in this way.

‡In the caravan version, the trip was 50 days journey by caravan, and 100 stades was the standard distance covered in one day. I have not seen an explicit reference to an ancient source which refers to the caravan method, but I have not searched very hard. The method is mentioned by Fischer [1975], among others. Dreyer [1905, p. 176] gives a reference to the pacing method. It is possible, so far as I know, that the caravan method is a modern idea with no classical justification, as the so-called "eclipse of Caesar" seems to be [Newton, 1970, pp. 70ff].

*I wish I had said this, but I cannot claim the credit. Keightley [in press] uses it about a controversial point in early Chinese chronology.

values and assessed the accuracy by comparing it with a modern determination. Most writers have overlooked the fact that we should not assess the accuracy of a measurement by comparing it with the "correct" result.

Let me give an example. The obliquity of the ecliptic in the year -114, carried to the second of arc, is 23° 42' 35", according to Newcomb's theory (Equation III.1); let us accept this value as correct for the sake of argument. The estimate made from the star data discussed in the preceding section, also carried to the second, is 23° 43' 06" [Newton, 1974], for the same year. These values differ by only 31". However, this is not the accuracy of the estimate made from the star data. Every measurement consists of two parts. One part can be called the central value, and the other part can be called the standard deviation. No measurement is considered complete in modern science unless a standard deviation, or its equivalent, accompanies the central value.

When we give a standard deviation, we recognize the fact that every measurement contains some error, and that we will not usually get the same value from different measurements. That is, the central value that we actually obtain, call it A, is to some extent a matter of chance. From the circumstances of the measurement, we estimate the standard deviation, call it σ, and we write the result in the form $A \pm \sigma$. What we mean is that the correct value lies between $A - \sigma$ and $A + \sigma$, with a probability that we assess at about two chances out of three.† Put another way, we believe that the error in the value A has only one chance in about three of being greater than σ.

In the case of the obliquity measurement for the year -114, the value assigned to σ is 3' 19". This, and not 31", is what we should take as the error in the estimated obliquity.

Let us now try to estimate σ for Eratosthenes' measurement of the circumference of the earth. Three individual measurements went into the final result, and we should look at each one individually. These are the latitude of Syene, the latitude of Alexandria, and the distance between them. We may ignore the fact that Alexandria is about 3° west of Syene rather than due north of it.

According to Eratosthenes, the latitude of Syene is equal to the obliquity, which was 23°.723 in his time. The correct latitude of Syene is about 24°.083, so that the error is about 0°.360, about 22'. The error in measuring the latitude of Alexandria should be about the same. We now need the error in the difference. If we have a value $A_1 \pm \sigma_1$ and a second value $A_2 \pm \sigma_2$, it is a standard result in statistics that the difference is $(A_1 - A_2) \pm \sqrt{(\sigma_1^2 + \sigma_2^2)}$. In this case, each σ is 0°.360, and the standard deviation that we should attach to the difference is 0°.5, almost

---

†This is not the precise definition of the standard deviation, but it will suffice in this approximate discussion. The exact definition is given in Section 5 of Appendix A.

-43-

exactly. That is, Eratosthenes' value for the difference in latitude should be written as $7°.2 \pm 0°.5$. The standard deviation of the difference in latitude is thus about 7 per cent.

Since we do not know how Eratosthenes found the distance between the two places, and since we do not even know what value he used for the distance, we have no way to estimate the corresponding accuracy. All we can say is that the final error in Eratosthenes' value of the circumference must be greater than the error in the difference of latitude alone. That is, the final standard deviation must be greater than 7 per cent.

This estimate of the error will be controversial, of course, and I do not wish to uphold rigorously the way in which it was derived. If Eratosthenes measured the latitude of Alexandria with a gnomon, which seems likely in view of his errors, Figure III.2 suggests that the apparent radius of the sun's disk is a plausible estimate of the error in the latitude of Alexandria. This is about $16'$. However, this figure does not apply to the measurement of the latitude of Syene. If the latitude were truly equal to the obliquity, the sun was directly overhead at noon on the day of the summer solstice. If the reader will redraw the figure for this circumstance, he will see that the penumbra lies equally around the base of the gnomon, and the error should be quite small, much less than the error made at Alexandria.

The fact that the error at Syene is greater than the error at Alexandria requires explanation. It does not seem possible to me that Eratosthenes actually made the measurement at Syene. He probably just used a traveller's statement that the sun was directly overhead at noon on the solstice, and thus took the latitude to equal his estimate of the obliquity, namely $23° 51' 20''$. Now, if we use $22'$ for the σ of the Syene measurement (and we cannot justify a smaller value), and if we use $16'$ for the measurement at Alexandria, we get $27'.2$ for the σ of the difference in latitude. This is an error of 6.3 percent, and it is a lower limit to the accuracy. If we guess 5 percent for the σ of the distance measurement, the overall accuracy is 8 percent. This, I believe, is about as good as we can do in estimating the accuracy of Eratosthenes' measurement. It is neither as good as many have claimed nor as bad as a few have claimed.

It looks as if Eratosthenes and his successors did not regard the measurement as one of particular accuracy, as we can see in three ways. First, it was obtained by combining round numbers. The circumference was 50 times the distance from Syene to Alexandria, and the distance was 5000 stades. It is not likely that measurement actually yielded these values. It is more likely that they are values rounded from measurements that were not considered very accurate,†

---

†If the figure of 5000 stades came from a caravan journey, it is the product of two round numbers, namely 50 days travel time and 100 stades per day. If it came from professional pacers, it is highly unlikely that their result was a multiple of 1000.

and the distance may simply have been a conventional value
used for conversational purposes. For example, Americans
often say that the United States is 3000 miles wide from
coast to coast, and a person might use this in casual conversation as the distance from Washington to San Francisco.
However, the distance is only about 2400 miles.

Second, the value of the circumference was soon changed
from 250 000 to 252 000 stades, either by Eratosthenes or by
someone close to him in time. The presumed reason [Dreyer,
1905, p. 175] was to get a convenient value, namely 700
stades, for the length of a degree of arc.

Third, there is a later measurement (and perhaps two)
that was obviously inaccurate. If Eratosthenes' value was
considered accurate, there would have been no reason to ad-
vocate a different and obviously crude value. The value in
question is due to Posidonius, a philosopher who did most of
his writing on Rhodes between -100 and -50, approximately.
He is somewhat later than Hipparchus and over a century
later than Eratosthenes. All of his work is lost except for
fragments. We learn of his estimate of the earth's size
only from what others said about it; see Dreyer [1905, pp.
175ff] for the sources.

According to Posidonius, the star Canopus (α Carinae)
just grazes the horizon at Rhodes, while it reaches an elevation angle of $7\frac{1}{2}$ degrees at Alexandria; this is 1/48 of a
full circle. The distance from Rhodes to Alexandria, again
according to Posidonius, is 5000 stades. Thus the circumference is 240 000 stades.

We do not know what stade was used† and so we cannot
put the result in modern terms. However, we know that the
result is a poor one, far inferior to Eratosthenes' value,
for two reasons. First, the distance from Alexandria to
Rhodes is over water, and it could not be measured with
nearly the accuracy of the land distance from Syene to Alexandria. An accuracy of 25 per cent would be good for an
over-the-water distance. Second, the difference in latitude,
namely $7\frac{1}{2}$ degrees, is quite poor; the correct value is about
$5\frac{1}{4}$ degrees.

At the time in question, the declination of Canopus was
about $-52°.7$. The latitude of Rhodes is about $36°.4$. Thus
the distance from Canopus to the zenith at Rhodes‡ was about
$89°.1$. That is, Canopus reached an elevation of about $0°.9$,
if we do not consider refraction. As actually seen, including an allowance for refraction, it appeared about $1°.4$
above the horizon, instead of being on the horizon as Posidonius stated.

The latitude of Alexandria is about $31°.2$, so that
Canopus rose about $6°.2$ above the horizon, including refraction, instead of $7°.5$ as Posidonius said. Posidonius should

† Dreyer says that the stade is the one of 157.5 meters, but
there is no basis for the statement known to me.

‡ When Canopus was in the meridian plane.

-45-

have gotten about 4°.8 for the difference in latitude,† instead of the difference of $7\frac{1}{2}$ degrees that he used.

Posidonius' error is hard to understand and it has excited much controversy, which Fischer [1975] summarizes. As a general matter, it does not seem likely to me that Posidonius observed any of the relevant quantities himself. It is more likely that he got his information from sources that he misunderstood or used carelessly. We have another instance of an extremely large error that he made [Dreyer, 1905, pp. 173-174]. In the course of arguing that the earth is not flat, Posidonius says that the star we call γ Draconis passes through the zenith in the town of Lysimachia in Thrace. The latitude of Lysimachia is about 40° 33', while the star was more than 53° north of the equator at the time in question. Here we have an error of about 13°, when an error of at most 30' is what we expect.

One explanation is that Posidonius merely intended the calculation as a lecture exercise. This is a stock explanation given in many histories of science in the face of an unreasonable error. It certainly does not apply to the relation of γ Draconis to Lysimachia, where the intent was to establish a scientific conclusion by means of observation. In my opinion, the explanation does not apply in any other instance where I have seen it invoked. In the instances that I know, the result had already been established by the use of genuine data. I do not see why a person would go to the trouble of making up a false example as an illustration when the original, using the same principles and based upon real data, was available.

This applies with force to the present situation. As a lecture illustration, Posidonius' method is identical with Eratosthenes'. In both cases, a difference in latitude is established, along with a distance, and the values are used in Equation III.2. Eratosthenes' numbers must surely have been available to Posidonius. Why, then, should he invoke a different set of numbers unless he thought he was doing something scientifically new, and not just giving an illustration with made-up numbers?

Ptolemy does not use the size of the earth in his Syntaxis, but he does use it in his Geographikes [Ptolemy, year unknown, Chapter I.7]. There he takes the circumference to be 180 000 stades, making 500 stades to the degree. There is a reasonable basis, however, for believing that his stade was larger than that used by Eratosthenes [Fischer, 1975], and that it was possibly the Royal Egyptian stade of about 210 meters. If so, this makes the circumference equal to about 37 800 kilometers, which is not bad.

---

†This differs from the correct value, which is about $5\frac{1}{4}$ degrees. Most of the difference comes from refraction; a small amount comes from the rounding that I have used in the calculations. Refraction is fairly serious at low elevation angles; this is a third reason why Posidonius' method is inferior to Eratosthenes'.

In sum, Greek astronomers by the year -200 knew the size of the earth with an accuracy of about 10 per cent. However, the accuracy of their result is not the important point; their accuracy was limited by the measuring tools available to them. The main point is that they were able to stand outside the earth and look at it with the eye of the imagination, which is not limited.

It does not seem that there was any improvement in knowing the size of the earth in the seven centuries between Eratosthenes and the close of Greek astronomy.

5.   The Measurement of Longitude

The equator furnishes a natural reference from which to measure latitude. There is no natural reference† from which to measure longitude, so that the meridian called 0° longitude must be chosen arbitrarily and by agreement of interested parties. At present, we take 0° to be the longitude of a particular point in the Greenwich Observatory. Ptolemy took it to be the longitude of the Canary Islands‡; he chose them because they were the westernmost point that he knew. Thus, in Ptolemy's geography, the longitudes of all known points were east of the reference.

If we exclude the use of artificial earth satellites, we have two basic ways for finding longitude. One way is to measure an east-west distance on the ground; by using the radius of the earth, this can be converted to a difference in longitude. The other way is to measure the local time of a single event that can be timed at two places; by using the rate of rotation of the earth, this can be converted to a difference in longitude.

In modern times, both methods have been used. The first can be applied with high accuracy for points that are directly connected by reasonably flat land, so that the distance can be surveyed with high accuracy. It is hard to apply to distances between continents, or to points separated by rugged terrain.

In a typical application of the second method, a radio signal is sent from, say, the Greenwich Observatory at noon, Greenwich time. The time of its arrival measured at, say, the Naval Observatory in Washington is H hours before noon, Washington time. Then the difference in longitude is 15H degrees, approximately.‡

†By this, I mean a reference position that is defined precisely by some physical property of the earth.

‡More accurately, the islands called the Fortunate Islands, which may include Madeira as well as the Canaries.

‡Of course, a number of corrections must be applied to this simple result. An obvious correction comes from the fact that Greenwich and Washington are not at the same latitude. Another comes from the time that the signal takes to travel between the places. We do not need to go into the exact values of the corrections involved. By Washington time, I mean the true time for the meridian of Washington, not the time of the zone in which Washington is located.

-47-

In Greek times, only one kind of event could be used in the timing method. That is the lunar eclipse. A lunar eclipse begins when the moon touches the earth's shadow and ends when the moon loses contact with it. Anyone who can see the eclipse sees it begin and end at the same times. If two observers measure and compare the times in their own local time bases, they can find the difference in the time bases and hence the difference in longitude.

Islamic astronomers used the method with moderate success; al-Biruni [1025] uses it in his geodetic study, for example. The Greeks were aware of the possibility, and Ptolemy [year unknown, Chapter I.4] refers to it. However, only one example occurs in ancient literature, so far as I know: The lunar eclipse of -330 September 20 was observed both in Carthage and in Babylonia, and the local times were recorded. Hence it should have been possible to find the difference in longitude between these places. Unfortunately, different ancient writers give different values of the times, and this lone example could not actually be used.

Thus Greek astronomers and geographers had to fall back upon the measurement of distances in order to find longitudes. In only a few places, such as Egypt, had there been much attempt at a systematic measurement of distances. Almost all distances had to be estimated from travellers' reports, sailing times, and the like. Thus it is not surprising to find that the extent of Asia was considerably exaggerated. Ptolemy assigns 180° of longitude from the Canary Islands to a point that is probably in central China; the correct value is about 125°. However, other ancient geographers used a different value and, indeed, any value that they gave was just a matter of speculation; Ptolemy had no real basis for the far-Asian longitudes that he assigned. Some geographers put eastern Asia only about 45° west of Europe.†

Surprisingly, Ptolemy did no better for the well-known Mediterranean regions. He thought that the northern coast of Africa was almost straight, and he gives the Mediterranean from Marseilles to Algeria a width of about 11° in latitude; this is too large by about 50 per cent.‡ He has 62° in longitude from Gibraltar to the northeastern corner of the Mediterranean. This is also too large by about 50 per cent; the correct value is about 41°.‡

In summary, Greek astronomers and geographers were able to measure longitude differences with an accuracy of about

---

†Some students of Columbus believe that he subscribed to such a belief, perhaps out of wishful thinking. The reader who is interested in Columbus's ideas and their background should probably start by studying Morison [1942].

‡However, it is about right for the difference in latitude between Marseilles and Alexandria. These are among the few points whose latitudes had been carefully measured (Section III.3).

‡Bunbury and Beazley [1911] give a good short summary of the main features of Ptolemy's geography.

50 per cent in the parts of the world that were fairly well known to them. They could only guess about poorly known places, of course. Their performance in measuring latitude was better, in spite of the example of northwest Africa, because latitude can be estimated by purely local observations of a fairly simple character, such as the maximum elevation that the sun obtains at the summer solstice.

6.   The Motion of the Earth

The most obvious astronomical motion is the diurnal rotation of the heavens with respect to the earth, which gives us the alternation of light and dark. The next most obvious one is probably the motion of the moon with respect to the sun, which gives the succession of lunar phases from new moon through first quarter to full moon through last quarter and back to the new moon. This is probably more obvious than the motion of the sun with respect to the stars, which gives rise to the succession of seasons.†

The succession of lunar phases can be explained by postulating that the moon revolves around the earth. This explanation was adopted at an early time in Greek astronomy and we still accept it today. The other motions were the subjects of lively controversy among Greek astronomers, who ended up by adopting positions opposite to those of modern astronomers.

The Greek astronomers recognized at an early stage that only the relative motion matters to astronomy. With regard to the diurnal motion, they recognized that one may say that the earth stands still while the heavens rotate around it, or that one may say that the heavens stand still while the earth rotates within it. Similarly, as they recognized, one may say that the sun stands still while the earth revolves around it once per year, or one may say that the earth stands still while the sun revolves around it once per year.‡ Either description saves the astronomical phenomena.

Before we look at Greek beliefs about the motion of the earth, let us look at a distinction that is important in modern science, the distinction between kinematics and dynamics. Kinematics is the subject whose field is the description of motion, while dynamics is the subject that studies the relation between motion and force. When we are interested in dynamics, we say that the earth revolves

---

†This statement about the seasons is not quite correct, but it served Greek astronomers for several centuries and it will serve us until we come to Chapter V.

‡Astronomers make a distinction between rotation and revolution. They use a rotation of a body to mean an angular motion that can be described as taking place about an axis through the center of mass of the body; thus the diurnal angular motion is a rotation. They use a revolution of a body to mean an angular motion that is described as taking place about some other axis; thus the annual motion is a revolution.

around the sun.† When we are interested only in kinematics, we usually say, even in modern astronomy, that the sun revolves around the earth. This emphasizes the principle of the relativity of motion, which is a principle as old as the Greeks. We are free to adopt the description that is simplest, and the explanation that is simplest depends upon what we are trying to do.‡

We saw in Section I.1 that many Greek philosophers and scientists made a distinction between astronomy and physics. The role of astronomy was simply to save the phenomena, simply to describe the appearances in the heavens. Since we can save the phenomena equally well by assuming that the earth is either at rest or in motion, the question of the earth's motion, either of rotation or revolution, was not a question for astronomy. It was a question either for physics or for philosophy; in modern times, it is a question for philosophy only.

Many Greek philosophers gave the impression in their writings that they understood thoroughly how the mind of the Deity worked, and they had no trouble in answering the question: The earth was completely at rest in the center of the universe.

The practioners of physics, who lacked an intimate knowledge of the Deity, had more difficulty in answering the question, but most of them came to the same answer. Their reasons were basically dynamical, but their understanding of dynamics was drastically different from ours. To us, the most fundamental principle in dynamics is that it takes a force to change a state of motion. To them, the most fundamental principle was that it took a force to maintain a state of motion. Thus, even if they went so far as to grant that the earth might be in motion, they could not really grant that houses, the trees, the air, the clouds, and so on, would share in this motion. To them, if the earth were rotating from west to east, in the manner that Milton describes at the head of this chapter,‡ everything else on earth would be

†More accurately, we say that everything in the solar system moves around the center of mass of the system. Since the sun is so massive, the center of mass is close to the center of the sun.

‡The expert in celestial mechanics will recognize that these remarks about the kinematics of the solar system are not rigorously accurate. There are some small effects, which amount to seconds of arc, that make it easier to say that the earth goes around the sun, even from the purely kinematic point of view. Since these effects can be seen only with the telescope, the discussion is correct if we deal only with naked eye observations.

‡In this quotation, Raphael is explaining celestial motions to Adam. He explains to Adam that the earth may be rotating from west to east, and also explains many other matters in celestial mechanics, such as the heliocentric theory, while cautioning Adam that he would do better to search after more important matters. Milton was apparently much interested in the Galileo controversy, and even visited Galileo in his nominal house arrest in 1638.

flying from east to west.

Ptolemy [ca. 142], for example, in his Chapter I.7, changes from astronomer to physicist and writes: "Certain ones . . . believe that no evidence contradicts, for example, that, the heaven being still, the earth turns on its axis once per day from west to east, or, if both turn, it is about the same axis and, as we have said, in a manner that is consistent with the relations that we observe between them. It is true that, considering only the appearances of the stars themselves, nothing keeps things from being thus, for simplicity; but such ones do not understand how, in view of what happens around us and in the air, their assumptions are ridiculous." He then goes on to explain the east-west motions in the air that would result.

However, Ptolemy does not stop here, and I am not sure that I understand the conclusion of his Chapter I.7. If we imagine that the atmosphere is somehow carried along with the earth, as I understand him to say, the bodies contained in the atmosphere would still be flying backward (westward). Finally, if we should imagine somehow that bodies in the atmosphere are carried along as if they made one body with it, neither the atmosphere nor the bodies would outstrip the other. This means that bodies which fly or are thrown would always remain in the same relative positions. This clearly does not happen, hence bodies cannot be entrained as one body with the atmosphere, and hence the earth cannot rotate. This is clearly a weak argument if it has been rendered correctly.

Whatever Ptolemy meant by this passage, he certainly teaches that the earth is a sphere, totally at rest, in the center of the universe.

The Greek astronomers coupled the motion of the earth with the question of whether the universe is finite or infinite. If the earth is at rest while the universe rotates around it, the universe must be finite. Otherwise, the parts of the universe at an infinite distance would have an infinite velocity, which is impossible. Hence, a stationary earth implies a finite universe. I cannot follow the argument. I do not see how one can accept a possible infinity in distance while denying a possible infinity in velocity.

The specific arguments that I have quoted directly concerned the rotation of the earth. By obvious extensions, they apply also to its revolution.

In sum, Greek astronomers, physicists, and philosophers had the imagination to develop the concept that the earth is in motion, but they ended up by rejecting it and teaching that the earth is absolutely at rest at the center of the universe and of the solar system. Some of the reasons advanced for this conclusion were physical and some were philosophical. We may suspect that the philosophical considerations were the dominant ones, and that the physical arguments were rationalizations of philosophical arguments, or even of theological ones.

THE FABRIC OF THE HEAVENS

> He his fabric of the heavens
> Hath left to their disputes - perhaps to move
> His laughter at their quaint opinions . . . .
>
> John Milton[†]

1.   The Shape and Texture of a Heavenly Fabric

It is not at all necessary to construct a fabric of the heavens in order to have a useful astronomy. The Babylonians, by about -300, had developed elaborate methods of calculating the positions of the sun, moon, and planets, and many of their methods of calculation have been preserved and published [Neugebauer, 1955]. The methods were not based upon any assumed structure or fabric of the heavens. They were purely mathematical in nature, and were rather like the traditional method of calculating tide tables.

If a person observes the tide at a particular place, he soon discovers that it oscillates through two full cycles during a lunar day.[‡] He can represent this oscillation by what is called a sinusoidal function, which he can calculate and compare with the measured tide. He will find that there is a considerable amount of tide left over after he allows for this sinusoidal function. He will probably observe next that the amount left over tends to oscillate twice per solar day, and he will calculate and subtract out an oscillation of this nature. He will still find that there is an amount left over, he will find an amplitude and frequency of another oscillation, and so on. After each oscillation that he allows for, he finds an ever-decreasing amount or residual (tide left over), which he represents by an oscillation of ever-decreasing size, if he has done his work correctly. He can keep this up until he gets tired, or until he can calculate the tide at any time as accurately as he wants to. He can do this without having any idea of the forces that raise the tide; that is, he can develop mathematical methods for calculating the tide without having any theory or conceptual structure of the tide.

The Babylonians did a similar thing in astronomy. They started by assuming that the moon, for example, moves uniformly through the heavens. When they allowed for this uniform motion, they found departures that can be as large as about 7°.5. The departures could be broken down into various oscillations of different periods, and, by adding up a number

---

[†]Paradise Lost, Book VIII, lines 76-78.

[‡]A lunar day means the interval from one time when the moon passes through the meridian of the observer to the next time it does so. The statement that there are two cycles of the tide per lunar day is true for most places, but not for all.

of oscillations, they could represent the motion as accurately as they wanted to, or as accurately as their observations allowed.

The Greek approach, in the times that interest us, was quite different. The Greek astronomers, like modern astronomers, postulated that the heavenly bodies obey certain laws or models that can be put into mathematical form. The detailed mathematical description of the motions then comes by working out the consequences of the basic laws, using the rules of mathematical analysis to pass from the mathematical statements of the laws to the mathematical representation of the motion.†

The first problem that the Greeks faced in building a fabric of the heavens was whether the solar system was centered upon the earth, the sun, or neither. As we saw in Section III.6, the Greek astronomers ended up by putting the earth at the center of the system, but this was not their only system. Oddly, in one of the earliest Greek systems of the heavens, neither the earth nor the sun was at the center. Instead, the earth, the sun, the moon, and the planets all revolved around a central fire. The earth revolved around the central fire once per day and kept the same face turned toward the fire as it did so. In this system, only the part of the earth that could not see the fire was habitable. The inhabitants of the side turned away from the fire saw all parts of the heavens in turn during a day, so that the postulated daily revolution produced the apparent diurnal rotation of the heavens.‡

There were several other Greek systems of the world. One of them was due to Aristarchus of Samos, who is credited with measuring the time of the summer solstice of -279. Since Aristarchus's work has been lost, we do not have all the details of his system, but it was probably a full heliocentric system. From what other Greek writers said about it, we can be sure that the earth performed an annual revolution around the sun while spinning daily on its axis in Aristarchus's system. However, his system found few people who mentioned it and even fewer who advocated it.

We may pass over other systems and go to the one that last occupied the field of intellectual combat among the Greeks. This is a geocentric system, which is often called

---

†I think it is fair to say that the basic laws in Greek astronomy were geometrical rather than physical. That is, Greek astronomers postulated that the heavenly bodies move in certain geometrical patterns instead of saying, as we do, that they move in response to certain physical forces.

‡This system, though strange in many ways, still causes us to admire the imagination that lay behind it, the imagination that allowed its makers to stand outside the system and to envisage the consequences of unorthodox points of view. A development of the Pythagorean school, it is described in considerable detail by Dreyer [1905], in his Chapter II.

the epicycle system. It will be the subject of the next
section.

2. The Epicycle System

The hardest part of understanding the epicycle system
is learning the vocabulary so that it comes naturally.
Otherwise, the system involves only simple concepts and is
not hard to understand.

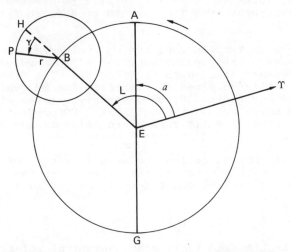

Figure IV.1   The simplest form of the epicycle system. E denotes the observer and the line
E ϓ is a reference line from which angular positions are measured. The circle with center at E,
that passes through ABG, is called the deferent. In most discussions, we take its radius BE to
be unity. B travels uniformly around the deferent in the direction of the arrow. P is the body
whose position is to be represented. At the same time that B is travelling uniformly around
the deferent, P travels uniformly around the circle whose center is B. This circle, whose
radius is PB, is called the epicycle. The line BH is a construction line used in discussion; it is
the extension of the deferent radius BE.

The basic parts of the system, which is essentially
just a geometrical pattern, are shown in Figure IV.1. The
epicycle system is used to describe the motion of a body in a
plane, and it must start with a reference line. This reference is the line Eϓ in the figure.† E is the position of
the observer. The circle that passes through points A, B,
and G is called the deferent circle, or simply the deferent,
and its center is at E. Greek astronomers used the sexagesimal notation and they found it convenient to take the
radius BE of the deferent to be 60. We work mainly in

---

†The symbol ϓ is the drawing of a pair of ram's horns, and
as such it is the symbol for the constellation Aries. Later
it came to be the symbol for the vernal equinox also. Specifically, it denotes the position of the sun at the instant
of the vernal equinox. Thus it is the line of intersection
of the ecliptic plane with the equatorial plane. For the
moment, however, I am using ϓ simply to denote any convenient reference.

decimal arithmetic, and we shall take the radius to be unity, unless we are reviewing a Greek discussion of the system. The point B moves uniformly around the deferent in the direction shown.

The system also involves a second circle whose center is B. P is the body whose position we want. At the same time that B moves uniformly around the deferent, P moves uniformly around the second circle, which is called the epicycle. The radius of the epicycle will be denoted by r; r is one of the parameters that are to be deduced from observations.

A modern writer who uses the epicycle system would probably measure the angular position of the radius PB from a fixed reference line, such as EΥ or EA. The Greeks, however, customarily measured it from the line BH, which is the extension of the radius BE; BH obviously rotates at the same rate that B moves around the deferent. PB may rotate in either direction, and its rate of rotation is another parameter that must be found from observations.

The angle BEΥ is called the mean longitude of P, because the angular position of B, on the average, is the same as that of P. BEΥ will be denoted by L, and it is a linear function of time:

$$L = L_0 + n(t - t_0).$$ (IV.1)

In this, t is time and t_0 is some convenient reference epoch, n is called the mean motion (mean angular velocity), and L_0 is called the mean longitude at the epoch. L_0 and n must be found from observations.

The Greeks called the angle HBP, measured from HB to PB in the direction in which P moves around the epicycle, by the name anomaly. We shall use γ to denote the anomaly. It is also a linear function of time:

$$\gamma = \gamma_0 + \gamma'(t - t_0).$$ (IV.2)

The quantities γ_0 and γ' will not be given names for the moment. They are obviously parameters that must also be found from observations. Altogether, then, there are five parameters that must be found from observations. They are L_0, n, γ_0, γ', and r.†

The Greek astronomers used the epicycle system in two different ways which, from the modern viewpoint, fill two quite different functions. In the first way, the epicycle radius PB always remained parallel to some fixed line AG, with P always being, say, directly above point B in the figure. P is then obviously farthest from E when B is at the point A on the deferent, so the direction of AE is

---

†If we refer the line PB to some reference line other than HB, we have some other linear equation in place of Equation IV.2. We do not use γ_0 and γ' if we do this, but we have two other parameters to take their places.

called apogee. The angle AE♈, taken positive in the counterclockwise direction, is called the longitude of apogee; it
will be denoted by a. P is closest to E when B coincides
with G, so EG is the direction of perigee.

The anomaly (angle PBH) now increases in the direction
opposite to the sense in which B moves, and in size it is
equal to angle BEA. We can thus identify angle BEA with the
anomaly γ. Clearly we have

$$L = a + \gamma. \tag{IV.3}$$

Here we are measuring the anomaly from apogee, which was the
Greek custom. Measuring it from perigee is more common in
modern astronomy.

We must now introduce more terminology. The angular
position of P from the line E♈ will be called the longitude
λ. This would be the angle PE♈ in Figure IV.1 if we had
drawn it in. We also introduce a quantity e_c that we call
the equation of the center. This term originated in medieval
times. Ptolemy calls it the προσθαφαιρεσις. The translators
of Ptolemy that I have seen do not translate this term; they
merely spell it using the letters of their own alphabet. e_c
is the difference between the longitude λ and the mean longitude L:

$$\lambda = L + e_c. \tag{IV.4}$$

In the epicycle system that we are discussing, we can easily
find e_c by using the law of sines for a plane triangle:

$$e_c = \tan^{-1}[-r \sin \gamma/(1 + r \cos \gamma)]. \tag{IV.5}$$

To see the significance of this, let us develop e_c as
a series of powers of r. Let us then drop all powers above
the second and add the result to L in Equation IV.4. We
find

$$\lambda = L - r \sin \gamma + \tfrac{1}{2}r^2 \sin 2\gamma, \tag{IV.6}$$

if angles are in radians rather than degrees.† Now the orbit
of a planet round the sun is, to high accuracy, an ellipse
with the sun at one focus, as shown in Figure IV.2. Here S
is a focus of the ellipse APG, AG is its major axis, and S♈
is the reference direction. G is the point of closest approach, and the angle GS♈ is called ω, the argument of perihelion.‡ Let the distance CG be unity and let the distance
CS be called the eccentricity e of the ellipse. The angle
PS♈ is the longitude λ of P.

Any text on celestial mechanics will show that λ can be

---

†See Section 1 of Appendix A for the meaning of a radian.

‡If the word planet is changed to sun, if sun is changed
to earth, and if perihelion is changed to perigee, this
discussion also applies to the apparent orbit of the sun
around the earth.

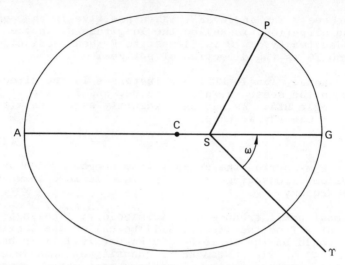

Figure IV.2   An elliptical orbit. S, the center of gravitation, is
at one focus of the ellipse, and G is the point of closest approach
to S. S ♈ is the reference direction, and GS ♈ is called ω, the longitude of perigee (or perihelion). Angle PS ♈ is the longitude of P.
If the distance CG is taken as unity, the distance CS is the
eccentricity.

written as†

$$\lambda = M + \omega + 2e \sin M + (5/4)e^2 \sin 2M, \qquad (IV.7)$$

in which M is a certain linear function of time. We cannot
directly compare Equations IV.6 and IV.7 because anomaly is
measured from perigee in Equation IV.7 and from apogee in
Equation IV.6.   In order to compare the equations, we let
$\omega = \pi + a$.‡   Then M + ω becomes M + π + a.   In view of
Equation IV.3, we can now identify M + π with the anomaly
γ, which is also a linear function of time.   When we make
these changes, Equation IV.7 becomes

$$\lambda = L - 2e \sin \gamma + (5/4)e^2 \sin 2\gamma. \qquad (IV.8)$$

If we identify r in Equation IV.6 with 2e in Equations
IV.7 or IV.8, the terms proportional to e are identical.
The terms proportional to e^2 differ by $(3/4)e^2 \sin 2\gamma$, which
has a maximum value of $(3/4)e^2$. For most objects in the
solar system, an angle of this size is too small to see with
the naked eye.   Thus Equation IV.6 gives the same longitude
as an ellipse, to the accuracy of observation that was usually available to Greek astronomers.

However, the epicycle model cannot agree with an

---

†Moulton [1914] is readily available and gives all the information about celestial mechanics that we need in this
work.

‡Angles are still in radians here, and π radians is the same
as 180°, or half of a circle.

ellipse in both longitude and distance. It is clear from Figure IV.1 that the distance to P varies from 1 - r to 1 + r in the epicycle model, while Figure IV.2 shows that the distance varies from 1 - e to 1 + e in the ellipse. If we equate r to 2e, the epicycle gives twice as much variation in distance as the ellipse does. If we equate the variations in distance by setting r equal to e, the epicycle then gives only half the correct variation in longitude.

Several writers including myself have said that Greek astronomers could not study the distances to the heavenly bodies with useful accuracy, and thus they did not realize this basic defect in their most important geometrical model. The statement about distances is correct with regard to the sun. However, in the course of writing this work, I have realized that it does not apply to the moon and the planets. As we shall see in later chapters, Greek astronomers could measure accurately the variation in the distance of a planet, but they could not measure its average distance. They could measure accurately both the average distance of the moon and its variation. Thus they had the information needed to show that the epicycle model is basically defective for both the moon and the planets.

The variations in longitude and distance are reconciled by the ellipse and by several other models. Greek astronomers developed one such model, which I shall describe in Section IV.5, and they applied it to all the planets except Mercury. Oddly, they never applied it to the moon, for which they had the most information.

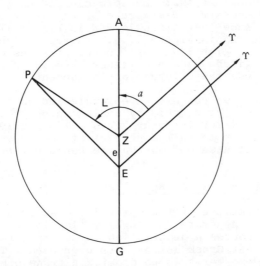

Figure IV.3   The eccentric. P is the object to be observed.
It moves uniformly around the circle APG whose center is Z.
The observer is not at Z but at the point E. The distance ZE is
called the eccentric distance or the eccentricity e. The direction
from Z to ♈ or from E to ♈ is the reference direction; and the
angle PE ♈ is the longitude λ of the object P.

At the close of Section IV.1, I used the epicycle system to denote a whole complex of systems, which involve elements beyond the epicycle and deferent. It would stagger the reader to introduce all of these in one chapter, and I shall not do so. However, it is best at this point to introduce two complications. We cannot even begin the detailed study of Greek astronomy without mastering the first of these, and we must master the second before we study the Greek treatment of the planets. I shall take up the first of these before I take up the second way in which the Greek astronomers used the epicycle model of Figure IV.1.

3. The Eccentric

The eccentric, in the Greek meaning of the term, is shown in Figure IV.3. In this figure, we have a deferent circle APG without an epicycle. The object P to be observed moves uniformly around the deferent circle, whose radius is unity† and whose center is Z. The observer's position E is no longer at the center of the deferent. Instead, it is offset by a distance ZE that we shall call the eccentric distance or the eccentricity e. Because of a paucity of familiar letters, I use the same letter that I used for the eccentricity of an ellipse in the preceding section. As before, the direction to ♈ is the reference direction. The angle PZ♈ is now the mean longitude L, which is a linear function of time. The angle PE♈ is the longitude λ. A is clearly the position at which P has the greatest distance, and the angle AZ♈ is the longitude of apogee, a.

Now go back to Figure IV.1, imagine that P is always directly above B, and imagine that the triangle PBE has been drawn in. Suppose that the epicycle radius r of Figure IV.1 equals the eccentric distance e of Figure IV.3. It is elementary to show that triangle PBE in Figure IV.1 is congruent to triangle PZE in Figure IV.3. Therefore the epicycle and the eccentric give identical representations of the motion, both in longitude and in distance, and it is indifferent whether we use the epicycle or the eccentric.

In fact, we can use an epicycle and an eccentric in the same picture. We do this by adding an epicycle circle drawn around the point P in Figure IV.3. If we take the parameters of the epicycle to be independent of the eccentric, this gives a more complicated motion than we could get by either device alone.

4. The Second Use of Epicycles

The first use of the epicycle, discussed in Section IV.2 above, is to generate an approximation to an elliptical orbit; this is of course a description in modern terms and not a description that Greek astronomers ever used. The second use, also in modern terms, is to provide a transformation from heliocentric coordinates to geocentric ones.

---

†Greek astronomers took the radius to be 60.

Suppose for the moment that the orbits of all of the planets around the sun, including the earth's, were exact circles. We of course never see the heliocentric positions of the other planets; we must view them from earth. Under the condition stated, we can picture the sun as going around the earth in a circle. Another planet, Venus say, goes around the sun in a circle. Then the sun becomes point B in Figure IV.1, and the orbit of the sun around the earth is the deferent circle. The heliocentric orbit of Venus is then the epicycle. If the orbits involved were circular, the representation of Figure IV.1 would be exact.

Point B would now move around the deferent circle in a year, and point P would move around the epicycle in 0.615 years when its position is measured from a direction fixed in space; 0.615 years is the heliocentric period of Venus. Put another way, B moves 0.98565 degrees per day with respect to the direction E♈, and P moves 1.60217 degrees per day with respect to the direction E♈. However, the Greek astronomers measured the motion of P with respect to the moving point H, which moves at the same rate as the sun. Thus what the Greeks called the anomaly γ increases by 1.60217 - 0.98565 = 0.61652 degrees per day.†

If we kept the same picture for the outer planets, the epicycle circle would be larger than the deferent circle. This would be legal, but it is not what the Greeks did. They tacitly took the epicycle always to be smaller than the deferent. The deferent and epicycle are still equivalent to real orbits, however. To see this, say for Mars, imagine for the moment that E is Mars while B is the sun. That is, the deferent is the orbit of the sun as seen from Mars, and the epicycle is then the orbit of the earth around the sun. This would be an Areocentric picture of the earth's orbit, with E as Mars and P as earth. To get the position of Mars as seen from earth, we add 180° to both angle BE♈ and angle PBH.

In this picture, point B moves around the circle in a Martian year, and hence it moves at the rate of 0.52407 degrees per day with respect to the reference line E♈. Point P moves around its circle at the rate of 0.98565 degrees per day with respect to E♈. Thus what Greek astronomers called the anomaly of Mars, and what a Martian astronomer would have called the anomaly of earth, increases by 0.98565 - 0.52407 = 0.46158 degrees per day.

In real life, all the planetary orbits are ellipses rather than circles, and this makes the planets deviate from the simple model of Figure IV.1 by amounts large enough to be seen with the naked eye. The Greek astronomers knew that Figure IV.1 was not adequate, although they did not word the matter in terms of ellipses, of course. The only Greek improvements over Figure IV.1 that we know of for the planets are those given by Ptolemy [ca. 142]. If there were others, they have not come down to us in the surviving literature.

---

†These numbers come from modern theories of planetary motion and are not exactly the ones used by the Greeks.

The variant of the epicyclic system that Ptolemy used for Venus, Mars, Jupiter, and Saturn is usually called the equant model, which will be described in the next section.

5. The Equant Model

The point E in Figure IV.1 serves three functions that can be identified. It gives the position of the observer, it is the center of the deferent circle, and it is the center of uniform rotation. In the model of Figure IV.3, the function of giving the position of the observer was moved to point E, but the other two functions were still left to the

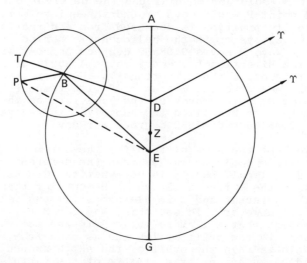

Figure IV.4   The equant model. The deferent circle ABG has its center at Z, and the radius BZ (not drawn) was taken as 60 by Greek astronomers; we take it as 1. Point B always stays on the deferent, but its rotation is not uniform about Z. Instead, it is uniform about D. That is, the angle BDΥ is a linear function of time. The observer is at E. A planet P moves uniformly around the circle whose center is B, and the angle PBT is called the anomaly γ. The angle PEΥ is the longitude of the planet. In practice, Greek astronomers assumed that the distances DZ ($= e_2$) and ZE ($= e_1$) are equal. When this assumption is made, the model is called the equant model, and point D is called the equant to point E.

point Z. In the model of Figure IV.4, the three functions are assigned to the three different points D, Z, and E. E is still the position of the observer; that is, it denotes the earth. Z is the center of the deferent circle ABG. B always stays on the deferent, but its rotation is uniform about D, not Z. That is, the angle BDΥ, which is called the mean longitude L of the planet, is a linear function of time.

P, as before, moves uniformly around the epicycle whose center is B. The anomaly γ is measured from the line BT, which is the extension of DB. That is, the anomaly is the angle PBT.

The planet is at apogee, that is, it is farthest from the earth, when point P is on the extension of the line EB. Similarly, the planet is at perigee when P is directly between B and E. However, when Ptolemy refers to the apogee or perigee of a planet, he does not mean these configurations. By apogee, he means the point A, which is the point on the deferent that is farthest from the earth E, and the angle AD♈ is the longitude a of apogee. Similarly, perigee in his usage means the point G.

To start with, there is no obvious relation between the distances ZE and DZ, so I shall give them different names. I shall call ZE the first eccentricity e_1 and I shall call DZ the second eccentricity e_2. Ptolemy uses observations to prove that $e_1 = e_2$, but, as I shall show in Section XI.5, his proof is seriously wanting, even by the standards of his own time. When the equality $e_1 = e_2$ is imposed upon the model, it is called the equant model, and the point D is called the equant to E, or simply the equant.

Ptolemy applies the equant model to Venus and the outer planets but not to Mercury. He uses a more complex model for Mercury, and I shall defer its description to Section X.1. However, we can easily see here why he needs a different model for Mercury.

We shall see in Section XI.1 that the use of two eccentricities gives us a better approximation to an elliptic orbit† than we can get from the single eccentric model of Figure IV.3. Therefore the deferent gives us a good description of the orbit of the sun; we remember that this is the orbit of the sun around the earth when the planet is Venus and the orbit of the sun around the planet when the planet is an outer one (Mars, Jupiter, or Saturn). The epicycle is then a heliocentric orbit. It is the heliocentric orbit of Venus itself when the model is used for Venus, and it is the heliocentric orbit of the earth when the model is used for the outer planets.

It happens that Venus and the earth have by far the smallest eccentricities of any of the planets. Thus a circular epicycle can give a fairly good result for Venus and for the outer planets. If we tried to use the same model for Mercury, however, the epicycle would have to represent the heliocentric orbit of Mercury, and it happens that Mercury has by far the largest eccentricity of all the planets. Thus the model of Figure IV.4 cannot give a good result for Mercury.

We need to summarize the notation that relates to Figure IV.4; we have already had to use most of it. The angle BD♈ is the mean longitude of the planet, which I denote by L with a subscript to indicate the planet. For the moment, when we are dealing only with a general planet, I shall use

---

†In particular, if the parameters are chosen so that the greatest oscillations in longitude are correctly represented, the greatest oscillations in distance are also correctly represented. The epicycle and a single eccentric cannot do this. See Section IV.2.

Lp for this angle. The angle AD⍦ is called a, the longitude of apogee. The angle TBP is the anomaly γ in the Greek usage, as we have already said. The distances ZE and DZ are the first and second eccentricities e_1 and e_2, and they have already been named. The distance PB is r, the radius of the epicycle. Now Lp and γ are linear functions of time, to wit:

$$L_p = L_0 + n(t - t_0),$$

$$\gamma = \gamma_0 + \gamma'(t - t_0).$$

(IV.9)

These have already been given in Equations IV.1 and IV.2, and are repeated for convenience.

Altogether we have the parameters a, e_1, e_2, r, L_0, n, γ_0, and γ', which are eight in number. However, as we shall see in Section IV.7, there are two relations among the eight, so that there are only six independent parameters. Two of these six are e_1 and e_2 and, if we require $e_1 = e_2$ as Ptolemy does, there are only five independent parameters for the equant model. These five must be found from the analysis of observations.

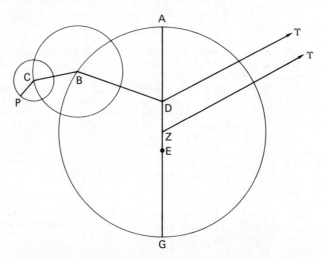

Figure IV.5   The double epicycle model. This consists of the model of Figure IV.4 with another epicycle added. Instead of having the planet P move around the circle whose center is B, a point C now moves uniformly around this circle. The planet P now moves uniformly around the second epicycle whose center is C. The model was never used in Greek times, although it is simpler than some which were. Models similar to it were introduced in late medieval times.

6.   A Model That Was Never Used

There is a modification of Figure IV.4 that seems obvious to me, but, so far as I know, it was never tried by the Greek astronomers. The modification is shown in Figure IV.5. It differs from Figure IV.4 only by adding a second epicycle, and we may call it the double epicycle model. In this model, the planet P does not lie on the circle with B as center. Instead, a point C moves uniformly around this circle. The planet P moves uniformly around a second epicycle whose center is C. Obviously, we do not have to have the equant model as the base for a second epicycle. We could just as well add an epicycle to the model of Figure IV.1, for example.

While this model was never used by Greek astronomers, at least not in any sources known to us, the second epicycle was introduced by Islamic astronomers around 1300 or 1350, and it was also used by Copernicus [1543, Chapter IV.3]. It is not known whether Copernicus knew of the earlier Islamic usage or whether he developed the second epicycle independently.

Although Greek astronomers did not use a second epicycle, the double epicycle model is both simpler and more accurate than some models which they did use. The remark has particular force with regard to the models that Ptolemy used for the moon and for Mercury, which will be described in Sections VII.3 and X.1. Further, the same model can be used for all the planets, instead of having to have a separate model for Mercury.

The model may seem obvious because we are conditioned to the idea of elliptical heliocentric orbits for the planets. From this viewpoint, we can easily see the functions of the various elements in Figure IV.5. Point B represents the sun. The deferent is the geocentric orbit of the sun if we are dealing with Mercury or Venus, and it is the planetocentric orbit of the sun if we are dealing with an outer planet. The use of the two eccentricities DZ and ZE allows us to describe the solar orbit with considerable accuracy. P must then move around B in an elliptical orbit, and use of the double epicycle gives us a reasonable approximation to this orbit.

Clearly, the direction of the radius CP is to be the direction of the appropriate aphelion as seen from the sun. We have seen that an epicycle cannot represent both longitude and radial distance on an elliptical orbit with high accuracy. Since the radius CP is sometimes "seen" from point E, we are interested in both the radial distance of P from B and the angular position of P as seen from B. This means that the length of CP must represent a compromise between accurate representation of angle and distance. Hence, the length of CP that will best fit the data probably lies between the eccentricity of the heliocentric orbit, using eccentricity in its modern meaning, and twice this value.

7.   Heliocentric Indications Known to the Greeks

I have already mentioned in Section IV.1 that Aristarchus of Samos, probably between -300 and -250, developed a system in which the earth made an annual revolution around the sun, and which was probably a fully heliocentric system. It is something of a puzzle that Greek astronomers did not adopt this system. There were two strong indications known to them, one from physics and one purely from astronomy, that the heliocentric system would save the phenomena in a simpler fashion

-65-

than the geocentric one. There were also some weaker indications that lead to the same conclusion.

The easier point to see is the one that comes from physics Aristarchus [-280], in his only surviving work, studied the sizes of the sun and moon, and I shall study his work in detail in Section VIII.2 and Appendix B. He concluded that the sun has about 300 times the volume of the earth,† so that the sun is presumably much more massive than the earth.

Greek physicists did not have a clear picture of inertia, but they were certainly aware that it is easier to move a small object than a large one. Thus, if the sun is 300 times as large as the earth, it is much more sensible to say that the earth goes around the sun than that the sun goes around the earth. This may have been the consideration that led Aristarchus to formulate a heliocentric system.

The second strong indication of a heliocentric system comes from the equant model of Figure IV.4, or even from the simpler model of Figure IV.1. Consider first an interior planet (Mercury or Venus) in Figure IV.4. The angle BD♈ is the mean longitude, call it L_p, of the planet. Now B is the position of the sun as seen from E. Thus angle BD♈ is also the mean longitude, call it $L_\odot$, of the sun. That is,

$$L_\odot = L_p \qquad \text{for Mercury or Venus.} \qquad (IV.10)$$

When Figure IV.4 is applied to an outer planet, the line from B to P is the line from the mean sun to the earth. That is, the angle between BP and D♈ is the mean longitude $L_\odot$ of the sun. This angle is composed of two parts. The first is angle BD♈, the mean longitude L_p of the planet. The second is angle TBP, which is the anomaly γ. That is,

$$L_\odot = L_p + \gamma \qquad \text{for an outer planet.} \qquad (IV.11)$$

Equations IV.10 and IV.11 follow immediately from a heliocentric theory, but there is no reason for them in a geocentric one. However, the Greek astronomers were well acquainted with these relations. Ptolemy, for example, uses them explicitly in Chapter IX.3 of the Syntaxis where he constructs some tables of planetary motion, but he does not comment upon their origin. He simply uses them as established relations that do not need comment.

In sum, Equations IV.10 and IV.11 between them represent altogether five relations. In a geocentric theory, they are just accidental relations that are found empirically. They result immediately and simply from a heliocentric theory, which thus represents a considerable simplification.

I shall describe two weaker indications in later parts of this work. In Section XI.6 we shall see a statement about a particular configuration of an outer planet and the earth.

---

†Ptolemy, in Chapter V.16 of the Syntaxis, finds a ratio of about 170. In this respect, as in so many others, there was no improvement in Greek astronomy after a very early period. The correct ratio is about 1 300 000.

This statement plays an important part in finding the parameters of the outer planets, and it has to be proved as a theorem which is not very obvious in the geocentric theory. In the heliocentric theory, in contrast, it is a mere tautology. In Section XII.4 we shall see the complex forms that the theories of the latitudes of the various planets took in the geocentric theory. In contrast, the latitudes obey a simple and obvious theory in the heliocentric picture and this theory is the same for all the planets.

We shall probably never know why the Greek astronomers rejected the heliocentric system, but we may speculate that the reason came from philosophy or even from theology.† To most Greek philosophers, the earth was the most important object in the universe, because it was the abode of man. To them, it was unthinkable that the earth, the center of the philosophical universe, should have any motion. Even to astronomers, this philosophical repugnance to the motion of the earth may have outweighed the simplification that it gave to astronomical theory.

Some Greek astronomers held a theory intermediate between the geocentric and heliocentric ones. In this theory, Mercury and Venus revolved around the sun, while the sun, moon, Mars, Jupiter, and Saturn revolved around the earth.‡ The revolution of Mars, Jupiter, and Saturn is to be interpreted in the sense that Figure IV.4, or some similar figure, represents their motion. Actually, though it is rarely pointed out, this theory is essentially the same as the geocentric theory, in the final Greek form that Ptolemy gave it.

Near the beginning of his Chapter IX.1, in which he starts the study of the planets, Ptolemy discusses the relative sizes of the various spheres. The largest is that of the stars, and the next are those of Saturn, Jupiter, and Mars, in descending order. The smallest is that of the moon. The remaining three are those of Mercury, Venus, and the sun. Some astronomers, he says, have placed the spheres of Mercury and Venus outside the sun, but he, along with most astronomers, puts Mercury and Venus inside the sun. This, as he points out, divides the planets which can have any angular distance from the sun and those which can never move far from the sun, and it is thus the most probable arrangement. However, as he further points out, there is no way to tell about the order of the spheres of the planets, because their distances can be found only from parallax, and no planet shows any measurable parallax.‡

---

†I mean Greek theology, not Christian, Jewish, or Muslim. Plutarch [ca. 90, Section VI] remarks that Aristarchus had been accused of impiety in suggesting that the earth revolves around the sun.

‡Dreyer [1905, Chapter VI] discusses this theory and some of its adherents.

‡Parallax refers to the way in which the direction to an object depends upon the position from which it is viewed. See Section VI.1.

When Ptolemy mentions the spheres of the sun, moon, and planets, he may mean spheres having the same radii as the deferents of those bodies.* The deferents of Mars, Jupiter, and Saturn are then larger than the deferent of the sun. For Mercury and Venus, he recognizes that B, the center of the epicycle in Figure IV.4, is always in the same direction as the sun, but he does not require B to be the sun, as the theory just mentioned would do. However, nothing that he says in the Syntaxis keeps B from being identical with the sun.

Further, when Mercury or Venus is actually nearest to the earth, the distance is (Figure IV.4) EG minus the epicycle radius BP. Observations determine the ratio of these distances, while the fact that neither Mercury nor Venus has a measurable parallax sets a minimum value to their difference. In consequence, in the Ptolemaic theory as stated in the Syntaxis, Mercury and Venus necessarily go around the sun in the sense that their orbits enclose the sun. The theory further allows, but does not require, Mercury and Venus to go around the sun in the sense that their epicycles have the sun as center.

During the course of a year, we do not see any changes in the arrangements of the stars with respect to each other, nor do we see any parallax of the stars.† This means one of two things: Either the earth does not revolve around the sun, or the distance to the stars is far greater than the distance to the sun. Ptolemy estimated the solar distance as about 1200 times the radius of the earth. Apparently most Greek astronomers found it unlikely that the stars could be enormously farther away yet, and thus they may have rejected the heliocentric hypothesis because they could not measure a stellar parallax.

8. The Physical Reality of Epicycles and Deferents

In a useful but oversimple distinction, let us speak of mathematical reality and physical reality. When we speak of the mathematical reality of epicycles, deferents, and the other paraphernalia of the epicycle system, we mean that there are certain curves in space along which certain motions take place. When we speak of their physical reality, we mean that there are tangible objects in space, and that these tangible objects control the observed celestial motions. We then have the interesting question: To the Greek thinkers, did the epicycles and so on have physical reality, did they have only mathematical reality, or did they have neither?

---

*He may also be thinking of two spheres for each planet, one lying at its least distance and one lying at its greatest distance. See the next section.

†We can measure stellar parallax with careful telescopic observations. The largest parallax known is for the star called Proxima Centauri, a star of about the 11th magnitude. Its parallax is about $0''.76$. See Section VI.1.

We cannot expect that all Greek astronomers, for example, had identical beliefs, and so we cannot give a rigorous answer to the question. However, I suspect that there was a majority opinion, or at least a plurality one. The answers that I am about to give were held by many people, but it should be recognized that there were probably many people with different beliefs. When I say "Greek astronomers believed" such and so, I mean only that many of them did; it would be tedious to keep noting exceptions.

To start with, Greek astronomers believed in the mathematical reality of the system. Since the astronomers used the system to calculate celestial positions, this statement seems almost a tautology. However, it is necessary to make it, because many modern writers on Greek astronomy say that the Greek astronomers did not accept even the mathematical reality of the system. Neugebauer [1957, pp. 195-196], for example, writes that Ptolemy "could not have doubted that the actual geocentric distances of the moon were very different from what his model required." To give another example, Dreyer [1905, p. 195] writes that Ptolemy's theory "could not claim to give the actual place of the moon in space, since it very grossly exaggerated the variation of the distance of the moon from the earth." These writers have apparently overlooked Ptolemy's observations dated 135 October 1 and 139 February 9. Ptolemy uses these observations, as well as some observations of lunar eclipses, to show that the distance to the moon is exactly what his model required. I shall analyze these observations in Sections VII.2 and VIII.5.

Thus Ptolemy, at least, claimed mathematical reality for his lunar system, and I have no doubt that other astronomers also believed in the mathematical reality of the epicycles and so on. There is no way to test a belief in mathematical reality for the planetary systems, because there was no direct way of measuring the planetary distances. However, if they thought that their devices had mathematical reality for the moon, there was no reason for them to doubt the same reality for the planets.

Many philosophers believed in the physical reality of the system as well as in its mathematical reality. They believed that the elements of the astronomical system were produced by physical, tangible, and impenetrable but transparent spheres. The idea of such spheres goes back at least to the homocentric spheres of Eudoxus, who wrote around -375. These were various spheres which rotated about various axes, carrying various astronomical elements with them in their rotations.† The system of Eudoxus had some success, until it was superseded, in the opinion of most workers, by the epicycle system. Aristotle was its best-known adherent. So far as we can judge from what they wrote, the advocates of this system thought that the spheres were real objects.

When the epicycle system was introduced, many philosophers apparently felt uneasy about it unless it could be reconciled with physically real mechanisms. Dreyer [1905,

---

†Dreyer [1905, Chapter IV] describes this system in considerable detail.

p. 160] describes one way of producing this reconciliation. In this reconciliation, the epicycle becomes the equator of a small sphere confined between two other spheres. One sphere would have a radius equal to ZB (Figure IV.4) minus the radius of the epicycle, and the other would have a radius equal to ZB plus the epicycle radius. The rotation of the two latter spheres would cause the epicyclic sphere, if we may call it that, to roll between them. This produces physically the motion that is described mathematically by the epicycle system.

I did not notice any passage in the Syntaxis which indicates that Ptolemy believed in the physical reality, as opposed to the mathematical reality, of his system. To be sure, there are a few passages, such as the one near the beginning of Chapter IX.1 that I have already mentioned, in which Ptolemy refers to the spheres of the planets, but in these places he may be using "sphere" merely as a convenient way to talk about the average distance to a planet or other body without attributing physical existence to it.

However, Goldstein [1967] has recently discovered both Arabic and Hebrew translations of a work that is now lost in the original Greek. The work is attributed to Ptolemy in the translations and there seems to be no reason to doubt the correctness of the attribution. It was certainly written after the Syntaxis, because it refers to the Syntaxis by name and it refers to quantitative results found there.

In this work, Ptolemy associates two spheres with each planet, including the sun and moon. The radius of one sphere equals the least distance of the planet from the earth and the radius of the other equals its greatest distance. These spheres are in motion, and their motions produce the observed planetary motions. Further, he assumes that the outer sphere of one planet just touches the inner sphere of the next planet, so that there is no empty space. From these assumptions, he derives the sizes of all the planetary spheres.

This system works remarkably well. It is worth spending some time on it because it bears on an important point in Ptolemy's solar and lunar theories, which I shall take up in Section VIII.7. Before we go to the quantitative details, we should note a feature of Ptolemy's attitude toward the system. In the last paragraph of Section 4,† after he says that the outer sphere of one planet just touches the inner sphere of the next planet in his system, Ptolemy says: "This arrangement is most plausible, for it is not conceivable that there should be in Nature a vacuum,‡ or any meaningless and useless thing. The distances of the spheres that we have mentioned are in agreement with our hypotheses. But if there is space or emptiness between the spheres then it is clear that the distances cannot be smaller, at any rate, than those mentioned." Thus Ptolemy leaves open the possibility that there is vacant space between one outer

---

†As the work is arranged in Goldstein's English translation, which I am using.

‡If so, what fills the space between the inner and outer spheres associated with a particular planet?

sphere and the next inner sphere, which would make a larger
universe than the one he is about to derive. However, I
think he clearly implies that there is no such vacant space
in his opinion, and that his scheme gives the true size of
the universe.

Ptolemy's scheme of sizes goes as follows: Dropping
fractions, he says, he has shown in the Syntaxis that the
least and greatest lunar distances are 33 and 64, respectively
(the earth radius is the unit of distance in all this discussion). Hence 64 is also the least distance of Mercury. He
has shown that the ratio of distances for Mercury is about
34/88, which gives 166 for Mercury's greatest distance. Similarly, the ratio for Venus is about 16/104, giving 1079 for
its greatest distance. Here there is a problem, because he
has shown that the least distance of the sun is 1160 rather
than 1079. Leaving this problem aside for the moment, the
sun's greatest distance is 1260, that of Mars is 8820, that
of Jupiter is 14 187, and that of Saturn is 19 865. This is
also the distance of the sphere that carries the fixed stars,
and hence it is the radius of the universe.

With regard to the gap between Venus and the sun, Ptolemy says, in Section 3 of his newly discovered work, that
here is a discrepancy which he cannot account for, because
the observations have led him inescapably to the distances
and ratios that he has stated. He speculates that he might
have made a slight error in the lunar distance. If he increases it slightly, he will thus increase the greatest distance calculated for Venus. At the same time, because of the
way he found the lunar distance, he will decrease the solar
distance, thus removing the discrepancy; I shall show this
connection between the lunar and solar distances in Section
VIII.7.

Ptolemy should have noted that the discrepancy between
Venus and the sun is a consequence of his arithmetic rather
than of errors in the lunar distance. To start with, the
greatest lunar distance according to the Syntaxis is 64;10
in sexagesimal notation rather than 64, as we shall see in
Section VIII.5. Passing over Mercury for a moment, the
ratio of the greatest and least distances for Venus (Section
XI.3) is 104;25 to 15;35, which comes out to be 6.700 535.
However, Ptolemy alters this to 104/16, which equals 6.5.
This change alone produces a change of about 34 in the greatest distance of Venus.

As we shall see in Section X.3, the ratio of the greatest to the least distances of Mercury that is found in the
Syntaxis is 91;30 to 33;4, which equals 2.767 137.† However,
Ptolemy says that the ratio he has found is 88 to 34, which
equals 2.588 235. It is easy to see how Ptolemy obtained
the values that he uses for Venus, but there is no explana-

---

†In stating these values, Goldstein [1967, p. 10] attributes
them to W. Hartner in a paper that I have not consulted.
In this paper, which was written in 1964, Hartner predicted
the existence of the work under discussion, and Hartner's
prediction led Goldstein to make his discovery.

tion of the values that he uses for Mercury. We can find no
reason why he should change 91;30 to 88, nor why he should
change 33;4 to 34; neither is a reasonable rounding or a
plausible copying error.

Use of the values found in the Syntaxis solves Ptolemy's
discrepancy, as we see from Table IV.1. This table gives
the ratios of the greatest to the least distances of Mercury

TABLE IV.1

SOME PLANETARY DISTANCES IN PTOLEMY'S SYSTEM

| Body | Ratio of greatest to least distance, from the Syntaxis | Greatest distance | |
		Derived from the Syntaxis	Stated by Ptolemy
Moon	---------	64.167	64
Mercury	2.767 137	177.558	166
Venus	6.700 535	1189.733	1079

and Venus that we have just found. Starting from the greatest lunar distance of 64;10 = 64.167, the table then gives
the greatest distances of Mercury and Venus that we derive
from these ratios. These distances are compared with those
stated by Ptolemy.†

Thus the greatest distance of Venus turns out to be
1190 rather than 1079. This is actually greater than the
least distance of the sun, which is 1160 in Ptolemy's scheme.
However, the greatest distance to Venus can be reduced to
1160 by unimportant changes in planetary parameters. It is
clear that all the arithmetic must be done with great care
if Ptolemy's scheme is to make sense, and it is also clear
that Ptolemy does not realize this fact.

On the basis of this newly discovered work, I think it
is safe to say that Ptolemy believed in the physical reality
of his system, in the sense that I defined this term above.
However, we should note that Ptolemy's scheme of distances
conflicts with his statement in Chapter IX.1 of the Syntaxis,
where he says that the planets have no measurable parallax,
which means that their distances are too great to measure.‡
Since Mercury in this new scheme is sometimes as close to
the earth as the moon is, and since the moon's distance can
be measured, the distance of Mercury can also be measured by
the same methods. Of course, this conflict does not contradict the idea that Ptolemy wrote both works. A person often
changes some of his ideas as he studies a subject in increasing depth.

---

†Goldstein notes that the distances stated by Ptolemy are not
consistent with the results found in the Syntaxis, but he
does not study the distances implied by the Syntaxis results.

‡We saw in the preceding section that this condition requires
the epicycles of Mercury and Venus to enclose the sun.

It is probably fair to say that belief in physically real celestial spheres was the orthodox but not unanimous belief in medieval Europe. When Brahe proved that the comet of 1585 was at planetary distances [Pannekoek, 1961, pp. 215-216], he did two important things. He proved that comets are not atmospheric or sub-lunar phenomena as Aristotle had taught; instead they belong in the realm of celestial phenomena. He also proved that an insubstantial thing like a comet can pass right through the celestial spheres, and thus he finally destroyed belief in their substantiality.

CHAPTER V

THE SUN AND RELATED PROBLEMS

1. The Seasons

Either the simple epicycle model of Figure IV.1 or the simple eccentric model of Figure IV.3 is capable of representing the sun's position with an accuracy that is better than 1' of arc. The models give exactly the same position if the parameters are chosen in the proper fashion, and it is arbitrary which model is used. The later Greek astronomers chose to use the eccentric for the sun, reserving the epicycle for more complex problems.

The parameters that are to be found are the eccentric distance e, the direction of apogee a, the mean motion n, which is equivalent to the length of the year, and the solar longitude at some specific epoch. The data for finding e and a are the lengths of the seasons.

The declination of a celestial body is the angular amount by which it is north or south of the equator. If the body is north, the declination is considered to be positive, and if it is south, the declination is considered negative. At the vernal equinox (Section I.3), the sun's declination is zero; at the summer solstice, it is a maximum and is equal to the obliquity ε; at the autumnal equinox, it is zero again; and at the winter solstice, it is a minimum and equal to $-\varepsilon$. The longitude of the sun changes by 90° between each of these events.

An observer who is north of the Tropic of Cancer can find the times of these events by following the elevation angle or the zenith angle of the sun, as in Section III.3. The zenith angle is to be measured when the sun is in the meridian plane (due south of the observer). When the zenith angle Z is a minimum, the time is the summer solstice, and

$$Z = \phi - \varepsilon. \tag{V.1}$$

When Z is a maximum, the time is the winter solstice and†

$$Z = \phi + \varepsilon. \tag{V.2}$$

In Equations V.1 and V.2, ϕ is the observer's latitude and ε is the obliquity. As we saw in Section III.3, ϕ is the average of the two values of Z (half of their sum) and ε is half of their difference.

---

†It is not likely that either solstice will come just at local noon, when the sun is in the meridian, but it is simple to allow for this. To give a simple example, suppose that Z has the same value on June 21 and June 22 of some year, and that it does not get this small on any day near this time. Then it is obvious that the solstice came at midnight between June 21 and 22 in that year.

When Z is halfway between the values given by Equations
V.1 and V.2, that is, when Z is equal to the latitude, the
time is either the vernal or the autumnal equinox; it is the
vernal equinox if Z is moving toward the minimum. Thus, with
no preceding knowledge of his latitude, of the obliquity, or
of the equinoxes and solstices, the observer can find all of
them from these simple measurements.

Refraction and other systematic effects cause some error
in these results, aside from accidental errors of observation.
At places such as Alexandria, the systematic error in Z at
the summer solstice is typically a few seconds of arc, and
the error at the winter solstice is typically about 2' of arc.
The error in ϕ and ε is thus about 1', and the error in the
time of an equinox is about 15 minutes [Newton, 1972]. If
there are no errors except the systematic ones, the correct
time is about 15 minutes earlier than the measured one for
the vernal equinox and about 15 minutes later for the autumnal equinox. There are no systematic errors of importance
in finding the times of the solstices.

TABLE V.1

GREEK DETERMINATIONS OF THE SEASONS,
COMPARED WITH VALUES FROM MODERN TABLES

Observer	Length in Days			
	Spring	Summer	Autumn	Winter
Euctemon[a]	95	92	89	89
Hipparchus[a]	$94\frac{1}{2}$	$92\frac{1}{2}$	$88\frac{1}{8}$	$90\frac{1}{8}$
Modern[b]	94.1	92.3	88.7	90.2

[a]According to Geminus [ca. -100].
[b]Calculated for the year -430 from the theory of
Newcomb [1895].

There were probably many Greek determinations of the
lengths of the seasons, but the only two that have come down
to us are those preserved by Geminus [ca. -100]. The earlier
determination is one that Geminus attributes to Euctemon.
Euctemon is believed to have worked closely with the astronomer Meton, who made a famous observation of the summer solstice in Athens in -431, and it may be that the measurements
attributed to Euctemon were part of their cooperative effort.†

†Geminus says that the seasons about to be given are those
according to both Euctemon and Callippus. Callippus worked
in Cyzicus, in Asia Minor, about a century after the time of
Meton and Euctemon. So far as I know, he is the first person who adopted $365\frac{1}{4}$ days (the Julian year) for the length
of the year. Since Callippus used the same lengths of the
seasons as Euctemon, it is probable that he adopted Euctemon'
seasons without making independent measurements, and I do
not count his seasons as the result of an independent determination.

The second determination was made by Hipparchus, who, we know, worked mainly on Rhodes. His known observations date from -161 to -127. The measurements of Euctemon and Hipparchus are summarized in Table V.1.

In the table, spring is counted as the interval from the vernal equinox to the summer solstice, and so on. Because of the small perturbations due to the gravitation of other planets upon the earth, the lengths of the seasons change slowly with time. The lengths labelled "Modern" in Table V.1 are calculated from the theory of Newcomb [1895], for the year -430. To the accuracy that we need, these are also the lengths in the time of Hipparchus.

In Chapter III.4 of the Syntaxis, Ptolemy confirms the values attributed to Hipparchus. There he says that Hipparchus found $94\frac{1}{2}$ days for the length of spring and $92\frac{1}{2}$ days for the length of summer. Later, he calculates that autumn and winter have $88\frac{1}{8}$ and $90\frac{1}{8}$ days, respectively, and he says that these agree with Hipparchus.

It is interesting to study the accuracy of the measured values, and I shall do so presently. First, however, I want to use them in deriving the parameters of the sun's orbit. Four parameters are to be found, namely a, e, γ_0, and γ'. For these purposes, we find it convenient to write Equations IV.3, IV.4, and IV.5 in the form

$$\lambda = a + \gamma - \tan^{-1}[(e \sin \gamma)/(1 + e \cos \gamma)]. \quad (V.3)$$

In this, the eccentric distance e has replaced the epicycle radius r, but the mathematical relations are unchanged. γ is a linear function of time:

$$\gamma = \gamma_0 + \gamma'(t - t_0). \qquad (V.4)$$

Let us take the time to be zero at the time of some specific vernal equinox, say that of the year Y, and let us set $t_0 = 0$ in Equation V.4. By definition, $\lambda = 0$ and $\gamma = \gamma_0$ when $t = 0$. Let N denote the sum of the lengths of the seasons; N = 365 days for the seasons attributed to Euctemon and N = $365\frac{1}{4}$ days for those attributed to Hipparchus.† At the equinox of year Y + 1, $\lambda = 360°$ and $\gamma = \gamma_0 + 360°$. Thus $\gamma' = 360/365$ if we are using the data of Euctemon, and $\gamma' = 360/365\frac{1}{4}$ if we are using the data of Hipparchus. We now have three parameters to find.

The lengths of the four seasons give us values of λ at five times. These times are the vernal equinox of year Y, the following summer solstice, autumnal equinox, winter solstice, and the vernal equinox of year Y + 1. Thus we start with five equations of the form of Equation V.3, at these five times. We found γ' by subtracting the equation

---

†Both Euctemon and Hipparchus knew more accurate values for the length of the year. In the considerations of this section, it is more important to maintain consistency with Table V.1 than it is to use the most accurate available value of the length of the year.

for t = 0 from the equation for the time one year later. Thus the equation for the time one year later has been eliminated, and we now have four equations for the three parameters a, e, and γ_0.

We cannot solve four independent equations for three unknowns. Since the four equations come from measurements of physical quantities, a modern scientist would use statistical methods to find the values of the three unknowns that give the best fit to the four equations. The Greek astronomers did not have these statistical methods, which have been developed only in the past century or so. They usually proceeded only by using just as many equations as unknowns.† Ptolemy, and probably Hipparchus before him, chose to omit the equation for the winter solstice. This is probably the reason why Ptolemy states only calculated values for the lengths of autumn and winter.

In the remaining three equations, it is possible to find a closed form for a and e, but it is necessary to solve for γ_0 by successive approximation. The procedures for doing this are straightforward but tedious, and there is no reason for giving the details. The value of γ_0 is needed for constructing ephemeris tables of the sun, but the parameters that interest us most are a and e. The results are‡:

$$a = 65\tfrac{1}{2} \text{ degrees}, \qquad e = 1/24 = 0.04167. \qquad \text{(V.5)}$$

We know now that the values of both a and e change slowly because of the gravitational perturbations of the other planets upon the orbital motion of the earth. The value of a calculated from the theory of Newcomb [1895], for the time of Hipparchus, is $65°.98$; thus Hipparchus did a good job of finding the position of apogee. The eccentricity of the solar orbit at the same time is 0.01755. To high accuracy, the value of e in Equation V.5 is twice the eccentricity if we use the term in its modern sense. Thus the value to compare with e in Equation V.5 is 0.03510. This comparison is not so good.

It is fairly simple to compare the times of the equinoxes that correspond to e = 0.04167 and to e = 0.03510. When we do so, we find that Hipparchus has the vernal equinox too early by about 7 hours and the autumnal equinox too late by the same amount. When the sun is near an equinox, its declination changes by almost exactly 1′ per hour. Hipparchus's errors mean that the sun was still about 7′ south of the equator when Hipparchus thought it was on the equator.

---

†Greek scientists had some idea of the merit of measuring a quantity several times and using some kind of average. Their average was not necessarily what we mean by the average; it might just be a number somewhere between the largest and smallest values. So far as I know, they did not extend this idea to several variables, perhaps because of the severe computational problems that an extension would have posed.

‡These are the values that Ptolemy finds in Chapter III.4 of the Syntaxis, using Hipparchus's data. I find $a = 65°.40$ and $e = 0.04137$ from the same data, in excellent agreement.

This means that his measurement of latitude was too large by 7', a surprisingly large error that cannot be explained by refraction. Further, according to Ptolemy [ca. 142, Chapter III.1], Hipparchus did not use a gnomon for finding the equinoxes.† Perhaps his instruments were not made or aligned with the accuracy that later astronomers could achieve.

We can make an independent estimate of the error that Hipparchus made in his latitude. A number of star declinations are among the few of his observations that have survived [Hipparchus, ca. -135], and Fotheringham [1918] has analyzed them. Fotheringham finds that Hipparchus's declinations are too large by 0°.073 (= 4'.4) on the average, which means that his value of the latitude was too large by the same amount.‡ This differs significantly from the error implied by his equinoxes. The inconsistency in his instruments suggests errors of several minutes in their construction or alignment, in addition to direct errors of observation.

In Chapter III.1 of the Syntaxis, Ptolemy says that he measured the times of the autumnal equinox in 139, of the vernal equinox in 140, and of the summer solstice in 140. Since he knows the length of the year with high accuracy, he can calculate the time of the autumnal equinox in 140. From the measurements, he finds that spring is $94\frac{1}{2}$ days long, that summer is $92\frac{1}{2}$ days long, and that autumn plus winter is $178\frac{1}{4}$ days. These, he points out in Chapter III.4 of the Syntaxis, are the same values that Hipparchus had found almost three centuries earlier (see Table V.1), and this proves that the apogee line of the sun's orbit always maintains the same position with respect to the equinoctial and solstitial points.

Actually, as we know from Newcomb's theory, the solar apogee moves by about 1°.72 per century with respect to the equinoctial and solstitial points. There are about 275 years from Hipparchus's measurements to the year 140, and the solar apogee moved nearly 5° in that time. That is, apogee was now at about $70\frac{1}{2}$ degrees rather than $65\frac{1}{2}$ degrees as it was in Hipparchus's time. Measurements with the accuracy that Ptolemy implies for his data would have shown this motion, and we must ask why Ptolemy did not find it.

The answer is simple and tragic. As I shall show beyond a reasonable doubt in Section V.4 below, the equinox

---

† He used a metal band curved into the form of a circle, placed with the circle parallel to the plane of the equator. Thus the instrument is sometimes called an equatorial ring. At the instant of an equinox, the northern part of the inner face of the ring is exactly shadowed by the southern part of the ring. At any other time, this part of the inside of the ring has light on either its upper or lower edge. If the inner face of the ring is suitably calibrated, it is possible to interpolate between observations made in the daytime in order to find the time of an equinox that occurred at night.

‡ On the basis of a more extensive analysis, I have found [Newton, 1974] that his declinations are too large by only 0°.049, approximately 2'.9.

and solstice measurements that Ptolemy claims to have made
were not made at all, but were fabricated from the very
theory which he claimed he was testing. That is, they were
calculated using Hipparchus's solar theory, and it was then
claimed that they were careful measurements which showed the
validity of that theory.

There is no reasonable question, I think, that this
situation means the occurrence of a scientific fraud. From
its mere existence, however, we cannot convict Ptolemy himself of the fraud. It is possible, for example, that Ptolemy had an assistant who was charged with making the detailed
measurements at the times and using the procedures that
Ptolemy specified. It is then possible that this assistant,
either from laziness or for other reasons, calculated the
data instead of observing them and thus abused Ptolemy's
confidence. I shall take up the question of who was the
author of this fraud later in this work. For the time being,
we must simply accept that Greek astronomy has been tragi-
cally blackened by this fraud and by its perpetuation in the
most famous work of Greek astronomy.

We can now study the accuracy of the data in Table V.1.

2. The Accuracy of the Equinoxes and Solstices

In Table V.1, we should first note that the total length
of the seasons attributed to Euctemon is 365 days while the
total attributed to Hipparchus is $365\frac{1}{4}$ days. In neither case
does the total represent the accuracy with which the length
of the year was known. Let us look first at the data of
Euctemon.

In the year -431, according to several ancient sources,
Meton introduced into Athens a calendar in which the length
of the year is 365 + (1/4) + (1/76) days. This calendar may
have been used for astronomical purposes, but it is doubtful that it was ever used as the civil calendar of Athens.†
Since Euctemon worked closely with Meton, he undoubtedly
knew that the year is appreciably longer than 365 days.
Therefore we may be sure that the seasons attributed to
Euctemon in Table V.1 have been rounded to whole days.

The situation with regard to Hipparchus's data is similar. Using data that will be presented in Section V.3,
Hipparchus concluded that the length of the year is 365 +
(1/4) - (1/300) days. However, he probably saw no way to
partition the fraction 1/300 among the seasons, and simply
let the seasons add up to $365\frac{1}{4}$ days.

---

†There is a large and controversial body of writing about the
calendars of Athens. In spite of the effort that has been
expended, we are not always able to give the Julian equivalent of a date given in an Athenian calendar. Meritt
[1961] summarizes much of the work that has been done on
the Athenian calendars, and there are of course later papers
on the subject. I have discussed in detail some of the conclusions given by Meritt and others [Newton, 1976, particularly Sections II.4, V.3, and VIII.4].

The root-mean-square error of the seasons according to
Euctemon is 0.78 days, while that according to Hipparchus is
0.38 days, about half the size. The length of a season is
found by subtracting two observed times, which are the observed times when the season began and ended. Statistically,
the root-mean-square error in a difference is $\sqrt{2}$ times the
root-mean-square error of the numbers subtracted. Thus the
root-mean-square error in an equinox or solstice is about
0.55 days for Euctemon and 0.27 days for Hipparchus.†

However, the latter calculation is based upon the expectation that equinoxes and solstices should have equal
accuracy. Since the measurements of equinoxes and solstices
involve different phenomena, there is no basis for such an
expectation. When we measure the time of a solstice, we
measure the time when the meridian elevation angle of the
sun has an extreme value, a maximum in the summer and a minimum in the winter. Near the extreme, the angle changes
slowly and finding the extreme requires the discrimination
of small changes. This means that the time is not well determined.

When we measure the time of an equinox, on the other
hand, we measure the time when the meridian elevation takes
on a specified value. Further, when the elevation is near
the specified value, it is changing most rapidly, and the
time of an equinox is well determined.

For these reasons, many writers on ancient astronomy
have said that the accuracy of a Greek equinox is much better
than the accuracy of a solstice. This confuses the concepts
of precision and accuracy. In order to make the difference
clear, suppose we measure the length of a line that is about
50 centimeters long, and that we do so with a meter stick
graduated in millimeters. We can easily estimate the length
to a tenth of a graduation, although there is a slight uncertainty in doing so. In a series of carefully made measurements, we expect a reasonable fraction to differ from the
mean by 0.1 millimeters. We expect few if any to differ by
0.2 millimeters or more. This scatter about the mean is
what we mean by precision; in the example, the precision is
about 0.1 millimeters.

However, if there are systematic factors which affect
all of a series of measurements, individual measurements may
not be within 0.1 or 0.2 millimeters of the correct value.
Suppose that our meter stick is actually 101 centimeters
long. Then each individual measurement of our line is in
error by 0.5 centimeters (5 millimeters), on the average.

---

†Euctemon's two largest errors are those for winter and
 spring, which are almost equal in size and opposite in
 sign. If we changed his vernal equinox to a time 1 day
 later, Euctemon's errors would actually be smaller than
 Hipparchus's. It is indeed unlikely that Euctemon would
 have made an error of a day in the vernal equinox without
 a corresponding error in the autumnal equinox. This suggests strongly that there has been a scribal error of 1
 day in transmitting his vernal equinox to us.

This is the accuracy of a measurement in the supposed experiment.

It is approximately correct to say that the precision
tells us the size of the random errors of measurement, while
the accuracy tells us the resultant of both the random errors and any systematic errors that may be present. The
accuracy can never be better than the precision and, in most
real situations, it is considerably worse.

Consider the measurement of a solstice again. We have
only to find the time when the meridian elevation of the sun
has an extreme. We do not have to measure the value at the
extreme position, and the time found is independent of errors
in construction or alignment, at least within reasonable
amounts. In fact, to the accuracy that concerns us, I cannot
think of any kind of error that affects a solstice in a systematic fashion. Here we have one of the rare occasions in
which the accuracy is virtually identical with the precision.
This is not so for the time of an equinox.

In order to find the time of an equinox, we must start
by measuring the elevation angles of the sun at the solstices
which we did not have to do to find the solstices themselves.
We must then construct the position which bisects these two
angles, and finally we must measure the times when the sun
passes these positions. Errors in measuring the solstitial
angles, and in constructing the bisector, make systematic
errors in the equinoxes, and the accuracy of an equinox is
thus inferior to its precision.

TABLE V.2

GREEK DETERMINATIONS OF SOLSTICE-TO-SOLSTICE
AND EQUINOX-TO-EQUINOX INTERVALS

Observer	Interval, days			
	SS to WS	WS to SS	VE to AE	AE to VE
Euctemon	181	184	187	178
Hipparchus	$180\frac{5}{8}$	$184\frac{5}{8}$	187	$178\frac{1}{4}$
Modern[a]	181.0	184.3	186.4	178.9

[a]Because of rounding, these add up to a length of the
year that is slightly too large.

In order to estimate the errors in equinoxes and solstices, we must study them independently. By adding the
appropriate seasons, we can find intervals between solstices
only and between equinoxes only.† The results are summarized
in Table V.2, which is based upon the data of Table V.1. In
the table, SS and WS mean the summer and winter solstices

†Most of the analysis that follows immediately is found in
Newton [1976, Section V.3]. Since this work will be cited
frequently, I shall henceforth designate it as APO.

while VE and AE mean the vernal and autumnal equinoxes. With
these abbreviations defined, I believe that the table is
self-explanatory.†

Look first at the solstitial intervals. Euctemon's
errors in them are 0.0 and 0.3 days, respectively, while
Hipparchus's errors are 0.375 and 0.325 days. Euctemon's
errors are actually smaller, but this may be in part an
accidental effect of the rounding that has occurred with
Euctemon's data. Certainly Euctemon's measurements are no
worse than Hipparchus's in spite of being three centuries
earlier. For safety, I shall assume that Euctemon's and
Hipparchus's solstitial errors are the same and equal to
0.35 days, which is Hipparchus's average error. Since the
interval is found by subtraction of two measured times, the
expected error in a single measurement of a solstice time is
$0.35/\sqrt{2} = 0.25$ days, or 6 hours.‡

There is confirmation of this estimate. Aristarchus of
Samos measured the summer solstice of -279 and found that it
came at 18 hours, local time, on -279 June 26.‡ Hipparchus
measured the solstice of -134 and found that it came at noon,
local time in Rhodes, on -134 June 26. Using the best estimates that we can make from the modern solar theory [APO,
Section V.3], we find that Aristarchus's time is too early
by about 7 hours and that Hipparchus's time is too late by
about 8 hours. These agree as closely as we can expect with
the estimate just made. In the rest of the study, I shall
use 7 hours as the expected error in a Greek measurement of
a solstice.

We should note explicitly that Euctemon achieved this
accuracy around the year -430. There was no improvement in
accuracy during the millenium that remained to Greek astronomy.

Now look at the intervals between equinoxes. Euctemon's
errors are larger than Hipparchus's, but not by a great amount.
We shall not be far wrong if we say that the error in an
equinoctial interval, for either observer, is 0.625 days, or
15 hours. The errors in equinoxes, as we saw earlier, are
probably dominated by systematic errors rather than random
errors, and the errors at vernal and autumnal equinoxes have
a strong tendency to be equal and opposite. To get the error in an individual equinox, then, we should divide by 2
rather than by $\sqrt{2}$. Thus the error in one Greek measurement

---

†As I wrote in Section III.4, we should estimate the accuracy of a measurement by analyzing the measurement process.
We cannot do this here, so we can estimate the accuracy
only by comparing measured values with "correct" ones.

‡It is possible that Hipparchus calculated the winter solstice instead of measuring it. Allowing for this does not
change the estimate of 6 hours by an important amount.

‡Syntaxis, Chapter III.1. Ptolemy does not explicitly state
this value nor the one that follows, but the information he
gives allows us to reconstruct the measurements accurately
and unambiguously.

of an equinox is about $7\frac{1}{2}$ hours. This is slightly larger than the error in a solstice.

Rather by accident, then, the accuracies of Greek solstices and equinoxes are about equal to each other, and the accuracy of each is about 7 hours. Euctemon had already achieved this accuracy by about -430, and there was no further improvement during the whole course of Greek astronomy. This accuracy strikes me as surprisingly poor, particularly for the equinoxes. Reasonable care should have given better accuracy.

3. The Length of the Year

We saw in Section I.3 that there are two different kinds of year. We tend to think first of the year as the time that the sun takes to complete a circuit of the heavens. We can measure this kind of year by following the progress of the sun among the stars, perhaps by noting the stars that rise just at sunset or that set just at sunrise, or by other ways of connecting the position of the sun with the stars. This kind of year is called the sidereal year.

The second kind of year is connected with the return of the seasons. The time from one equinox or solstice to the next one of the same kind is called the tropical year. If we base our calendar upon this kind of year, we will always plant and reap our crops, clean out our swimming pools, have the World Series, and perform other rites of the seasons, at about the same point in the calendar every year. The year that measures the return of the seasons is obviously more important in our lives than the year that merely says where the sun is among the stars, so much so that "year" without qualification always means the tropical year.

The length of the year was an important topic to early astronomers, and much of the literature in Greek astronomy is connected with it. It may be that the sidereal year is easier to measure than the tropical year. Although Euctemon, and presumably his colleague Meton, had fairly accurate values for the lengths of the seasons, the year that they used seems to have been the sidereal year. At least the length of the year in Meton's calendar (Section V.2 above) was 1/76 days longer than $365\frac{1}{4}$ days, and this is closer to the sidereal year than to the tropical year.

We know now that the sidereal year is about 365.256 days while the tropical year is about 365.242 days; the difference is about 20 minutes per year. In three centuries, say, the accumulated difference between time kept by the different kinds of year is about 6000 minutes, or about 4 days. This difference is readily observable. For awhile after Meton and Euctemon, the difference seems to have affected the adopted length of the year without producing a realization that there were two different years. Thus Callippus, about a century after Meton and Euctemon, reduced the year to $365\frac{1}{4}$ days but, so far as we know, he did not recognize the two types of year. This recognition was first made explicit by Hipparchus.

According to Ptolemy's account in the Syntaxis [Chapter III.1], Hipparchus was much concerned about measuring the length of the year and determining whether it was constant or not. Ptolemy quotes six measurements of the autumnal equinox that Hipparchus made between -161 and -142. The first three, made in -161, -158, and -157, are not consistent with the last three, which were made in -146, -145, and -142; Hipparchus was probably still refining his instruments and procedures in these early years. The one in -145 is $365\frac{1}{4}$ days after the one in -146. However, the one in -142, instead of being $4 \times 365\frac{1}{4} = 1461$ days later than the one in -146, is 1461 days less 6 hours later.

Ptolemy also gives fourteen measurements of the vernal equinox made by Hipparchus for years between -145 and -127. The last one of these, which came at 18 hours on -127 March 23, is the last observation by Hipparchus that is known. All of this long series of vernal equinoxes is exactly consistent with a year of $365\frac{1}{4}$ days.† Ptolemy says that Hipparchus found nothing in his equinoctial observations to make him doubt that the year is constant, but that some observations of eclipses may have left him feeling somewhat unsure about the matter.

Hipparchus may not have used any of his equinox observations to deduce the length of the year, however. If Ptolemy's statements are correct, Hipparchus did not trust the equinox times fully, even though they suggested a constant length of the year. Ptolemy quotes Hipparchus as saying that one edge of the equatorial ring that he used sometimes lighted up, went dark, and lighted up again a few hours later during one equinox. This made Hipparchus fear that instruments to measure the equinoxes might not be stable enough to give dependable results.‡ He may also have been

†Hipparchus rounds all of the equinox times to the nearest quarter of a day. That is, he merely says that an equinox came at midnight, in the morning, at noon, or in the evening. Since the year is slightly less than $365\frac{1}{4}$ days, a series of measurements stated to the quarter of a day can be consistent with an interval of $365\frac{1}{4}$ days for many years, but it must finally show an inconsistency. There are reasons to believe that the inconsistency shown in the autumnal equinoxes between -145 and -142 is real, and not an error of observation [Newton, 1970, Section II.1].

‡Pannekoek [1961, p. 125], for example, points out that this can be an effect of refraction. One sequence of events that can give this result is the following: Suppose that the sun is a few minutes of arc south of the equator at sunrise on the day of the vernal equinox. Refraction makes it appear higher in the heavens than it really is, and this can make it illuminate the northern edge of the ring. As the sun rises in the sky, refraction decreases rapidly and the illumination goes back to the southern edge. Finally, in a few hours, the sun truly crosses the equator and the upper edge is lighted again. However, solar heating can cause geometrical instability of the sort that Hipparchus feared. I do not know whether we have enough details about the instruments and their mountings to exclude this possibility.

suspicious of the equinoxes for the reasons outlined in the preceding section.

Whatever his reasons may have been, Hipparchus finally deduced the length of the year only from the summer solstice of -279, measured by Aristarchus, and that of -134, measured by himself, according to what Ptolemy tells us in Chapter III.1 of the Syntaxis. The times of these solstices were given in the preceding section. These solstices are 145 years apart, and the interval between them was 12 hours less than it would have been if the year were exactly $365\frac{1}{4}$ days. In 290 years, then, a number that Hipparchus rounded to 300 years, the recurrence of the solstices would fall short by 1 day. In other words, the length of the year is 365 + (1/4) - (1/300) days. This is about 5 minutes less than $365\frac{1}{4}$ days; the true value is about 11 minutes less.

Hipparchus thus found the length of the year with an accuracy of about 6 minutes, about 1 part in 10^5. This accuracy was never exceeded in Greek astronomy, nor was it exceeded anywhere for 1000 years.[†]

This is a good place to summarize Hipparchus's accomplishments with respect to the sun. There are two aspects, the short term and the long term. In the short term, he had the positions of the sun's apogee and perigee located accurately, but he exaggerated the eccentricity of the sun's orbit. This caused a periodic error, with a period of a year, which peaked at about 20' of arc; this is the amount that the sun moves in about 8 hours. For the long term, he had an error of about 6 minutes in the length of the year. Since 8 hours is about 480 minutes, the long term error equalled the maximum short term error in about 80 years after his own time, and the long term error dominated the results from his theory thereafter. Excellent though it was, his theory needed replacement after, say, 50 years, but it was not replaced for 1000 years.

Hipparchus's measurement of the length of the year had a consequence of high importance. It left no question that the tropical year (equinox to equinox) is shorter than the sidereal year (star to star). That is, the sun travels a shorter distance between successive passages through, say, the vernal equinox than it does between successive passages of the same star. This can happen only if the equinox is slowly travelling westward through the stars to meet the

---

[†]So far as I know, Islamic astronomers were the first to find a better length of the tropical year. There is a belief that the Mayans, at about the same time, had a calendar that challenges the Gregorian calendar in accuracy. This belief depends upon the assumption that certain inscriptions are calendrical in nature. Thompson [1974, p. 96], for one, believes that the inscriptions do not refer to the calendar, if I understand him correctly. He believes that they refer to civil events such as the accessions of rulers.

sun. This motion is the precession of the equinoxes[†] that was described in Section I.3.

There are two ways to measure the rate of precession of the equinoxes. One way is to measure the difference between the ordinary (tropical) year and the sidereal year. The other is to measure the rate at which the coordinates of the stars change. I shall discuss the second method in Chapter IX.

TABLE V.3

PTOLEMY'S ALLEGED EQUINOX AND
SOLSTICE OBSERVATIONS

Reported time		Correct time[a]	
Day	Hour[b]	Day	Hour[b]
132 Sep 25	14	132 Sep 24	9.9
139 Sep 26	07	139 Sep 25	2.6
140 Mar 22	13	140 Mar 21	9.4
140 Jun 25	02	140 Jun 23	14.0

[a]As calculated from modern tables.
[b]Local time at Alexandria.

4.   Ptolemy's Alleged Observations of the Equinoxes and Solstices

It is now time to take up the observations of the equinoxes and solstices that Ptolemy [ca. 142, Chapters III.1 and III.7] claims to have made. Three of them have already been mentioned in Section V.1 above, and all of them are summarized in Table V.3. The table first gives the days and hours of the alleged observations, in local time at Alexandria. It then gives the correct time, meaning the time as calculated from the modern theory of the sun. In calculating these times, I first used the theory of Newcomb [1895], which assumes that the sun has no secular acceleration. I then assumed, on the basis of earlier studies [APO] that the sun has an acceleration of 3 seconds of arc per century per century due to tidal friction and other non-gravitational effects. This is probably correct within 0.5 second per century per century. The total error in the calculated times in Table V.3 is probably no more than half an hour. The calculated times are also given in local time at Alexandria.

The errors in Ptolemy's times are enormous. The three equinoxes are all too late by about 28 hours, while the

---

[†]The plural is often used in this phrase because both equinoxes travel together. The solstices also move, but this is rarely mentioned.

solstice is too late by 36 hours.†

We saw in Section V.2 that Euctemon, Aristarchus, and Hipparchus, many centuries before Ptolemy, could measure equinox and solstice times with errors of about 7 hours, a quarter of Ptolemy's errors or less. Yet Ptolemy's times should have been more accurate. In Chapter III.1 of the Syntaxis, where Ptolemy presents his alleged measurements, he says that he made them using the instruments described at the beginning of his book. The descriptions occur in Chapter I.12 of the Syntaxis.

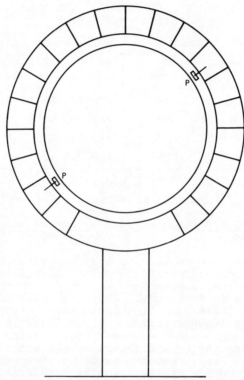

Figure V.1   Ptolemy's meridian circle. A metal ring is contained between the two largest circles. In the figure, this ring is marked at intervals of 15°, to suggest that it carries a graduated circle. Ptolemy says that his circle was graduated in fractions of a degree. An inner metal ring, which can rotate with respect to the outer one, carries two metal plates PP, which in turn carry needles that point to the graduations. The whole is mounted in the meridian plane. In use, the inner circle is rotated until one plate P exactly shadows the other, and the elevation angle of the sun is then read by means of the needles.

---

†However, calculations made in the way just described agree well with large bodies of data that are centuries before the time of Ptolemy. This proves that there are no significant errors in the calculations or in our understanding of the calendar.

The first instrument must have been something like
Figure V.1. The two largest circles represent a large metal
ring mounted on a pedestal. I have shown marks on this ring
at intervals of 15° to suggest that the ring carries a graduated circle. Ptolemy says that each degree was marked on
his instrument, and that each degree had as many subdivisions
as it could accommodate. However, he does not tell us the
number of subdivisions. A smaller ring can rotate within the
graduated circle. This ring carries two small plates marked
PP in the figure. The plates are just alike, are diametri-
cally opposed to each other, and each carries a needle whose
position is to be read on the graduated circle.

By means of plumb bobs, meridian lines, and perhaps
other aids, the graduated circle is lined up with the meridian plane and the zero mark is lined up with the zenith.†

To use the instrument, which can only be used at times
very close to noon, the inner circle is rotated until one
plate P is exactly shadowed by the other. This gives what
is called technically a null reading, and it is capable of
high precision. The position of the sun is then found by
reading the positions of the needles with respect to the
graduated circle.‡

Ptolemy says that the second instrument is easier to
use, and it is certainly easier to describe. It consists of
a quadrant of a circle carefully oriented to lie in the
meridian plane. The quadrant is again divided, according to
Ptolemy, into degrees and their subdivisions, and again he
does not specify the subdivisions. A small pin is placed at
the center of the circle and its noontime shadow is cast onto
the graduated circle. The position of the shadow is then
read. While this instrument is easier to use, it probably
does not have the precision of the first instrument.‡

If either of these instruments was made and used with
the care that Ptolemy implies, the accuracy of the elevation angles of the sun should have been around 2' of arc,
it seems to me. This means that the accuracy of the equinoxes should have been about two hours. The accuracy of a
solstice is harder to assess, but I shall assume, on the
basis of Section V.2, that it is comparable to the accuracy
of an equinox.

---

†When I say zero mark, I mean the reference mark that Ptolemy
chose to have point toward the zenith. I do not know whether
he called this 0°, 90°, or possibly some other value. Since
Ptolemy refers to measuring the angle from the vertical, he
probably called it 0°.

‡Ptolemy does not say specifically that he reads both needles,
but this is his implication, it seems to me. He does say
specifically that both plates have needles.

‡There are several reasons for this. An obvious one is that
the position must be read "on the fly" with the second instrument. With the first, the plates are turned to the
correct alignment by means of the shadow. The angles can
then be read at leisure.

Although many have tried, no one has succeeded in advancing a source of observational error that can account for Ptolemy's errors. One point that cannot be overcome is that all of the errors are in the same direction. Now the basic quantity that is measured by Ptolemy's instruments is the elevation angle of the sun at noon. If a systematic error in the elevation angle makes a vernal equinox too early, say, it makes an autumnal equinox too late and it does not affect the time of a solstice.

If we assume that Ptolemy's times were indeed observed, as he claims that they were, we must then assume that his errors are accidental errors of observation. To be conservative while posing a simple computational problem, let us assume that each error in Table V.3 amounts to ten standard deviations. If we further assume that the errors obey what is called the normal law of error, we can calculate the probability that four successive errors each amount to ten standard deviations in the same direction. This probability is 1 out of 10^{92} [Newton, 1973].†

I do not mean for this probability to be taken literally. In any actual situation, there is always some question about the accuracy with which the normal law of error applies. In this situation, there is no reason to assume that it applies with enough accuracy to give quantitative meaning to a probability like 1 in 10^{92}. What this calculation of probability does tell us is that the probability is small far beyond any ordinary experience. For simplicity, I shall neglect this small probability and say that there is no chance that Ptolemy's errors were accidental errors of observation. Where, then, did they come from?

TABLE V.4

HOW PTOLEMY'S EQUINOX AND SOLSTICE OBSERVATIONS
WERE FABRICATED

| Starting observation | | Number of | Fabricated time | | Reported |
Day	Hour	years	Day	Hour	hour
-146 Sep 27	00	278	132 Sep 25	13.8	14
-146 Sep 27	00	285	139 Sep 26	07.2	07
-145 Mar 24	06	285	140 Mar 22	13.2	13
-431 Jun 27	06	571	140 Jun 25	02.3	02

The answer comes from experience in teaching laboratory courses in the physical sciences at the introductory level. In these courses, the student is instructed to perform an experiment or make an observation that will verify some law or the value of some important constant that is in fact well known. As every teacher of such courses knows, what many mediocre students do is to calculate backward from what is

---

†I shall cite this reference as Part I in the rest of this work.

to be proved to the data needed to do it. They then pretend
that these are the data they found.

This answer is suggested by Ptolemy's own words. He
emphasizes that his alleged measurements, which are summarized in Table V.3, prove that Hipparchus's length of the
year is highly accurate. Because of this emphasis, I decided to see whether Ptolemy's data had been found by starting from what they proved and calculating back to the data.
The relevant calculations are summarized in Table V.4.

One way in which Ptolemy verified the length of the year
was by comparing one of his autumnal equinoxes with the one
measured by Hipparchus in -146, which came at the midnight
(00 hours) beginning -146 September 27. Hipparchus's observation is the one labelled "starting observation" in the
first row of Table V.4. Ptolemy claims [Syntaxis, Chapter
III.7] that the autumnal equinox 278 years later was "one
that he measured with the greatest care" (μια των ακριβεζατα
ληφθεισων). Hipparchus's length of the year is equal to
365.246 666 667 days. When we add 278 multiples of this to
the starting epoch, keeping a precision of 0.1 hours in the
calculations, we find the "fabricated time" given in the
first row of Table V.4. This is 132 September 25, at 13.8
hours. Ptolemy claims that he found the equinox on this
date, at 14 hours.

This particular observation is not one that Ptolemy
used to "verify" Hipparchus's length of the year. It is the
one that he used to find the position of the sun at the
fundamental epoch of all his ephemerides, which is noon on
-746 February 26. He "verifies"the length of the year in
Chapter III.1 of the Syntaxis. First he compares Hipparchus's equinox of -146 September 27 with the one that came
285 years later. In the year 139, he says, "we observed
again with sureness the autumnal equinox" (...ημεις ετηρησαμεν
ασφαλεζατα παλιν την μετοπωρινην ισημεριαν....), and he
found that it came at 07 hours on 139 September 26. The
fabricated time for this equinox in Table V.4 is 07.2 hours
on this date. He also "verifies" the length of the year by
using Hipparchus's vernal equinox of -145 and the one that
he claims to have measured at 13 hours on 140 March 22; the
fabricated hour is 13.2 .

When he uses the summer solstice, he does not use one
that Hipparchus measured. Instead, he uses the one attributed
to Meton and Euctemon, at 06 hours on -431 June 27. He compares it with the one that he measured "with great care"† at
02 hours on 140 June 25, and again finds Hipparchus's length
of the year.‡

†Ptolemy "doth protest too much, methinks" (Hamlet, Act III,
Scene 2, line 240) about the accuracy of his observations
and the care with which he made them.

‡Although the calculation of the fabricated times is tedious,
I recommend strongly that the reader calculate at least one
of them. Perhaps because of its very tediousness, I find
that the completed calculation has a great intellectual im-
pact.

In every instance, when the time called the fabricated time in Table V.4 is rounded to the even hour, it agrees exactly with the time that Ptolemy claims to have observed. Thus we have two hypotheses about Ptolemy's alleged observations. The hypothesis that they were in fact the result of observation proves unable to account for the results. On the other hand, the hypothesis that they were fabricated gives the results exactly and in a straightforward manner.

In calculating the probability that the times were observed, I assumed that each error equalled ten standard deviations, which is equivalent to assuming that the standard deviation of an observation is about 3 hours. This assumption is reasonable, but it does not matter for present purposes whether it is correct or not. The reader may make any assumption that he likes about the standard deviation and the other properties of the measurement process. Whatever assumptions he makes, he cannot explain the errors in Ptolemy's times by the hypothesis that they were obtained from observation. The reason is that the times agree exactly with the values obtained by calculation. It is inconceivable that this happened for all of the equinox or solstice observations that Ptolemy claims to have made, no matter what the sources of error are.

After he quotes Hipparchus's first observation of the vernal equinox, Ptolemy adds: ".. and he states that the ring† at Alexandria was equally lighted on its two faces at the fifth hour, so that the two observations of the same equinox disagreed by 5 hours, .." The fifth hour here means the fifth hour of the day. If Ptolemy has quoted Hipparchus correctly, Hipparchus took this to be the end of the fifth hour, or 11 hours from midnight.‡ The date is -145 March 24.

When Ptolemy fabricates his own observations, he starts from Hipparchus's observation on this date, as Table V.4 shows; this observation was at 06 hours on -145 March 24. It is ironic that the Alexandria observation is probably more accurate. The best estimate I can make of the vernal equinox of -145 is that it came at 10 hours on March 24, in either Rhodes or Alexandria, whose local times differ by only a few minutes. Thus Hipparchus's error on this date was about 4 hours, while the error in Alexandria was only about 1 hour.

The publishing history of the calculations summarized in Table V.4 is somewhat complicated. I made them, probably in the spring of 1968, in the process of writing Chapter II of an earlier book [Newton, 1970]. After I finished this book, I wrote a survey article [Newton, 1969] in which I included the calculations. Because of the accidents of publishing schedules, the article, though written later, was published first. When I was nearly through with the book,

---

†This apparently means the equatorial ring that I mentioned earlier. "He" refers to Hipparchus.

‡I have overlooked this observation in my earlier writing. It was called to my attention by reading Muller [1975, p. 7.7].

I discovered that the calculations relating to the equinoxes, but not to the solstice, were contained in a doctoral dissertation [Britton, 1967] that, so far as I know, is still unpublished.

In this same period, I studied histories of ancient astronomy and discovered that some writers, notably Delambre [1817], suspected that Ptolemy's solar data had been fabricated. On p. xxvi of volume 1, for example, Delambre says: ".. there would never have been any doubt in that regard† if these equinoxes, compared to modern values, had not given the year a length that it is impossible to admit. All is explained if these equinoxes were calculations that had been presented as real observations." However, arguments of this sort, at least those that I saw, including Delambre's, were based upon the large size of Ptolemy's errors, and none gave the telling argument contained in Table V.4. An argument based on the size of the errors can be countered by demonstrating the possibility of large errors, but this does not affect the argument of Table V.4.

Later, when I turned my attention to medieval astronomy, I read Delambre's history of medieval astronomy [Delambre, 1819]. To my astonishment, I found that Delambre, on p. lxviii, gave the calculations relating to the solstice and to one of the equinoxes. He used the introduction to his study of medieval astronomy in order to answer comments that had been made about his study of ancient astronomy. For this reason, I did not discover his calculations in my original study of ancient astronomy. So far as I know, Delambre's was the first publication of any of the calculations, and he was the first person who produced the unanswerable argument that Ptolemy's alleged equinoxes and solstices were fabricated. I do not know of any publication of the argument between Delambre's and mine.

In some circumstances, the fabrication of information can have an innocent explanation.‡ Under the circumstances that prevail here, I can see no explanation except deliberate deceit. There is much emphasis on the care with which the measurements were made, and the measurements were used to establish the values of certain astronomical parameters, parameters which Ptolemy tells us are to be found from observation, not from sitting and thinking about what the gods intended. However, it does not follow that Ptolemy was himself responsible for the deceit.

It is possible that Ptolemy had an assistant who was supposed to work only under Ptolemy's orders and whom Ptolemy thus did not feel required to acknowledge. If there were such an assistant, he may have supplied Ptolemy with fabricated data in place of genuine measurements, and Ptolemy

---

†That is, with regard to the authenticity of Ptolemy's claimed observations.

‡We shall see an example in the next section. In this example, it is not claimed that the information was obtained by careful observation; the information is merely presented as useful.

may have used the fabrications in good faith. This idea receives some support from Ptolemy's use of "we". "We" made the observations of the equinoxes and solstices. "We" may be a way of saying that Ptolemy and someone else were responsible for the observations. However, it may be only the editorial "we", and Ptolemy uses the plural form in places where it definitely is the editorial form. In other places, such as in stating the theorem of Menelaos,[†] he uses the first person singular, so he is not rigorous in choosing between the singular and plural.

For the moment, we can draw no conclusions about the author of the fabrications shown in Table V.4. We can decide upon the author only when we have more evidence.

Since the error in Ptolemy's equinox and solstice times is somewhat more than a day, the error in his solar theory is somewhat more than a degree, for epochs in his own time.

5. The Fabricated Solstice of -431 (Meton's Solstice)

We have remarked before that a calendar was introduced into Athens in the year -431, and that the length of the year in this calendar was 365 + (1/4) + (1/76) days. Some sources attribute this calendar to Meton, some to Euctemon, and some to both.

An inscription called the Milesian parapegm says that the calendar was inaugurated with the summer solstice of the year that we call -431 (Diels and Rehm, 1904).[‡] The inscription also says that this solstice came on the 13th day of the last month of the Athenian calendar, and that this was the 21st day of the 7th month of the Egyptian calendar. The parapegm also refers to the summer solstice in the year -108.[♯] For this reason, we may safely assume that the inscription was prepared some time close to the year -108.

The historian Diodorus, writing about a century later, also refers to the solstice of -431 and gives the Athenian date but not the Egyptian one.[*] We do not know whether Diodorus is an independent authority or whether he derived his information, directly or indirectly, from the parapegm.

Ptolemy [Syntaxis, Chapter III.1] gives information that is not in the parapegm; he gives the hour of the day. He says that the solstice in -431, as measured by Meton and Euctemon, came at 06 hours on the 21st day of the 7th Egyptian month, but he does not give the Athenian date. We have no difficulty in understanding the Egyptian calendar, and the time that Ptolemy states is 06 hours on -431 June 27.

[†]See Section II.4.

[‡]Meritt [1961, p. 4] also quotes the relevant parts of this inscription.

[♯]There is a slight disagreement in the scholarly literature about the year -108. All scholars agree that the year is close to -108, and this is all that we need here.

[*]Meritt [1961, p. 4] also gives the quotation from Diodorus.

This is the time used for the starting epoch in the fabrication of the alleged solstice of 140 June 25 in Table V.4.

The solstice actually came at about 10 hours on -431 June 28, so that the time which Ptolemy gives is about 28 hours too early.[†] We saw in Section V.2 that Euctemon's solstices actually have a standard error of about 7 hours. The probability that a solstice is in error by as much as 28 hours (four standard deviations) is only 1 in about 16 000.

Again fabrication provides a simple and exact explanation of an error that is almost impossible if it had really come from observation. We have seen (Section V.2) that Hipparchus observed a summer solstice at noon on -134 June 26, and that he based his year upon this observation. His year is 365.246 666 667 days. If we subtract 297 multiples of this from noon on -134 June 26, we get 5.8 hours on -431 June 27. When this is rounded to the even hour, it agrees exactly with the time that Ptolemy states.

There is in fact further confirmation that the solstice of -431 has been fabricated. We cannot translate the early Athenian calendar exactly into our calendar, but we do know many things about it. We know that the Athenian months were based upon the phases of the moon. The date -431 June 27 is wrong for the 13th day of a lunar month, for any plausible theory of the relation between the moon and the month. Further, it has proved impossible to reconcile -431 June 27 with what we do know about the Athenian calendar, if we assume that it came on the 13th of the month. Thus everything that we know tells us that the date -431 June 27 is wrong.

There is no doubt beyond the most frivolous one, I think, that the solstice attributed to Meton (and/or Euctemon) has been fabricated. The questions are: Who did the fabricating and why did he do it? In considering the questions, we must note that the parapegm gives the date but only Ptolemy gives the hour. Hence we must consider the possibility that the date and the hour were fabricated separately. We should take up the date first.

I have written extensively on this subject in Sections II.4, V.3, and VIII.4 of APO. It would take us too far afield to review all of the evidence. The following, I believe, explains all of the known facts about the date in a simple and plausible manner: Some persons in Miletus, about the year -108, decided to dedicate an inscription to Meton and his calendar. They knew the year of his famous solstice, and they knew that it came on the 13th day of the Athenian month, but they had already lost the knowledge of how to read the Athenian calendar. Only about 20 years before, Hipparchus had developed a theory of the sun that was quite accurate for its time and whose accuracy the Milesians had no reason to suspect. Since they did not know what else to do, the persons who prepared the parapegm calculated the solstice from Hipparchus's theory. Their reasons were innocent, and they would

[†]The discussion of this section closely follows Section VIII.4 of APO.

doubtless be distressed if they knew of all the trouble
that their fabrication has caused later scholars.†

I see no way to decide about the fabrication of the
hour. It is probable that the hour was calculated at the
same time as the date, and it is possible that the combination of date and hour survived and was transmitted to Ptolemy in a source that is now lost. It is also possible that
Ptolemy (or his anonymous assistant) fabricated the hour
independently in the manner that we have just described.

Since the correct solstice time is about 10 hours on
-431 June 28, and since there is only 1 chance in 3 that the
error exceeds 7 hours, the probability is rather high that
the solstice time observed by Meton came on June 28. However, the Athenian day began at sunset, only about 10 hours
after the correct time. Thus a somewhat large but still
plausible error could have put the observed time after sunset, and hence on the Athenian day that corresponds to June
29. There is a much smaller chance, almost a negligible one,
that the observed time could have been before sunset on the
preceding day, which would let it correspond to June 27.

Thus we should make the 13th day of the Athenian month
correspond to June 28, or possibly to June 29, but almost
surely not to June 27. A specialist in the Athenian calendar,
with this as his starting point, should be able to decide
quickly between June 28 and June 29, and thence to construct
a consistent theory of the Athenian calendar.

6. Ptolemy's Alleged Observations of the Obliquity and of
    the Latitude of Alexandria

In Chapter I.12 of the Syntaxis, Ptolemy describes two
instruments, which should have high accuracy, and which he
claims that he used in measuring the elevation angle of the
sun at noon.‡ In the same chapter, he says that he used
these instruments to measure the zenith angles at the two
solstices. Every time, he says, the difference was between
47 2/3 and 47 3/4 degrees. This implies that the difference
is $47°$ plus $42\frac{1}{2}$ minutes. However, he says, the difference
is the same as 11/83 of a full circle, which is the value
that Eratosthenes found and that Hipparchus used.

Now 11/83 of a circle is slightly more than $47° 42' 39''$,
whereas Ptolemy's data imply $47° 42' 30''$.‡ I think it should

---

†The assumption that the solstice was fabricated after Hipparchus's death explains the fact that Hipparchus did not
use it. If the alleged observation had been accepted as a
valid observation in Hipparchus's time, it is highly likely
that he would have used it in his study of the length of
the year. To him, it would have looked like excellent confirmation of the constancy of the year, and this was a subject that he studied intensively.

‡See Section V.4 and Figure V.1 above.

‡al-Biruni [1025, p. 59] is the earliest writer I have seen
who points out this discrepancy.

not disturb us to have Ptolemy say that the two results are the same. By the standards of the time, agreement within 9″ of arc would have been regarded as an excellent confirmation. What disturbs us is that Ptolemy came at all close to Eratosthenes' value.

In the time of Ptolemy, the correct value of the difference, which is equal to twice the obliquity ε, was 47° 21′ 27″ [Part I], 21′ less than the value Ptolemy claims. With the instruments that Ptolemy describes, there is no discernible source of bias other than a small amount of refraction. If Ptolemy had been silent on the matter, I would have been willing to allow several minutes of arc, say 5′, as the standard deviation for a single measurement of 2ε, but Ptolemy does not allow such a large value. He says that the difference always lay within a range of 5′. This means that the standard deviation can be no more than about 2′; for simplicity, let us say that the error in Ptolemy's value is 10 standard deviations. (See Section 5 of Appendix A.)

The language implies that he measured 2ε several times; it is reasonable to take 4 as the number of times. Thus, as with the equinoxes and solstices, we have four measurements all with errors of 10 standard deviations and in the same direction. The probability that this could have happened by chance is 1 out of 10^{92}, a truly negligible probability.

On the other hand, Ptolemy's value agrees far beyond the attainable level of accuracy[†] with the already famous value measured by Eratosthenes almost four centuries earlier. There is no question, I think, that Ptolemy's alleged measurements of the obliquity were fabricated. The obliquity that results from the fabrication is 23° 51′ 20″.

There are two elements of heavy irony in this situation. The first comes from the fact that the obliquity has been steadily decreasing since well before the time of Eratosthenes. In his time, the value of 2ε was about 47° 27′ 9″, almost 6′ larger than in Ptolemy's time. If Ptolemy had made an honest measurement of 2ε, he would have found that it was decreasing, and he would have made an astronomical discovery of the first rank.

The second concerns an Islamic mathematician and astronomer named Abu Sahl al-Kuhi, who observed in Baghdad around 990. Abu Sahl measured the obliquity and the latitude of Baghdad [al-Biruni, 1025, pp. 69-70]. He found ε = 23° 51′ 20″ and a latitude of 33° 41′ 20″. The correct latitude is about 33° 20′; Abu Sahl's error is the enormous one of 21′. By his time, the value of ε had decreased to about 23° 35′, which means that Abu Sahl's error in 2ε was about 32′. That is, in measuring the elevation angle of the sun, Abu Sahl made errors equalling the apparent diameter of the sun. We may flatly say that this is impossible, and there is no doubt that Abu Sahl's claimed set of measurements is a hoax.

[†]With 4 measurements, each having a standard deviation of 2′, the attainable level of accuracy is $2'/\sqrt{4} = 1'$. Ptolemy's agreement is within 9″.

The irony is that he committed this hoax to confirm Ptolemy's value, which was already a hoax.

In Chapter I.12, immediately after he gives his alleged data on the obliquity, Ptolemy points out that the same data give the latitude of the point of observation. However, he does not give the latitude of his position here, presumably because it is not relevant to what he is doing. He does not even name the observing position. He supplies both items of information in Chapter V.12, where they are relevant to the immediate topic. His observations were made in Alexandria, he says, and he says that the latitude there is 30° 58'.

According to the Times Atlas [1955], the latitude of Alexandria is 31° 13'. Ptolemy and the Times Atlas may not have meant exactly the same thing by Alexandria; a city does have some spread in latitude. Some scholars think that Ptolemy worked at the temple of Canopus, which was a short distance from Alexandria at the place now called Abu Qir, but which could reasonably have been included under the general place name of Alexandria. If this is correct, Ptolemy's latitude was 31° 19'.

Since Ptolemy did not make the observations of the obliquity, he naturally did not make the observation of latitude that inevitably accompanies the obliquity. Conceivably, he could have made another set of measurements, from which he gave us the latitude while concealing the accompanying obliquity. However, his error makes this unlikely. The error in the latitude is at least 15', about equal to the apparent radius of the sun. An error this large is impossible with the instruments that Ptolemy describes.

We can make a formal estimate of probability [Part I]. Since the measurement of latitude is sensitive to systematic errors in the instruments, we can ignore Ptolemy's statement about the repeatability of the measurements. I think that 5' is a generous estimate of the standard deviation of the systematic errors. The error is either 15' or 21', and thus it is at least 3 standard deviations. The probability that an error is this large by chance is less than 0.003.

If Ptolemy fabricated the latitude, there must have been an earlier measurement that he accepted as standard. No such measurement has come down to us, and we can only guess its origin. However, I think we can make an educated guess with a high degree of plausibility [Part I].

We believe that Eratosthenes measured the obliquity and obtained the value of 23° 51' 20" that we have been discussing. Since he measured the obliquity, he necessarily measured the latitude of the observing point also. We also know that he had a value for the latitude of Alexandria, because he used it in estimating the size of the earth (Section III.4). Hence it is likely that the value which he used for the latitude of Alexandria is a value that he had measured. We may well guess, from what we now know about Ptolemy, that his fabricated latitude of Alexandria was simply a pre-existing value, and thus we may guess that Ptolemy's value of 30° 58' was Eratosthenes' measured value. Since Eratosthenes' work is lost, we must ask if this value is consistent with what we know about his results.

Eratosthenes (Section III.4 above) took the latitude of Syene to be equal to the obliquity, namely 23° 51' 20", and he used 1/50 of a circle as the difference in latitude between there and Alexandria; we may feel sure that the denominator has been rounded. If he measured 30° 58' for the latitude of Alexandria,[†] the difference is 7° 6' 40", which is about 1/50.6 of a circle. Since Eratosthenes was obviously not striving for high accuracy, I think it is quite reasonable to assume that he would have rounded this to 1/50.

With the latitude, we cannot achieve the high level of confidence that we could with Ptolemy's alleged measurements of equinoxes, solstices, and the obliquity. Nonetheless, there is no reasonable doubt that his claimed measurement of latitude was fabricated. It is likely that his claimed measurement was in fact the one made by Eratosthenes, but we can never know with confidence, unless new documentary material turns up.

Ptolemy says that Hipparchus also used Eratosthenes' value of the obliquity, implying that he did not measure it independently. Now we know from Ptolemy that Hipparchus measured the time of a summer solstice and, according to Geminus, he might have measured the time of a winter solstice also.[‡] We might think for a moment that Hipparchus measured the times of these events without measuring the corresponding elevation angles of the sun. However, Hipparchus also measured the times of equinoxes, and he could not have measured these unless he had measured his latitude. It is conceivable that he measured his latitude solely from star observations and not from solar ones. However, since he certainly made almost all of the solar observations needed to find his latitude, it would be surprising if he did not make them all. If he did, he must have also found the obliquity. If he found the obliquity, it would be surprising if he found the same value as Eratosthenes.

Against this, we must set the fact that his equinoxes are in error by about 7 hours. The size and the sign of his errors indicate that his value of the obliquity was too large by about 7' of arc. This is not far from the error he would have made if he used Eratosthenes' value of the obliquity.

It is possible that Ptolemy has deceived us about what Hipparchus did; we shall find in later chapters that he frequently deceives us about the work of other astronomers. However, if we accept what Ptolemy says, we can reconcile all the known facts by this set of events: Hipparchus measured the zenith distance of the sun at the summer solstice

---

[†]An error of 15' in the latitude is reasonable for Eratosthenes, who presumably used a gnomon (Figure III.2 above). It is unreasonable with the advanced methods described by Ptolemy.

[‡]See Section V.1 above.

and from this he set the plane of his equatorial instrument
by using Eratosthenes' value of the obliquity. He then took
the equinoxes to be the times when the sun crossed the plane
found this way. He did not observe the winter solstice, but
calculated it from the parameters that he found for the summer solstice and the two equinoxes. While this accounts for
the evidence, it is still surprising that Hipparchus did not
observe the winter solstice and the obliquity independently.

Let us return to Ptolemy's alleged measurement of the
obliquity for two final points. Britton [1969] has analyzed
thoroughly the errors in measuring the obliquity by means of
the meridian instruments that Ptolemy describes. He finds
that Ptolemy would have made the error he did if he consistently read the instruments at a time that is about 30 minutes from noon rather than at noon. Since he has found a
possible source of error that is consistent with Ptolemy's
result, Britton concludes that Ptolemy's measurement is
genuine and not fabricated, as Delambre had alleged.

If a person consistently reads the zenith angle at the
same hour every time, if this hour is not noon, and if the
person uses the angle as if it had been measured at noon, he
will certainly make an error in inferring the obliquity. I
have not checked Britton's analysis of the error in detail,
but his basic methods are correct and I am willing to accept
his analysis. Unfortunately, his analysis is irrelevant to
the question of whether Ptolemy's observations are genuine
or fraudulent.

The basis of this statement is one that I have already
mentioned. The telling point is not the size of the error
in the observation, or whether there is a possible source
for an error of this size. The telling point is the precise
agreement of the "observed" value with a preassigned result.
Britton's analysis shifts the error from the reading of
angles to the measurement of the hour without affecting the
telling point. For the sake of illustration, let me assume
that the time error needed to account for Ptolemy's error in
the obliquity is 30 minutes, within a margin of about 2 minutes.† Since Ptolemy, by using a sun dial or the equivalent,
could certainly determine the time of noon within a few minutes, it is unlikely that he would make an error of 30 minutes in finding noon. If he did make an error of this size
for some reason we do not know, it is unlikely that he would
make the same error every year, as his claims about the constancy of his results demand. Even if his errors should be
consistent for some reason we do not know, the error in the
obliquity is a sensitive function of the time error, and it

---

†Britton points out that the error does not have to be the
same at the summer and winter solstices. For example, if
the error were 0 minutes at one solstice, it would have to
be somewhat more than 40 minutes at the other, since the
error in the obliquity is nearly a quadratic function of
the time error. At a given solstice, however, the error
must be highly constant from one year to the next. For
simplicity, I shall talk in terms of the same error at each
solstice.

is highly unlikely that he would make exactly the error
that is needed to make his result agree with Eratosthenes'.†

As an aside, Ptolemy's statement is the only basis for
assuming that Eratosthenes measured 2ε to be 11/83 of a
circle, so far as I know. If this is correct, we must admit
the following possibility: Ptolemy did measure the value of
2ε that he claims, namely $47° 42' 30''$. In order to add to
the impressiveness of his result, he claimed that it agreed
closely with a value that the famous Eratosthenes had measured four centuries before. In order to be still more impressive, he stated Eratosthenes' result in the archaic form
of 11/83 of a circle instead of in a "modern" form using
degrees, minutes, and seconds.

If this is what happened, it is still fraudulent. The
only question is whether the fraudulent measurement is the
one that Ptolemy claims to have made himself, whether it is
the one that he attributes to Eratosthenes, or whether both
are fraudulent. I shall assume in the rest of this work
that Ptolemy's measurement is fraudulent. This leaves open
the question about the measurement attributed to Eratosthenes.
It is not highly important for present purposes whether this
assumption is the correct choice or not. Fraud is involved
in Ptolemy's claims about the obliquity in any case, and the
remaining writing will be simpler if we make a specific
choice.

I have dwelt upon Britton's analysis in considerable
detail in order to emphasize an important point that was
mentioned earlier. It is fruitless to look for sources of
error that can possibly explain the size of the errors that
Ptolemy makes. Even if there are such sources, they cannot
explain how Ptolemy, by accident, over and over again, happened to make just those errors which allow his erroneous
theories to agree with preassigned values, namely values
that were accepted from the work of earlier astronomers.

Britton makes another point. Ptolemy describes the
instruments and methods that he used to find the obliquity
in considerable detail. It is not plausible that he would

---

†If I understand his conclusions correctly, Britton decides
that the following is the most likely explanation: Ptolemy
decided to make his measurements at 30 minutes from noon
every time at both solstices, with the intention of correcting these readings to give the correct reading at noon.
He applied the correction correctly at the winter solstice
but he applied it with the wrong sign at the summer solstice. I see no reason why he should choose this complicated procedure instead of simply making the readings at
noon. If Ptolemy did do this, he must have known the theory
involved in the correction, and it is surprising that he
would make the same sign error at every summer solstice.
Even if we grant all this, we are still confronted with the
unanswerable argument. It is highly unlikely that Ptolemy,
by accident, would choose to make his readings at just that
time which makes his obliquity agree almost exactly with
that of Eratosthenes.

follow such an elaborate procedure if he fabricated the results. Therefore the fact that Ptolemy did describe his instruments and methods establishes a strong presumption that his observations are genuine.

Now an astronomer or other scientist who makes valid observations naturally tends to tell us how he made them. What does a person do who has fabricated some observations and who wants us to accept them as genuine? If a genuine observer usually describes his instruments, a fraud who wants his results to be accepted will be careful to do the same thing. In other words, we expect to find the instruments and methods described whether the observations are genuine or fabricated. The fact that Ptolemy describes his instruments tells us nothing about the genuiness of his observations.

Thus, in spite of Britton's arguments, our earlier conclusion stands. With odds of more than 300 to 1, Ptolemy's measurement of the obliquity is fabricated.

CHAPTER VI

THE LONGITUDE OF THE FULL MOON

1.   Parallax

Parallax is the difference in direction to an object as seen from two different points. In astronomy, we almost always take one of the points to be the center of the earth and the other to be a point on the surface where astronomical observations are made. Clearly, if the object viewed is far away compared with the size of the earth, the parallax is small. For this reason, we can neglect parallax when we are dealing with the sun, the planets, or the stars, to the accuracy of observations made with the naked eye. However, the moon is close enough that we must take parallax into account from the beginning of the lunar theory.†

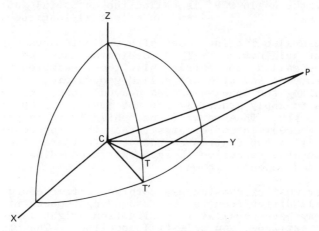

Figure VI.1   Parallax in astronomy. C is the center of the earth and T is the location of an observer on its surface. A plane through T and the axis CZ cuts the XY plane in line CT'. The angular position of T is specified by the coordinate angles XCT'and TCT'. The angular position of an exterior point P is specified by locating a point P' (not shown) analogous to T' and giving the coordinate angles XCP' and PCP'. Suppose we draw another set of coordinate axes centered at the observer T, and locate P in the new system by the same construction. Unless T is on the line CP, the coordinate angles centered on T differ from those centered on C. This is the phenomenon of parallax.

---

†The kind of range finder that is commonly used on cameras depends upon parallax. The range finder in effect measures the amount of parallax between two points having a known relation to each other, and finds the range (distance) from the measured parallax. In a similar way, astronomers can use parallax to find distance.

The general phenomenon is illustrated in Figure VI.1. The figure shows coordinate axes XYZ. We take X to be the direction to the vernal equinox, as seen from the center C of the earth; the directions of Y and Z will be mentioned later. T is an observer on the surface of the earth, and we shall take his distance from C to be unity. We specify his position on the surface by means of two angular coordinates. To find these, we draw a plane through his position and the Z-axis; this is the plane passing through Z, C, T, and the point labelled T'. One of the coordinates is the angle TCT', and the other is the angle T'CX.

We can imagine the angles analogous to TCT' and T'CX being drawn for the point P. They would be labelled PCP' and P'CX if they had been drawn in. If Z is the North Pole, so that the XY plane is the plane of the equator, angle PCP' is called the declination and angle P'CX is called the right ascension; declination is analogous to latitude and right ascension to longitude. If the XY plane is the plane of the ecliptic, the angle PCP' is called the celestial latitude and the angle P'CX is called the celestial longitude.†

We can also imagine a set of coordinate axes parallel to XYZ but originating at T. Coordinates measured in this system are called topocentric while those measured in the XYZ system centered at C are called geocentric. In deriving or tabulating the ephemeris of a heavenly body, we want the ephemeris to apply generally and not just for a specific point on earth. Thus we want the ephemeris to be put in terms of geocentric coordinates rather than topocentric ones. However, it is topocentric coordinates that are observed. We must be able to derive the geocentric coordinates from the observed topocentric ones.

Under most circumstances, both geocentric coordinate angles will differ from the corresponding topocentric ones. Thus we may have parallax in declination, right ascension, celestial latitude, and celestial longitude. That is, when we specify the parallax, we must also specify the coordinate to which it applies.

We may always draw a plane through C, T, and P, and, under some important circumstances, we are interested only in parallax in that plane; there is clearly no parallax in a coordinate normal to that plane. This situation is shown in Figure VI.2, in which the X'Y' plane is the plane through C, T, and P. There is only one angular coordinate in this plane. The angle PCX' is the geocentric coordinate angle λ of the point P. The angle λ_T is the corresponding topocentric coordinate of P. Finally, the angle TCX' is the coordinate λ_O that locates T with respect to C. The distance CP is denoted by R, and the radius CT is 1.

It is simple to work out the exact behavior of parallax

---

†Usually there is no danger of confusing celestial latitude and longitude with their counterparts defined for points on the earth. When there is no danger, I shall omit the adjective celestial.

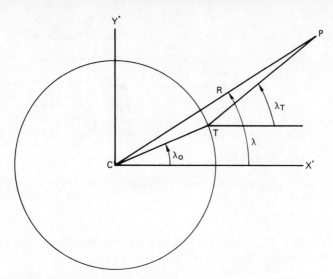

Figure VI.2   The particular case of parallax in a plane. The
plane of the figure is the plane passing through points C, P, and T
of Figure VI.1, and X' and Y' are coordinate axes in this plane. λ
is the coordinate angle of P as seen from C, λ_T is the coordinate
angle of P as seen from T, and λ_0 is the coordinate angle of the
observer T as seen from C. The difference $\lambda_T - \lambda$ is the parallax
of P in this particular case. The difference $\lambda_T - \lambda$ is equal to
the angle CPT, so that the parallax of P is also equal to the
angle between C and T as seen from the exterior point P.

in the plane case. The geocentric rectangular coordinates
X_T' and Y_T' of T and X_P' and Y_P' of P are

$$X_T' = \cos \lambda_0, \qquad Y_T' = \sin \lambda_0,$$
$$X_P' = R \cos \lambda, \qquad Y_P' = R \sin \lambda.$$

Hence one coordinate of P as seen from T is $R \cos \lambda - \cos \lambda_0$
and the other coordinate is $R \sin \lambda - \sin \lambda_0$. The angle λ_T
is then found from

$$\tan \lambda_T = (R \sin \lambda - \sin \lambda_0)/(R \cos \lambda - \cos \lambda_0).$$

What we want is the difference between λ_T and λ. By
using the formula from trigonometry for tan (A - B) in terms
of tan A and tan B, we readily find

$$\tan(\lambda_T - \lambda) = [\sin (\lambda - \lambda_0)]/[R - \cos (\lambda - \lambda_0)].$$

$$(VI.1)$$

I have derived Equation VI.1 because I want to point out
explicitly how the parallax $\lambda_T - \lambda$ depends upon R and upon
the difference in direction $\lambda - \lambda_0$, which is the difference
in geocentric direction between the observed point P and the
observer's point T. We can derive equations like Equation
VI.1 for the general case of Figure VI.1, but the process is
considerably more tedious. Since we shall not need to use

the general case, I shall not give the results.†

We also want the maximum value that the parallax can have. The maximum occurs when CT and TP in Figure VI.2 are perpendicular, so that the line of sight TP is tangent to the earth. That is, the line TP is a horizontal line when the parallax is a maximum. For this reason, the maximum value is often called the horizontal parallax. The maximum will be denoted by the symbol Π, and it is easy to find that

$$\sin \Pi = 1/R. \qquad \text{(VI.2)}$$

If R is large, the sine and the angle are equal if Π is in radians, and

$$\Pi = 1/R \qquad \text{(VI.3)}$$

when R is large. Thus measuring the horizontal parallax Π is equivalent to measuring the distance R.

When R is about 60, as it is for the moon, Equation VI.3 is usually not accurate enough; we must use Equation VI.2 instead. For the sun and the planets, we may use Equation VI.3 in all but the most precise telescopic work.

I mentioned the parallax of the stars briefly in Section IV.7. We have no need to deal with it quantitatively in this work, but it is desirable to say a little about it in order to clear up a possible misunderstanding. In Equation VI.3, the unity in the numerator is the radius of the earth, the distance of the observer from the center. When we are dealing with the stars, this is not the important distance. In Figure VI.2, if C becomes the sun, T becomes the earth, and the circle centered on C becomes the earth's orbit. As T moves around the orbit during the course of a year, there is a parallax effect of the stars. The maximum parallax for a particular star is still given by Equation VI.3, but the unity in the numerator is now the average radius of the earth's orbit, which is about 23 000 times the radius of the terrestrial globe. Even so, the greatest parallax known for any star is less than 1″ of arc. This means that the distance to the nearest star is more than 200 000 times the radius of the earth's orbit.

2. Evection and Variation

Let the symbol M denote the mean anomaly of the moon in its modern sense; it is the angle from the position of perigee to the mean moon. The equation of the center $e_{c\,\mathbb{C}}$ for the moon‡ contains more than 1500 terms in the modern theory of the moon. If we keep only the four largest terms, we have

---

†They can be found in Section 2F of the Explanatory Supplement [1961].

‡See Section IV.2 for the definition of this term.

$$e_{c\text{\textflorin}} = 6°.29 \sin M + 0°.22 \sin 2M$$
$$+ 1°.27 \sin (2D - M) + 0°.66 \sin 2D. \tag{VI.4}$$

In this, D is the quantity called the mean elongation of the moon. That is, if $L_\odot$ and $L_\text{\textflorin}$ are the mean longitudes of the sun and moon,

$$D = L_\text{\textflorin} - L_\odot. \tag{VI.5}$$

The first two terms in Equation VI.4 arise from the eccentricity of the lunar orbit; there are also steadily decreasing terms in sin 3M and so on.†

Rigorously, the new moon occurs when the difference between the longitude $\lambda_\odot$ of the sun and the longitude $\lambda_\text{\textflorin}$ of the moon is 0°, the first quarter occurs when the difference is 90°, and so on. With reasonable accuracy, however, we may say that D is 0° at the new moon, 90° at the first quarter, and so on. For brevity, I shall adopt the latter usage.

At the new moon or full moon, then, 2D is either 0° or 360°, which amounts to the same thing as 0° when we calculate a trigonometric function. Thus, at the syzygies, Equation VI.4 reduces to

$$e_{c\text{\textflorin}} = 5°.02 \sin M + 0°.22 \sin 2M. \tag{VI.6}$$

At the first or last quarter, when D is either 90° or 270°, Equation VI.4 reduces to

$$e_{c\text{\textflorin}} = 7°.56 \sin M + 0°.22 \sin 2M. \tag{VI.7}$$

We should note that the term 0°.66 sin 2D is zero at all the principal phases and that it makes no contribution to either Equation VI.6 or VI.7.

The difference between Equations VI.6 and VI.7 comes from the term 1°.27 sin (2D - M). It subtracts directly from the leading term in Equation VI.4 at the syzygies (new moon or full moon) and it adds directly to it at the quarters. We know now that this effect comes from the gravitation of the sun acting on the moon, and the derivation of this effect was one of the great triumphs of gravitational theory. The effect itself, however, was already known to Hipparchus, although in terms different from those we have used.

The word evection is sometimes used to mean the quantity 1°.27 sin (2D - M) in Equation VI.4, it is sometimes used to mean the fact that the coefficients in Equations VI.6 and VI.7 are different, and it is sometimes used to mean the difference between the coefficients. For the present, I shall use evection with the last meaning. That is, I shall say that the evection equals 2°.54 according to modern theory.

---

†The reader should refer back to the discussion of the epicycle model in Section IV.2.

When it is used in the lunar theory, the word variation refers to the term $0°.66 \sin 2D$ in Equation VI.4. As we have noted, it equals zero at the syzygies and the quarters. It has its greatest effect when D is an odd multiple of $45°$, which happens at the octant phases, midway between the syzygies and the quarters. It also comes from the gravitation of the sun on the moon.†

Ptolemy builds up a model with properties that correspond to the terms involving $\sin M$ and $\sin 2M$ in Equation VI.4, as well as to the evection. He builds up the model in stages, and I shall follow the order of his development. At first, he deals only with the position of the full moon, which can be represented by the simple epicycle model of Figure IV.1 in Section IV.2. The reader will remember that the model has five parameters that must be found from observation. These are the mean longitude L_0 at some specified epoch, the rate of change n of the mean longitude L, the mean anomaly γ_0 at the same epoch, the rate of change γ' of the mean anomaly γ, and the radius r of the epicycle.

3.   The Equation of Time

Before we take up the model of lunar motion, we must deal with the quantity called the equation of time. We say that the time is noon when the sun is in the meridian plane of a particular location.‡ The interval between successive noons is not the interval that the earth takes to make a full revolution in space, because the sun itself is moving in the same direction as the earth's rotation, as I mentioned in Section I.6. Since the sun does not move with uniform speed, the amount that the earth must rotate between successive noons is also not uniform. The deviation between true local time, in which the sun is in the meridian at the time called noon, and a uniform scale of time, is called the equation of time.

More specifically, when the sun is in the plane of the meridian on a particular day, we say that the time is apparent noon on that day. We may divide the time between apparent noon on day D and apparent noon on day D + 1 into 24 equal "apparent" hours. Time measured in this way is called apparent time. The uniform time scale obtained by averaging apparent time is called mean time. On the average, apparent and mean time are equal. The equation of time is the difference between mean and apparent time, and thus it must be zero on the average.

---

†More accurately, it and the other terms involving the position of the sun, including the evection, come from the fact that the gravitation of the sun on the moon is not the same size as its gravitation on the earth, since the earth and moon are not in the same place.

‡Here I am not talking about conventionalized zone time. I am talking about the measure of time that is unique to any given place, or at least to all places with the same longitude. This is called true local time.

The equation of time arises from two independent phenomena. The first comes from the eccentricity of the sun's orbit. The sun travels faster when it is at perigee than when it is at apogee, and thus a solar day is longer when the sun is at perigee. The maximum angle between the true and mean sun is about twice the eccentricity, if we use eccentricity in its modern meaning. For the sun, e is about 0.0175, and twice this is about 0.035 radians, or about 2°.0. Since it takes the earth 8 minutes to rotate 2°, the eccentricity gives a contribution to the equation of time that goes through a full cycle in a year and that has a maximum value of about 8 minutes.

The second phenomenon comes from the obliquity of the sun's orbit (the ecliptic). Even if the sun moved uniformly around the ecliptic, its motion when projected onto the equator would not be uniform, and it is the projection onto the equator that determines solar time. In particular, we see that the projection onto the equatorial plane moves more slowly than average when the sun is near the equinoxes and that it moves faster than average near the solstices. The oscillation in time produced by this effect goes through a full cycle between the vernal equinox and the autumnal one, and it goes through another full cycle between the autumnal equinox and the next vernal one. Thus it goes through two full cycles a year because of the obliquity, and the maximum effect is slightly more than 9 minutes.†

When the two effects are combined, we have a rather irregular oscillation with two cycles per year. The equation of time is about -14 minutes‡ in mid-February, rises to about +4 minutes in mid-May, falls to about -6 minutes at the end of July, and rises again to more than +16 minutes about the first of November. From there it falls again to the mid-February minimum.

The Greek astronomers, at least as early as Hipparchus, knew about the equation of time and could calculate it.⧧ Discovering the equation of time and developing the theory of it strikes me as one of the high points in Greek astronomy. In 16 minutes, which is the greatest value of the equation, only the moon moves far enough to matter in the context of naked eye observations. Thus the later Greek astronomers, so far as I know, used the equation of time regularly in connection with the lunar motion but never with solar or planetary motion.

---

†More precisely, it is $\frac{1}{4} \sin^2 \varepsilon$ in radians, which is about 0.04 radians. This is about 2°.3, which is equivalent to about 9.2 minutes of time.

‡This means that apparent time is 14 minutes earlier than mean time.

⧧At least, this seems to be the implication of Chapter V.3 of the Syntaxis, in which the equation of time is used in connection with an observation made by Hipparchus. It is possible, of course, that Ptolemy uses the equation in analyzing this observation but that Hipparchus himself had not done so.

Ptolemy describes how to calculate the equation of time in Chapter III.9 of the Syntaxis.

4. The Mean Motions of the Moon

The fact that the angular velocity of the moon goes through an oscillation, having a place where it is a minimum (apogee) and another where it is a maximum (perigee), has been known for a long time. The Babylonians [Neugebauer, 1955, volume 1] knew of this oscillation in velocity and had well-developed methods of including it when they calculated ephemerides of the moon. By following perigee and apogee over long periods of time, they also knew that the positions of the lunar perigee and apogee moved steadily through the heavens, and this knowledge passed to the Greek astronomers.

The Greek astronomers expressed this phenomenon by saying that the mean motion in longitude, which we call $n_{\mathbb{C}}$ here, is not the same as the mean motion in anomaly, which we call $\gamma_{\mathbb{C}}'$ here. By combining Babylonian observations with his own, Hipparchus discovered [Syntaxis, Chapter IV.2] that the interval of 126 007 days plus 1 hour contains 4267 synodic months, 4573 revolutions of the anomaly, and 4612 revolutions of the moon around the heavens, less $7\frac{1}{2}$ degrees.[†]

These figures lead to 29;31,50,8,9 days for the length of the synodic month, as many writers have pointed out. However, Ptolemy says that Hipparchus found from them that the synodic (ordinary) month has 29;31,50,8,20 days. As it happens, 29;31,50,8,20 days is the length of the month that Babylonian astronomers used [Neugebauer, 1955, volume 1, p. 78] in calculating the lunar ephemeris before the time of Hipparchus. Thus the length of the month that Ptolemy uses in his lunar theory has a Babylonian origin and cannot be due to Hipparchus as he claims.

I do not know how to explain this situation. Since Ptolemy says that Hipparchus used Babylonian data in establishing the length of the month, it may be that Hipparchus adopted a Babylonian value and that Ptolemy misunderstood what he had done. This, however, does not account for Ptolemy's statement that 126 007 days plus 1 hour equals 4267 months.[‡]

It does not matter whether one uses 29;31,50,8,9, or 29;31,50,8,20 days for the length of the month. The difference is about 3 parts in 10^8, which is beyond the accuracy of the data by at least an order of magnitude. Luckily for the historian of science, many parameters in ancient and medieval astronomy are given to a number of figures far beyond the number of significant ones. Suppose two parameters

---

[†]That is, the moon failed by $7\frac{1}{2}$ degrees to complete 4612 sidereal revolutions in the time stated.

[‡]It may be that Ptolemy has deliberately deceived us. He may have decided to use the Babylonian month for reasons that we do not know, and wanted to invoke the prestige of Hipparchus in support of it.

from, say, Greece and Babylon are given only to the level of observational accuracy. Agreement of the parameters does not tell us whether there is a cultural relation or not, because the agreement is required by the physical situation. However, if the parameters are given well beyond the level of possible accuracy, and if they agree as far as they are given, it is highly unlikely that the agreement happened by chance. That is, agreement at a non-significant level astronomically almost surely indicates contact at a significant level culturally.

Ptolemy now multiplies the mean motion of the sun in a day, which is 0;59,8,17,13,12,31 degrees if Hipparchus's length of the year is correct, by 29;31,50,8,20 days to get the angular distance that the sun moves in a mean month. He adds 360° to get the distance that the moon moves in the same time. Division by 29;31,50,8,20 days then gives the mean motion of the moon in a day. The result, which I have tested, is 13;10,34,58,33,30,30 degrees per day in longitude. Note that the length of the month was taken to the fourth sexagesimal position, which was about an order of magnitude beyond the accuracy of the data. Ptolemy gives the mean motion to the sixth position; the two added positions represent a factor of 3600, more than three orders of magnitude. Thus the mean motion that Ptolemy uses is more precise than the data warrant by more than four orders of magnitude.

To summarize this point, the value that Ptolemy finds for $n_{\mathbb{C}}$, the mean motion of the moon in longitude, is

$$n_{\mathbb{C}} = 13;10,34,58,33,30,30 = 13.176\ 382\ 215\ 2\ \text{degrees/day}.$$

$$(VI.8)$$

This is significant only to about 1 part in 10^7, and the fourth sexagesimal position has only partial significance.

Ptolemy then uses the finding that 126 007 days plus 1 hour contains 4573 full revolutions of the anomaly, and finds that the daily change $\gamma_{\mathbb{C}}'$ is 13;3,53,56,29,38,38 degrees per day. I cannot verify this value. On the basis of the data given, I find that $\gamma_{\mathbb{C}}'$ is 13;3,53,56,34,21,41 degrees per day. Pedersen [1974, p. 163] summarizes some arguments which indicate that all these parameters have a Babylonian origin. This may well be so, but it is puzzling that Ptolemy quotes data which do not lead to the parameters that he uses.

Whatever may be the explanation, Ptolemy does not adopt the value that he finds here. He will show, he says, from observations that he has made and that he will present later,† that the value of $n_{\mathbb{C}}$ in Equation VI.8 is quite accurate, but that the value found for $\gamma_{\mathbb{C}}'$ must be decreased by [Syntaxis, Chapter IV.3] 0;0,0,0,11,46,39 degrees per day. On this basis, he finally adopts

---

†He presents the data in Chapter IV.6 of the Syntaxis and derives the correction to $\gamma_{\mathbb{C}}'$ in Chapter IV.7.

$$\gamma_{\mathbb{C}}' = 13;3,53,56,17,51,59 = 13.064\ 982\ 860\ 0\ \text{degrees/day}$$

$$(\text{VI}.9)$$

As we shall see in Section VI.7, the change in $\gamma_{\mathbb{C}}'$ is unwarranted.

In the same chapter of the Syntaxis, Ptolemy also finds an angular velocity that describes the moon's behavior in latitude. It is too early to take up this value here, and I shall return to it in Section VI.9.

5. The Use of Lunar Eclipses in Studying the Lunar Motion

After he finds the mean motions given in the preceding section, Ptolemy finds a model that will account for the position of the moon at syzygy. After that, he will find the corrections needed at other phases because of the evection and other observed departures from the simple model. The model that he uses for the syzygy positions is the simple epicycle model of Figure IV.1.

We should notice one consequence of the mean motions derived in the preceding section. Since the value of $\gamma_{\mathbb{C}}'$ is slightly less than the value of $n_{\mathbb{C}}$, the radius r of the epicycle rotates slowly with respect to the equinox. The difference between the values in Equations VI.8 and VI.9 is 0.111 399 355 2 degrees per day. Thus the epicycle radius makes a full revolution in about 3232 days, about 8.85 years. It rotates in the counter-clockwise direction, which is the direction in which the moon itself moves.

In Chapter IV.1 of the Syntaxis, Ptolemy discusses the kind of observation that should be used in establishing the theory of the moon. We should not trust to luck by using all kinds of observations indiscriminately, he says. It is important that we restrict ourselves to the observations which are not only the oldest but which are also based upon lunar eclipses; by these alone can we find the true position of the moon. All other observations, whether they are based upon the position of the moon with respect to the stars, or made with the aid of instruments, or made during solar eclipses, can produce error because of parallax. Thus, in developing the theory of the lunar position at the syzygies, he will use only measurements of lunar eclipses, which are not affected by parallax.

It is correct that the measured times when lunar eclipses begin and end are not affected strongly by parallax. The point on the moon that first touches the earth's shadow can be seen by everyone for whom the moon is above the horizon and, if this event could be observed, there would be no effect of parallax at all. However, particularly for observations made with the naked eye, the beginning cannot be seen until an appreciable part of the moon is immersed in the shadow, and the time may now be slightly affected by parallax. This effect of parallax, though present, is certainly much less than the effect of parallax upon other kinds of measurement.

The point about the value of eclipses, if I have read it correctly, is of little importance, and it does not deserve the emphasis that Ptolemy puts on it. The effect of parallax upon a measurement can be calculated quite accurately and hence removed from account. The real question is then: What kind of lunar measurement can be made with the greatest accuracy? Is it, for example, the time when the moon begins to occult a star? Or is it the time when a lunar eclipse begins or ends? I am not sure of the answer, but I suspect that the time of an occultation is considerably more accurate than the time of a lunar eclipse. The edge of the earth's shadow is rather diffuse, because of refraction in the earth's atmosphere, and the beginning or end of an eclipse is not a well-defined event. The edge of the moon, on the other hand, is sharp, and the beginning or end of an occultation is well defined.

The time that is wanted from a lunar eclipse is the middle of the eclipse, when the longitudes of the sun and moon differ by $180°$.[†] The middle cannot be judged at all accurately with the eye. Ideally, then, what should be done is to measure the times of beginning and end and take the average as the time of the middle. Sometimes, particularly with the very old observations, only the beginning was measured, but this problem can be overcome. The duration of an eclipse can be calculated, and thus the time of the middle can be estimated from the time of the beginning or end alone.[‡] From the time of the middle, we calculate the longitude of the sun, and this gives us the longitude of the moon.

Each eclipse thus gives us the longitude of the moon at a measured time. Now the epicycle model of Figure IV.1 has five parameters that must be found from the analysis of observations. We have already found two of these, namely $n_{\mathbb{C}}$ and $\gamma_{\mathbb{C}}'$, leaving three to be found. These three are $L_{\mathbb{C}O}$, the mean longitude at some epoch, $\gamma_{\mathbb{C}O}$, the anomaly at some epoch, and $r_{\mathbb{C}}$, the radius of the epicycle.

In finding the longitude of the moon, it is the ratio of r to the radius of the deferent circle that matters. For the time being, we are free to choose the deferent radius arbitrarily. When I am discussing Ptolemy's results, I shall follow Ptolemy's choice and take the radius of the deferent to be 60.

In finding the three parameters just enumerated, Ptolemy says [Chapter IV.5] that he will follow the method that Hipparchus has already used. This is to use the middle times of three eclipses. This really means that he will use the position of the moon at three measured times since, as we have seen, this is the information that eclipses give us.

[†]This is not exact for an individual eclipse, but the error in it is small in the context of naked-eye observations. Further, the error in the statement is zero when averaged over all eclipses.

[‡]Note that Ptolemy is willing to let his results depend upon calculating the duration of an eclipse but not upon calculating the parallax.

Each position gives an equation connecting the three unknown parameters, and three equations are needed in order to solve for the unknowns.

Ptolemy finds the parameters in Chapter IV.6 of the Syntaxis by using a geometrical method that Neugebauer [1957, pp. 210ff] summarizes. I have checked his solution by a different method that depends but little upon geometry. In describing it, let me omit the subscript $\mathbb{C}$ which tells us that we are dealing with the moon, and let me use instead a numerical subscript that tells which eclipse we are dealing with. Let λ_1, L_1, and γ_1 be the true longitude, mean longitude, and anomaly, respectively, at the middle of the first eclipse, and let r be the radius of the epicycle. From Figure IV.1, it is straightforward to find the relation

$$\tan \lambda_1 = \frac{\sin L_1 + r \sin (L_1 - \gamma_1)}{\cos L_1 + r \cos (L_1 - \gamma_1)}. \qquad (VI.10)$$

We have similar relations for the second and third eclipses that involve λ_2, L_2, γ_2, λ_3, L_3, and γ_3. All the λ's are known from the observations. If we let t_1, t_2, and t_3 denote the times of the eclipses, $L_2 = L_1 + n_{\mathbb{C}}(t_2 - t_1)$, and we have similar relations for γ_2, L_3, and γ_3. Thus the L's and γ's for the second and third eclipses are written in terms of L_1 and γ_1, and we have three equations for the three unknowns L_1, γ_1, and r.†

TABLE VI.1

TWO TRIADS OF LUNAR ECLIPSES USED BY PTOLEMY

Date	Hour[a]	$\lambda_{\odot}$
-720 Mar 19	20 $\frac{2}{3}$	354 $\frac{1}{2}$
-719 Mar 8	23 $\frac{1}{6}$	343 $\frac{3}{4}$
-719 Sep 1	19 $\frac{2}{3}$	153 $\frac{1}{4}$
133 May 6	23 $\frac{1}{4}$	43 $\frac{1}{4}$
134 Oct 20	23	205 $\frac{1}{6}$
136 Mar 6	04	344 $\frac{1}{12}$

[a]In mean time at Alexandria.

---

†Since r occurs linearly in the equations, it can be eliminated readily, leaving two equations in two unknowns. These must be solved by iteration, but the process is straightforward and converges rapidly.

As I just said, Ptolemy writes in his Chapter IV.5 that
he will follow Hipparchus's method of finding the lunar
parameters by using triads of eclipses. He devotes the rest
of this chapter to proving that the eccentric model of Figure IV.3 and the epicycle model of Figure IV.1 give the same
quantitative results for the longitude; we saw this in Section IV.3 above. He first gives data on triads of eclipses
in his Chapter IV.6, where he uses two such triads. The
information about these triads is summarized in Table VI.1,
but the matter is considerably more complex than a simple
table can indicate.

Ptolemy first takes up three eclipses that were already
about as ancient to him as William the Conqueror is to us.
The first happened in the night between the dates that we
call -720 March 19 and 20. It began more than an hour of the
night after the moon rose at Babylon, and it was total. Since
the moon was full, we can equate moonrise to sunset. An hour
of the night in this context means a twelfth part of the
interval between sunset and sunrise. For a date so near the
equinox, Ptolemy says, we can take this to be equal to an
hour in its meaning of 1/24 of a day, so that the eclipse
began $4\frac{1}{2}$ hours before midnight. Since the eclipse was total,
its middle came 2 hours later, at $2\frac{1}{2}$ hours before midnight.
Finally, since the time difference between Babylon and Alexandria is 5/6 hours,† the time was 3 1/3 hours before midnight at Alexandria. This is the hour entered in Table VI.1.
The longitude of the sun at this time was $354\frac{1}{2}$ degrees, according to Ptolemy's solar theory.

The second eclipse is reported in a quite different
manner. It came during the night between -719 March 8 and 9,
and only the southern fourth of the moon was eclipsed. Since
the middle of the eclipse came at midnight at Babylon, it
came at 5/6 hours before midnight at Alexandria. This is the
time entered in Table VI.1, but Ptolemy makes one more correction to the time that I shall discuss in a moment. The
longitude of the sun was $343\frac{3}{4}$ degrees.

The third eclipse is reported in the same manner as the
first. It came in the night between -719 September 1 and 2.
The eclipse began after the moon rose, and the moon was
eclipsed by more than half on its northern side. This means,
according to Ptolemy, that the eclipse began at 0.5 hours of
the night. On this date, the length of the night was 11
ordinary hours, so the eclipse began 5 ordinary hours before
midnight. An eclipse of this magnitude lasts three hours, so
the middle was at $3\frac{1}{2}$ hours before midnight, Babylon time, or
4 1/3 hours before midnight, Alexandria time. This is the
time entered in the table; again there is another correction
yet to be noted. The longitude of the sun was $153\frac{1}{4}$ degrees.

---

†The correct interval is about 58 minutes. Since the important results depend more upon the intervals between eclipses
than upon the actual time of any eclipse, using 50 minutes
rather than 58 minutes has little consequence.

The further correction that Ptolemy makes to the times of the second and third eclipses comes from the equation of time. He says, after giving the above data, that the time interval between the first and second eclipses was 354 days plus $2\frac{1}{2}$ hours in apparent time, or 354 days plus (2 + 1/2 + 1/15) hours in mean time. As well as I can make out, Ptolemy leaves the time of the first eclipse unchanged, and changes only the time of the second eclipse to make it agree with the interval of 354 days plus (2 + 1/2 + 1/15) hours.

Similarly, he says that the interval between the second and third eclipses is 176 days plus $20\frac{1}{2}$ hours of apparent time, or 176 days plus 20 1/5 hours of mean time. He alters the third time accordingly.

It is clear that there is much uncertainty in assigning the times of the first and third eclipses. "More than an hour" does not necessarily mean $1\frac{1}{2}$ hours, as Ptolemy takes it; similarly, "after the moon rose" does not necessarily mean half an hour after moonrise. Of course, Ptolemy was in a better position than we are to know what the Babylonian conventions were. If they gave time only by saying "after H hours" or "after more than H hours", and if they never said, for example, "after less than H hours", we can do no better than to take "more than an hour" as $1\frac{1}{2}$ hours and to take "after moonrise" as $\frac{1}{2}$ hour. In doing so, we must recognize that the times are uncertain by about 15 minutes.

Ptolemy says that half of the first eclipse lasted 120 minutes and that half of the third eclipse lasted 90 minutes. Oppolzer [1887] lists these times as 110 minutes and 78 minutes, respectively. Thus Ptolemy has a systematic error in the times of the first and third eclipses. We do not know how the time of the second eclipse was found, and we do not know whether the same error is present in it. If all three eclipses contain the same error, it is of little consequence.

People do not have to be consistent in their habits, but they usually are fairly so. Thus I find it remarkable that the style of the second eclipse is so different from that of the first and third ones. I shall return to this matter later.

The next triad of eclipses that Ptolemy reports are some that he claims to have observed himself "with great accuracy"; these are also summarized in Table VI.1. The middle of the first came, he says, at 3/4 hours before midnight between 133 May 6 and 7,† and it was total. The longitude of the sun was $43\frac{1}{4}$ degrees. The middle of the second came 1 hour before the midnight between 134 October 20 and 21; it was three-fourths eclipsed on the northern side. The longitude of the sun was 205 1/6 degrees. The middle of the third came at 4 hours after midnight between 136 March 5 and 6; it was half eclipsed on the northern side. The longitude

---

†I hope it is clear that Ptolemy did not use our calendar. I have translated all his dates into the Julian calendar. It will be necessary to consider how he wrote dates only when there is a problem connected with a date.

of the sun was 344 1/12 degrees.†

Ptolemy then gives the time intervals. The first interval is 1 Egyptian year‡ plus 166 days plus 23 3/4 hours in apparent time, but 23 5/8 hours in mean time. The second interval is 1 Egyptian year plus 137 days plus 5 hours in apparent time, but $5\frac{1}{2}$ hours in mean time.

The longitude of the sun given in all these records is calculated rather than observed, but this is not fabrication under the circumstances. One of the premises in the eclipse method is that the position of the moon is to be taken from the calculated position of the sun.

Before I give the parameters that are found from these two triads of eclipses, let me present the other two triads that Ptolemy gives. They are found in Chapter IV.11 of the Syntaxis. In discussing them, it is useful to introduce a parameter E, called the maximum equation of the center. In the epicycle or eccentric model, the equation of the center is given by (Equation IV.5 in Section IV.2)

$$\tan e_c = -r \sin \gamma/(1 + r \cos \gamma).$$

In this, r is the epicycle radius or eccentric distance (more properly, the ratio of the radius or distance to the radius of the deferent), and γ is the anomaly measured from apogee. The maximum value of e_c comes when the radius of the epicycle or the eccentric, as the case may be, is perpendicular to the line of sight. Then

$$E = \sin^{-1} r. \tag{VI.11}$$

It comes when γ = E + 90°.

In Chapter IV.11 of the Syntaxis, Ptolemy says that the value of E that he has found from the first two triads of eclipses differs from what Hipparchus had found. Hipparchus had analyzed one triad of eclipses using the epicycle model and found r = 6;15 (if the deferent radius is 60) and E = 5;49 degrees. Apparently there was a slip of the pen here, because the value of E for this value of r is 5;59 degrees. Hipparchus analyzed another triad of eclipses using the eccentric model and found r = 4;46 and E = 4;34 degrees; the value of E should be 4;33 degrees, so Ptolemy has made a trifling error.

Ptolemy goes on to say that the difference does not arise from using different models, as some people have supposed, because the epicycle and eccentric models give identical results if they are based upon the same data. Further, there is no appreciable error in the data. The difference arises, he says, because Hipparchus made errors in his calculations, as he will now demonstrate.

---

†In Greek numerals, the fraction of a degree was written as ιβ′ in one text. In an earlier work [Part III], I misread this as 12′.

‡The Egyptian year had exactly 365 days.

The first triad of eclipses was observed in Babylon. The first eclipse in this triad was the eclipse of -382 December 23. There was only a small part of the moon that was eclipsed on its northeast side when half of an hour of the night remained, and the moon set still eclipsed. "Hour of the night" means a twelfth part of the interval between sunset and sunrise, as it did in the earlier Babylonian records. This sounds to me as if the stated time was the middle of the eclipse, but Ptolemy takes it to be the beginning, and he takes the half-duration to be 45 minutes.† As with the other eclipses, this value is presumably based upon a calculation. After he applies all of the corrections, including the equation of time, Ptolemy concludes that the middle of the eclipse came at $6\frac{1}{4}$ hours after midnight, Alexandria time. At this time, the longitude of the sun was 268;18 degrees.

The second eclipse was that of -381 June 18. The moon was eclipsed from the northeast side when the first hour of the night had already passed. Ptolemy takes this to mean that the eclipse began at $1\frac{1}{2}$ hours of the night. He does not give the magnitude of the eclipse, but he takes the half-duration to be 90 minutes.‡ After making the necessary corrections, he concludes that the middle of the eclipse came at 19 5/6 hours, mean time, when the longitude of the sun was 81;46 degrees.

The third eclipse was that of -381 December 12. It began from the northeast side during the fourth hour of the night; Ptolemy thus takes the beginning to be at $3\frac{1}{2}$ hours of the night, and he takes the half-duration to be 2 hours.‡ After making the necessary corrections, he concludes that the middle of the eclipse came at 21 5/6 hours, mean time, when the longitude of the sun was 257;30 degrees.

Ptolemy also gives the relevant intervals. The interval from the middle of the first eclipse to the middle of the second was, he says, 177 days plus 13 3/5 hours, and the sun advanced 173;28 degrees in longitude. Hipparchus, however, took the interval to be 177 days plus 13 3/4 hours, and he took the advance of the sun to be 173 degrees less an eighth of a degree.

The relevant data are summarized in Table VI.2. There I list, for each date, the hour and the longitude of the sun which Ptolemy says that Hipparchus used. I then list the hour and the longitude of the sun which Ptolemy says is correct. A few comments must be made. For the second eclipse, Ptolemy says that the hour was 19;50. However, if we add the interval that he states to the first time, we get the hour to be 19;51. The difference is unimportant. In preparing Table VI.2, I have preserved the stated interval,

---

†Oppolzer [1887] lists 52 minutes as the half-duration for this eclipse.

‡Oppolzer lists 75 minutes.

‡Oppolzer lists 110 minutes. The eclipse was total, as Ptolemy states.

since it is more important than the absolute value of the time.

TABLE VI.2

TWO TRIADS OF LUNAR ECLIPSES USED BY
HIPPARCHUS AND ANALYZED AGAIN BY PTOLEMY

Date[a]	Values that Ptolemy Attributes to Hipparchus		Ptolemy's Values		E, degrees	
					Values that Ptolemy Attributes to Hipparchus	My Solution
	Hour[b]	$\lambda_\odot$	Hour[b]	$\lambda_\odot$		
-382 Dec 23	06;15	268;18	06;15	268;18	5;49	5;18
-381 Jun 18	20	81;10	19;51	81;46		
-381 Dec 12	21;40	257;18	21;51	257;30		
-200 Sep 22	18;30	176;06	18;30	176;06	4;34	4;51
-199 Mar 19[c]	00;30	356;27	01;20	356;17		
-199 Sep 12	00;50	165;00	01;44	165;12		

[a]Greenwich time.

[b]Mean time at Alexandria.

[c]Eclipse was on -199 Mar 20 at Alexandria.

Ptolemy does not say whether Hipparchus made an error in the time and the position of the sun for the first eclipse or not. In order to see what happens, I assume in the table that Hipparchus used the same data for the first eclipse as Ptolemy. This gives 20 hours for the second eclipse and 81;10,30 for the position of the sun. For simplicity, I drop the 30″ in Table VI.2.

From the middle of the second to the middle of the third eclipse, Ptolemy says that the correct interval is 177 days plus 2 hours, and that the sun moved 175;44 degrees. Use of these values gives "Ptolemy's values" for the eclipse of -381 December 12 in the table. He says that Hipparchus took the interval to be 177 days plus 1 2/3 hours, and he took the advance of the sun to be 175;8 degrees.

I believe that there has been another slip in the last figure. If we use it, we must believe that Hipparchus made an error of more than 1 degree in the motion of the sun in less than a year. Since Hipparchus is the source for Ptolemy's own tables of the sun,† I find such an error incredible, and I assume that the correct value is 176;8 degrees.

†We should remember that the position of the sun is supposed to be calculated rather than observed in this method of using eclipses.

In the table, I have listed 81;10 for the second longitude of the sun according to Hipparchus. The difference
between this and the first longitude is 172;52, and it is
plausible that Ptolemy would have stated this as 173 degrees
minus an eighth. It is not plausible that Hipparchus would
have stated the longitude as 81;10,30. Hence I think that
it is reasonable to adopt the value 81;10.

I started with the assumption that Hipparchus and Ptolemy used the same data for the first eclipse, and I constructed the "values that Ptolemy attributes to Hipparchus"
by using this assumption and the intervals that Ptolemy
states. I was forced to this procedure because Ptolemy
does not give the actual times and solar positions that
Hipparchus used. The hours and longitudes found in this
way are all values that Hipparchus might reasonably have
stated, and this is not so for many assumptions about the
starting values. However, a variety of starting values do
lead to plausible later values, and the assumption is not
solidly established. In the use that I shall make of Table
VI.2, the differences are more important than the actual
values, and I shall stay with the assumption for the sake
of definiteness.

Let us turn to the second of Hipparchus's triads, the
one that he used with the eccentric model. These were observed at Alexandria, and the first was the eclipse of -200
September 22. The eclipse began half an hour before the
moon rose and ended at $2\frac{1}{2}$ hours of the night; hence the
middle was at 1 hour of the night.† After he makes the usual
corrections, Ptolemy concludes that the middle of the eclipse
came at $18\frac{1}{2}$ hours, mean time. It is obvious that the data
in this record are not the raw data; the beginning could not
have been observed if it came half an hour before the moon
rose. If this eclipse was actually observed, the middle and
the beginning must have been inferred from the end and from
the half-duration, and Ptolemy has suppressed the actual
data. The longitude of the sun was 176;06 degrees. I have
entered the time and position in Table VI.2, both as Ptolemy's values and the values that he attributes to Hipparchus. Here I am making the same assumption that I made for
the triad beginning on -382 December 23.

The second eclipse, which was on -199 March 19, began
at 5 1/3 hours of the night, and it was total. Hence Ptolemy takes the half-duration to be 2 hours; Oppolzer lists
108 minutes. Ptolemy concludes that the middle of the
eclipse came at 1 1/3 hours, mean time, and that the longitude of the sun was 356;17 degrees. Hence the difference
between the first and second eclipses were 178 days plus
6 5/6 hours in time and 180;11 degrees in longitude. Hipparchus, however, took the differences to be 178 days plus
6 hours in time and 180;20 degrees in solar longitude.

The third eclipse, which was on -199 September 12,
began when 6 2/3 hours of the night had passed, and it was

---

†Oppolzer lists 89 minutes as the half-duration of this
eclipse.

total. According to Hipparchus, the middle came when 8 1/3 hours of the night had passed, that is, at 2 1/3 hours of the night after midnight. Ptolemy finally concludes that the middle of the eclipse came at 1 3/4 hours, mean time, and that the longitude of the sun was 165;12 degrees. This makes the differences between the second and third eclipses equal to 176 days plus 2/5 hours in time† and 168;55 degrees in the longitude of the sun. Hipparchus took the differences to be 176 days plus a third of an hour in time and 168;33 degrees in the longitude of the sun.

The times and longitudes according to Ptolemy and Hipparchus are listed in Table VI.2. First look at the longitudes attributed to Hipparchus. If we use the intervals that Ptolemy says Hipparchus used, we have 356;26 for the second eclipse and 164;59 for the second. These are implausible values for Hipparchus to have stated, but they become plausible if we use 180;21 rather than 180;20 for the first difference. Since Ptolemy does not try to preserve accuracy to the minute, I think that he altered the first difference by 1 minute, and I think that Hipparchus's longitudes were probably those listed in the table.

Now look at the hours for the second and third eclipses. If we use the intervals which Ptolemy says that Hipparchus used, we get 00;30 and 00;50 hours, respectively; these are almost an hour too early. Further, Ptolemy says that Hipparchus put the middle of the third eclipse at 2 1/3 hours of the night after midnight, as we say above. This is compatible with taking 01;50 hours in mean time, but not with taking 00;50 hours. Thus I think that Hipparchus took the time interval between the first and second eclipses to be 178 days plus 7 hours, not plus 6 hours as Ptolemy says he did. However, in preparing the table, I have used the time intervals that Ptolemy states.

I cannot make much sense out of what Ptolemy says about Hipparchus's use of these eclipses. I find it hard to believe that Hipparchus made an error of about 36' in calculating the motion of the sun between -382 December 23 and -381 June 18. I also find it hard to believe that he made an error of almost an hour in the time interval between the eclipses of -200 September 22 and -199 March 19, although this may be a slip in writing, as I said above. What is worse, I cannot reconcile the data with the values of E, the maximum equation of the center, that are found from them.

In Table VI.2, I give the values of E which Ptolemy says that Hipparchus found and the values that I find by analyzing the data independently, using the method described in Section VI.5. He says that Hipparchus found E = 5;49 degrees from the first triad, while I find 5;18. He says that Hipparchus found E = 4;34 from the second triad, while I find 4;51. I shall return to the significance of this point in the next section.

---

†This puts the hour at 1;44 while Ptolemy finds 1;45 hours from analysis of the record. This means that the difference was 176 days plus 25 minutes, which Ptolemy changed to 2/5 hours.

The title of this section implies that all four triads of eclipses, both those in Table VI.1 and those in Table VI.2, were fabricated. I now turn to the proof of this assertion.

TABLE VI.3

PARAMETERS INFERRED FROM FOUR TRIADS
OF LUNAR ECLIPSES USED BY PTOLEMY

Years	Present Solution			Ptolemy's Solution		
	L_2	γ_2	E	L_2	γ_2	E
-720/-719	164;44	12;18	4;59	164;44	12;24	4;59
-720/-719[a]	164;45	13;01	4;50	---	---	---
133/136	29;31	64;38	5;00	29;30	64;38	5;00
-382/-381	263;59	28;25	5;02	263;58	27;37	5;01
-200/-199	181;07	109;39	5;00	181;07	109;28	5;01

[a] A solution with the time of one eclipse changed by 0.01 days in order to test the sensitivity to measurement error.

7. The Proof of Fabrication

From each of the first two triads of eclipses, Ptolemy derives the values of three parameters. The first parameter is L_2, the mean longitude of the moon at the time of the middle eclipse in each triad. The second parameter is γ_2, the anomaly of the moon at the same time. The third one is r, the radius of the lunar epicycle. In order to facilitate comparison with modern results, I have chosen to replace this by E, the maximum equation of the center, which is derived from r by means of Equation VI.11.

For the last two triads of eclipses, Ptolemy gives the mean longitude and the anomaly at the time of each eclipse. He also says that r, "according to our reckoning", is $5\frac{1}{4}$ if the deferent radius is 60. It is not clear whether he means that this is the value he has found by actual analysis of these triads, or whether it is what he has found from other data.

The parameters inferred from the four triads of eclipses are summarized in Table VI.3. For each triad, I give the values of L_2, γ_2, and E that I have found by an independent solution, as well as the values that Ptolemy finds. For the first triad, I also give the values that I find by altering the time of the second eclipse; I shall discuss the significance of this solution in a moment.

My solutions agree as closely as we can expect for the parameters L_2 and E in every case. The solutions for γ_2 agree exactly for the second triad (years 133 to 136), but they disagree by amounts ranging from 6' to 48' for the other triads. However, the longitude of the moon is rather

-122-

insensitive to the value of $\gamma_\mathbb{C}$, so, conversely, the value
inferred for γ_2 is highly sensitive to small details of the
computational procedure. Thus the differences in γ_2 are not
important.

The interesting thing about Table VI.3 is the close
agreement of the values of E for each triad. The first thing
we must do in interpreting this result is to study the effect of error in measuring the times. In order to study this
effect, I changed the time of the second eclipse in the first
triad by 0.01 days, or 14.4 minutes, while leaving the other
data unchanged. The result was to change the value of E by
9'. It also changed the value γ_2 by 43', showing the extreme
sensitivity of this parameter.

Even if the clocks used to find the times had no error,
the way in which most of the times are stated leaves an uncertainty of at least 15 minutes in each time. I think it
is reasonable to take 0.01 days as the standard deviation in
a measurement of the time; if anything, the correct value is
probably larger than this. Further, there is considerable
uncertainty in deciding when the eclipse began. To be conservative, however, I shall ignore this point. Now we change
E by 9' when we change an individual time by one standard
deviation. If E has the same sensitivity to each time, the
standard deviation of the error in E is $\sqrt{3}$ times the change
due to an individual error. That is, the standard deviation
in the value of E found from a triad of eclipses should be
$9'\sqrt{3}$; let us say 15' for convenience.

All of Ptolemy's values of E agree within 1', which is
1/15 of a standard deviation. The probability that two consecutive values of E should agree this closely is 1 chance
out of 40, and the probability that four consecutive values
should agree is 1 chance out of $(40)^3$, which is 1 chance out
of 64 000. This chance is so small that we can neglect it
for practical purposes.

We must conclude that the agreement of the values of E
in Table VI.3 could not have happened by the chance occurrence of measurement errors. It could have happened only if
three of the triads were fabricated to make them agree with
the remaining one, or if all four were fabricated.

Parenthetically, we should note how Ptolemy handles the
results, which he usually expressed by means of the epicycle
radius r. He takes the deferent radius to be 60. From the
first triad (years -720 and -719), he finds r = 5;13 and
from the second triad (years 133 to 136, the one that he
claims to have observed himself) he finds 5;14. In the rest
of the Syntaxis, including the discussion of the other triads, he rounds to 5;15 which he usually writes as $5\frac{1}{4}$.

The slight variation in E shown in Table VI.3 is consistent with the hypothesis of fabrication. In the fabrication, the data were fabricated using Ptolemy's theory of
the moon. Rounding of numbers necessarily occurred in the
process of calculating the fabricated data and, in any case,
the numbers had to be rounded so that they would look like

plausible results of measurement.† Then, when the fabricated data are solved for the parameters, there is necessarily a small amount of scatter in the final results.

In Part III, I produced strong evidence that the second
triad of eclipses was fabricated. The argument was based
upon the fact that the mean rates of motion of the sun and
moon used in Ptolemy's tables have some error, and eclipses
actually observed in his own time should show systematic departures from his solar and lunar tables. Nonetheless, if
we calculate the circumstances of the eclipses from his solar and lunar theories, the agreement is exact within the
scatter that we expect from the fabrication process.

The argument based upon the values of E leaves open the
possibility that one triad of eclipses is genuine. As a
possibly genuine one, we can eliminate the triad that Ptolemy
claims to have observed by the argument of Part III. We can
also eliminate the first and oldest triad. In order to do so,
we must ask the question: What is the correct value of E?
The answer is given by Equation VI.6. The maximum value of
$e_{c\,\mathbb{C}}$ from that equation is $5°.04$, slightly more than $5;02$
degrees. This is remarkably close to Ptolemy's value. However, the agreement must be an accident. If the measurements
of the first triad are highly accurate, the value of E should
not agree with the correct value.

The reason for this comes from the equation of the center for the moon. I have given the four largest terms in
this equation in Equation VI.4. The largest term that has
been omitted from Equation VI.4, $\delta e_{c\,\mathbb{C}}$, say, equals

$$\delta e_{c\,\mathbb{C}} = -\,0°.1856 \sin M_{\odot}. \qquad (VI.12)$$

In this, $M_{\odot}$ is the mean anomaly of the sun, using the term
mean anomaly in its modern sense. At the eclipses of -720
March 19 and -719 March 8, which are the first two eclipses
in the first triad, $M_{\odot}$ was slightly more than $90°$, and at
the eclipse of -719 September 1 it was slightly more than
$276°$. Thus $\delta e_{c\,\mathbb{C}}$ was close to $-0°.18$ for the eclipses in
March and close to $+0°.18$ for the eclipse in September.
When he analyzed this triad of eclipses, Ptolemy should not
have used the correct values of $\lambda_{\mathbb{C}}$, the longitude of the
moon, although he had no way to know this. He should have
replaced $\lambda_{\mathbb{C}}$ by $\lambda_{\mathbb{C}} + \delta e_{c\,\mathbb{C}}$.

I have repeated the analysis of this triad after making
this replacement. The results are

$$L_2 = 164;42, \qquad \gamma_2 = 9;57, \qquad E = 4;52,$$

all in degrees. If this triad of eclipses gives an accurate
value of E, it can only be because accidental errors of measurement happen to compensate for the effect of $\delta e_{c\,\mathbb{C}}$. In
this case, the value obtained agrees with a specific value,

---

†For example, Ptolemy could not say that an eclipse began 2
hours and 23 minutes after midnight. No one in or before
his time measured times that closely.

namely 5;02 degrees, within 3' or 0.2 standard deviations. The odds against this happening by chance are 11 to 1. This is large, but not astronomically so.

We should also ask whether either of the last two triads in Table VI.3 is genuine, bearing in mind that at most one can be. As far as the theory of errors is concerned, all we can say is that the preceding calculation applies approximately to either of the last two individually. However, Ptolemy treats both triads in exactly the same way, and this makes it likely that both are genuine or both are fabricated. Since at least one is fabricated, this makes it likely that both are.

For the moment, let us assume that all four triads are fabricated. If we assume this, we must immediately ask the question: Where did the fabricator get the value of E that he used in the fabrication? Answering this question is easy, and all the bits of the puzzle fall into place as soon as we have answered it.

Earlier [Part III], I wrote that the value that Ptolemy used for E came from Hipparchus, and Dreyer [1905, p. 164] says the same thing. Dreyer does not identify the source of his statement, but my source was the first part of Chapter IV.5 of the Syntaxis, where Ptolemy says that he will use the same method that Hipparchus used in taking three eclipses and finding the greatest difference from the mean motion.† Thus, when Ptolemy proceeded to give the circumstances of the eclipses of -720 and -719, I thought he meant that Hipparchus had used these eclipses, and that he was following Hipparchus here. On further reflection, I doubt this interpretation. Ptolemy does not introduce the eclipses of -720 and -719 until his Chapter IV.6, and when he does so, he does not mention Hipparchus in connection with them. He merely says that he is choosing three ancient eclipses that appear to have been well observed. When he does mention Hipparchus again, in Chapter IV.11, he emphasizes the difference between his results and those of Hipparchus.

Now let us look at the values of E from the eclipses that Hipparchus did use.‡ If we use the values of E that actually correspond to the data, the average value of E is 5;04,30 degrees. If we use the values which Ptolemy says that Hipparchus found, the average is 5;11,30. Either value suggests 5° as a convenient whole number. We shall see in later parts of this work that the fabricator had a predilection for whole numbers. Thus it is quite likely that he should have chosen E = 5° on the basis of Hipparchus's work, even though this may not have been a value that Hipparchus himself ever obtained. It is a coincidence that the round number is rather accurate. The fabricator was not always so lucky when he chose a round number.

We may thus assume that the fabrication was done

---

†By this I think he means what we have called the maximum equation of the center.

‡At least, he used them according to Ptolemy's statement.

TABLE VI.4

CALCULATED AND STATED VALUES OF
LONGITUDE FOR FOUR TRIADS OF ECLIPSES

Date	Calculated $\lambda_{\mathbb{C}} \pm 180°$		Stated $\lambda_\odot$		Calculated $\lambda_\odot$	
	°	′	°	′	°	′
-720 Mar 19	354	30	354	30	354	38
-719 Mar 8	343	45	343	45	343	45
-719 Sep 1	153	15	153	15	153	19
133 May 6	43	15	43	15	43	14
134 Oct 20	205	9	205	10	205	8
136 Mar 6	344	5	344	5	344	3
-382 Dec 23	268	16	268	18	268	16
-381 Jun 18	81	49	81	46	81	44
-381 Dec 12	257	28	257	30	257	29
-200 Sep 22	176	8	176	6	176	5
-199 Mar 19	356	16	356	17	356	16
-199 Sep 12	345	13	165	12	165	11

in order to yield the value E = 5°. When the fabrication
started, there was no need to get any particular values of
$L_{\mathbb{C}}$ and $\gamma_{\mathbb{C}}$ at any specific epoch. Hence, in the first triad
that was used, two eclipses could be genuine and only one
needed to be fabricated. In the oldest triad of eclipses,
which is also the first one that Ptolemy discusses, we remember that Ptolemy gives no details about how the middle
time of the second eclipse was found, but that he gives
elaborate details about the other eclipses in the triad.
Further, he states the mean longitude and anomaly for this
eclipse but not for the others. This suggests that only the
second eclipse in this triad was fabricated. The mean longitude and anomaly could be stated for this same eclipse, because they would then furnish the basis for Ptolemy's tables
of these quantities. They could not be stated for the other
eclipses, however: If these eclipses are genuine, the mean
longitude and anomaly for them are not expected to be consistent with the values for the fabricated eclipse and they
are not expected to be consistent with Ptolemy's tables.

This hypothesis is tested quantitatively in Table VI.4,
for all four triads of eclipses. The first column gives the
dates of the 12 eclipses. The second column gives the values
of the lunar longitude at the middle times of the eclipses,
as calculated from Ptolemy's theory. The longitude of the
moon is then changed by 180° so that it may be compared with

the longitude of the sun in the next column. I calculated
the lunar longitudes by using Equation IV.5 directly, instead of using Ptolemy's tables for the lunar equation of
the center. I tried using his tables before I did this, but
I found the needed interpolation to be ambiguous. That is,
for several eclipses, there was no way to decide how Ptolemy
might have done the interpolation.

The third column gives the longitudes of the sun which
Ptolemy states for the middle time of each eclipse. This
column should be compared with the preceding one. In every
case, the disagreement is one that could easily arise from
the computations themselves. The largest disagreement is 3',
for the eclipse of -381 June 18, and for this case a plausible interpolation in Ptolemy's table [Chapter IV.10 of the
Syntaxis] gives exact agreement.†

It is out of the question that this agreement happened
by chance. The average disagreement is exactly 1'; for simplicity, let us suppose that all disagreements are this size.
Suppose that the standard deviation in a measurement of the
time is only 15 minutes, which is probably an underestimate.
In 15 minutes, the moon moves about 8' with respect to the
sun, so the standard deviation of the difference between the
two columns should be 8'. Thus, the values in the second
column agree with pre-assigned values within one-eighth of a
standard deviation in each of twelve cases. The probability
that this could happen by accidental errors of measurement is
almost exactly 0.1 for a single case. The probability that
this could happen by chance for all 12 eclipses is one chance
out of 10^{12}.

Now look at the last column in Table VI.4, which gives
the lóngitude of the sun as calculated from Ptolemy's theory
for the middle time of each eclipse. For the solar longitudes, there is little ambiguity in the values that would be
found by interpolating in Ptolemy's table of the sun [Chapter
III.6 of the Syntaxis], so I found the values in the fourth
column by using that table. The agreement is exact for the
eclipse of -719 March 8, the second eclipse in the first
triad, and it is never more than 2' for any eclipse in the
last three triads. The disagreements for the last three
triads could easily arise from the computations themselves.
For the eclipses of -720 March 19 and -719 September 1, however, the disagreements are 8' and 4', respectively. I have
found no likely way to account for disagreements of this size

---

†The anomaly is 27;37 degrees for this eclipse. Ptolemy's
table for the equation of the center gives 1;53 degrees
when the anomaly is 24° and 2;19 when the anomaly is 30°.
If the fabricator found 2;12 by interpolation, he would
have found 261;46 degrees for the longitude of the moon,
in exact agreement with the longitude of the sun. The
value 2;12 is the result of a plausible interpolation done
by mental calculation.

from allowable variations in the method of calculation or interpolation.† This means that the times of these two eclipses do not agree with Ptolemy's tables of the sun and moon, although the times of the other ten eclipses do agree.

This is the result that we predicted before we prepared Table VI.4, and we may now guess what happened. The fabricator chose to fabricate the time of the eclipse of -719 March 8 so that the first triad of eclipses would yield a value of E close to 5°. However, he did not take the trouble to fabricate the times of the other two eclipses in this triad. Instead, he solved the first triad and found the mean longitude and anomaly of the moon at the (fabricated) time of the second eclipse, the one of -719 March 8. Next, he used these results to calculate the longitude of the moon at the genuine times of the first and third eclipses, and changed these calculated values by 180°. Finally, he asserted falsely that the longitude of the sun at these times, as calculated from the theory of the sun, was equal to the values obtained in this way. All other triads of eclipses were fabricated in their entirety, using the parameters found from the partly genuine and partly fabricated first triad.

This argument suggests that the stated times and magnitudes of the eclipses of -720 March 19 and -719 September 1 are genuine, with the only fabrication being in Ptolemy's statement about the solar longitude. However, there is other evidence, based upon the nature of the Babylonian calendar and the way Ptolemy uses it in these records, which suggests that these records are also fabricated. Because the necessary discussion of the Babylonian calendar is fairly long, I shall put this discussion in Appendix C, instead of breaking into the main line of argument here. If the evidence of Appendix C is correct, Ptolemy fabricated the times and magnitudes of the eclipses of -720 March 19 and -719 September 1 in a way that I have not yet discovered.

Since there are several suspicious circumstances about these eclipses, without any of the arguments being conclusive, I shall label these eclipses "may be fabricated". Unless their validity can be firmly established by evidence that has not come to light, it is certainly not safe to use these eclipses in any kind of research.

In sum, of the twelve eclipses in Table VI.4, the eclipses of -720 March 19 and -719 September 1 may be fabricated, and the other ten eclipses certainly are. Every triad, when considered as a unit, involves fabricated data. With regard to the last two triads, the ones that Hipparchus used, the conclusion applies only to the data that Ptolemy claims as correct. In the original form that Hipparchus used, the data were probably genuine. Unfortunately,

---

†For the eclipse of -719 September 1, it happens that the anomaly was 90°.2 from apogee. Thus there is no need to interpolate. To an accuracy that is much better than 1′, the equation of the center equals its maximum value, which is 2;23 degrees.

Ptolemy does not tell us the measured times that Hipparchus
used; he tells us only the intervals between the eclipses.
It was only as a speculation, made in order to have some
numbers to calculate with, that I assumed Hipparchus's time
to be the same as Ptolemy's for the first eclipse in each
triad.

In Chapter IV.7 of the Syntaxis, Ptolemy points out the
values he has found for L_2 and γ_2 from the oldest triad of
eclipses (Table VI.3 above) and from the triad that he claims
to have observed himself. The change in L_2 agrees exactly
with Hipparchus's data on the mean motion (see Section VI.4
above), he says, but the change in γ_2 disagrees by 17' of
arc. The time between the eclipses is 311 783 days plus
23 1/3 hours. Thus he changes Hipparchus's value of $\gamma_{\mathbb{C}}'$ by
0;0,0,0,11,46,39 degrees per day, as we have already related.
This change is of course totally illusory, since both eclipses were fabricated, and it is simply a numerical accident.
When the fabricated eclipses were solved for the parameters
L_2, γ_2, and r, rounding and other minor matters kept the solution from yielding exactly the parameters that Ptolemy
used in the fabrication. The value of L_2 is insensitive to
details of the calculations, so the change in longitude
agreed exactly with the fabrication. The value of γ_2 is
more sensitive, and the inverse solution yielded values
slightly different from those used in the fabrication. Thus
Ptolemy "found" that a small change was needed in $\gamma_{\mathbb{C}}'$. The
reader should notice that the original value was only significant to about 10 in the fourth sexagesimal position, and
that the change is about this size.

There is an interesting point about the eclipse of -200
September 22. Ptolemy says that the eclipse began half an
hour before the moon rose. As I noted, the data as Ptolemy
gives them could not possibly be observed data, and it is
hard to understand the record if the observation is genuine.
It is easy to understand if the record is fabricated. The
times were fudged and Ptolemy failed to notice that his way
of stating them is not consistent with observation.

There is also an interesting point about the triad of
eclipses in -382 and -381. It has been recognized for a
long time that some of the reported circumstances are hard
to understand if the eclipses were observed in Babylon, as
Ptolemy seems to say. Attempts to explain the records have
led to considerable controversy in the literature, some of
which I summarized earlier [Newton, 1970, pp. 140-142]. Now
we see that there is really no problem. The eclipses were
fabricated, and the puzzling features of the records result
from the fabrication.

Finally, there is an oddity about the eclipse of 136
March 6 that should be pointed out. In Part III, I concluded
that Ptolemy fabricated this eclipse along with the other
eclipses in the same triad. I did so by using the equivalent
of Table VI.4 for these three eclipses. When I calculated $\lambda_{\mathbb{C}}$
$\pm$ 180° for the eclipses, I used Ptolemy's full lunar theory

which will be described in the next chapter. For 136 March 6, this gave 344;12 degrees for $\lambda_{\mathbb{C}} \pm 180°$.

At the same time, I made an error in reading the longitude of the sun for this eclipse, as I noted above. In Ptolemy's text, the longitude is written as $\overline{\iota\delta}\ \iota\beta'$ degrees. I accidentally misread $\iota\beta'$ as 12' rather than as 1/12 degrees, which gave 344;12 degrees for the stated value of $\lambda_{\odot}$. This agreed exactly with the calculated value of $\lambda_{\mathbb{C}} \pm 180°$.

Technically, it is correct to use the full lunar theory for calculating the circumstances of eclipses. The theory developed in this chapter applies rigorously only when the mean elongation D of 0° or 180°, that is, when the mean moon is opposite to the mean sun. At the time of a lunar eclipse, the true moon is opposite to the true sun, and D is usually different from 180°. After I wrote Part III, however, I realized that Ptolemy bases his eclipse theory only upon the theory developed in this chapter, in spite of the technical error in doing so. In testing eclipse records for evidence of fabrication, therefore, one should also use only the theory of this chapter, and I used only this theory in preparing Table VI.4.

Changing from one theory to the other makes little difference in $\lambda_{\mathbb{C}} \pm 180°$ except for 136 March 6. For this date, it changes $\lambda_{\mathbb{C}} \pm 180°$ from 344;12 to 344;05 degrees, and the correct reading of $\lambda_{\odot}$ is also 344;05 degrees. Thus $\lambda_{\mathbb{C}} \pm 180°$ agreed exactly with $\lambda_{\odot}$ both in Table VI.4 and in Part III, even though the values of $\lambda_{\odot}$ used in two places were different.†

8. The Author of the Fraud

The fraud that we have found here is quite different from the fraud that we found in the preceding chapter. It was clear that the solar observations which Ptolemy claimed to have made were fraudulent, but there was no reason to

---

†This sounds as if the agreement found in Part III was a matter of luck, since the agreement depended upon the cancellation of two errors. However, a factor other than luck operated in this situation and it seems to operate in many similar situations. Suppose that I had made only one of the errors in Part III; for example, suppose that I had used the theory of this chapter instead of the general theory, while misreading $\lambda_{\odot}$ as 344;12 degrees. I would have found that $\lambda_{\odot}$ and $\lambda_{\mathbb{C}} \pm 180°$ disagreed by 7' for the eclipse of 136 March 6 but that they agreed closely for the other two eclipses in the triad. This discrepancy would have caused me to investigate the circumstances more closely. In doing so, I would undoubtedly have discovered the error in reading $\lambda_{\odot}$ for 136 March 6. In other words, significant errors stand a good chance of being detected, and errors that escape detection are likely to be unimportant.

suspect the validity of the observations which he quoted from the works of others.† Here, however, in addition to the fraudulent observations that Ptolemy claims to have made, we have a systematic falsification of history, and we have false statements about the results of calculations.

It is not too hard to imagine that Ptolemy instructed an assistant to observe, say, the equinoxes and solstices in Table V.3, and that the assistant deceived Ptolemy by fabricating them instead of observing them.‡ How do we account for the fabrication of already ancient records, however?

We must first imagine that Ptolemy did not consult the records himself but that he sent the same assistant to do so. This seems slightly unusual, but we can imagine reasons for the action. Instead of simply copying the data from the records, however, the assistant now fabricated the data. This is immediately harder to understand. The most likely reason for fabricating personal observation on the part of the assistant is laziness, but that does not apply here. It would have been much easier to copy the data than to fabricate them.‡ Hence the assistant must have had some other motive, and this makes the matter somewhat more unlikely. However, the assistant may have had motives that we do not know about, and that we might even find hard to imagine, and we still admit a possibility that an assistant was responsible for the fabricated ancient data. We must also admit that the possibility is less likely than it was when only fabricated current observations were involved.

We must also consider the writing of the <u>Syntaxis</u>. The writing seems to be designed as part of the fraud. For example, consider what Ptolemy writes about Hipparchus and the two triads of eclipses in -382/-381 and -200/-199. He says that Hipparchus got different values of E from the two triads, and that this was the result of Hipparchus's error in calculation. In saying this, he clearly does not realize (or he is trying to conceal his realization) that different triads must yield different values of E if the measurements and calculations are correct. He then gives the times of the observations and the calculated longitude of the sun at these times. Next he asserts that Hipparchus obtained certain changes in time and longitude which are incompatible with the recorded times and longitudes, and the errors that he imputes to Hipparchus are unlikely ones for Hipparchus to have made. Finally, he uses the changes in time and longitude which he claims are correct and gets values of E that agree almost exactly with each other and with E from two <u>other triads of</u> eclipses.

† I except the fabricated solstice of -431 that was attributed to Meton. This fabrication, at least in part, had an innocent explanation and dates from long before Ptolemy.

‡ Though dishonest, the assistant must have been faithful in one sense. If there were such a person, he worked for Ptolemy from 132 September 24 to 140 June 23 and probably longer.

‡ To be safe, the assistant would also have to replace the authentic record by a fabricated one in the archives.

At this point, we may try to say that Ptolemy's assistant supplied Ptolemy with the false information involved in this discussion. If we try to say this, we must also say that the hypothetical assistant did the observing, the search of the literature for old observations, and the analysis of both old and new observations. If the assistant did all this, what did Ptolemy do? Further, if we do say this, we still have not considered the actual writing; only Ptolemy could have decided what he wrote. We must now ask: Why did Ptolemy claim that Hipparchus made serious errors in analyzing the eclipses?

If Ptolemy felt impelled to refer to Hipparchus and these eclipses, he could have said that the epicycle (or eccentric) model may not account exactly for the longitude of the full moon, and that Hipparchus should perhaps have gotten different values of E from different triads; this statement has the merit of being true. He could also have said that measurement error can make the inferred values of E vary appreciably from one set of eclipses to another; this statement also has the merit of being true. However, Ptolemy does not choose to write either of these true statements, and he decides instead to make a false challenge to Hipparchus's results. This sounds like a decision that could only have been made by the person who actually wrote the Syntaxis.

By making suppositions that are elaborate enough, I am sure that we can construct a theory in which Ptolemy is the victim and not the author of the fraud. However, in the spirit of saving the phenomena, we should adopt the simplest hypothesis that is consistent with the circumstances. The simplest hypothesis by far is that the fraud was the deliberate work of Ptolemy.

I will agree that Ptolemy's guilt has probably not been proved beyond a reasonable doubt by the evidence presented so far. Before this work is finished, however, I believe that the reader will find the evidence to be overwhelming. For simplicity in the rest of the work, then, I shall usually write as if Ptolemy were the proved author of the fraud. On occasion, however, I shall continue to ask if the hypothetical assistant could have been the guilty person.

The assumption of Ptolemy's guilt requires the development of one corollary. Let me assume for the sake of argument that Ptolemy's guilt has been proved with regard to the lunar eclipses. Does this prove him guilty with regard to the solar observations?

In criminal law, as I understand it on the basis of extensive reading of mystery stories, guilt of one crime is not accepted as proof of guilt of a different crime. For example, suppose that there is a burglary on January 17 in one town and another on February 2 in a nearby town. Suppose that X is convicted of the burglary on February 2 in a court of law. This proves nothing about the identity of the burglar of January 17.

The reason is that there are many burglars who could have committed the crime of January 17. Except in the rare circumstance that some known evidence proves that both burglaries were done by the same person, proving one proves nothing about the other. Here, however, we have the rare circumstance. Unless we try to imagine that Ptolemy and his hypothetical assistant were both busily engaged in committing scientific fraud, each independently and without the knowledge of the other, there is only one guilty party here. Hence if we can prove that Ptolemy is guilty of some of the fraud, we have proved, with an extremely high probability, that he is guilty of it all.

9. The Moon's Node

So far, we have considered only the moon's motion in the plane of its orbit, and we have considered that aspect of its motion only at the full moon. Instead of going on to consider the motion in the orbital plane at other phases, I shall follow Ptolemy's order of development and take up the motion of the orbital plane itself.

We summarized the basic situation in Section I.4. The plane of the moon's orbit makes an angle of slightly more than 5° with the plane of the ecliptic. This angle, called the inclination, is not constant. Its mean value is about 5°.144, and the inclination oscillates about this value with a period of about 9.3 years and an amplitude on either side of the mean of about 0°.15. It also has other significant variations.

The line in which the plane of the moon's orbit intersects the plane of the ecliptic is called the line of nodes, and the moon crosses this line twice per month. When the moon crosses it going from the south to the north side, the direction from the earth to the moon is called the ascending node. When the moon crosses the line of nodes going from the north to the south side, the direction to the moon is called the descending node. The line of nodes moves in a westerly direction around the plane of the ecliptic, making a complete turn around the ecliptic in about 18.6 years. We want to be able to locate, say, the ascending node at any time. In this section, I shall describe how the Greek astronomers did this. Again following Ptolemy's order of presentation, I shall defer the Greek measurements of the inclination to a later section.

In studying the motion of the node, Ptolemy says that he will use [Chapters IV.2 and IV.9 of the Syntaxis] the method that Hipparchus has already used, as well as some of the same data. In fact, it is likely that the method was already understood at least as far back as Aristarchus, $1\frac{1}{2}$ centuries before Hipparchus and $4\frac{1}{2}$ centuries before Ptolemy. The method again involves the use of lunar eclipses. The necessary geometry is shown in Figure VI.3.

As the sun goes around the earth, the earth's shadow is always being thrown into space in the direction opposite the sun. The axis of the shadow obviously lies in the ecliptic. At the time of a full moon, let us imagine that a screen is put up at the same distance as the moon. On the screen, we would see a circular cross section of the shadow, and this cross section would have an apparent size, just as the disks of the sun and moon do. The size of the shadow varies with the distance from the earth. At a given point, the size of the shadow also depends upon the distance from the earth to the sun. This effect is small and Ptolemy neglects it, so far as I can judge from his discussion.

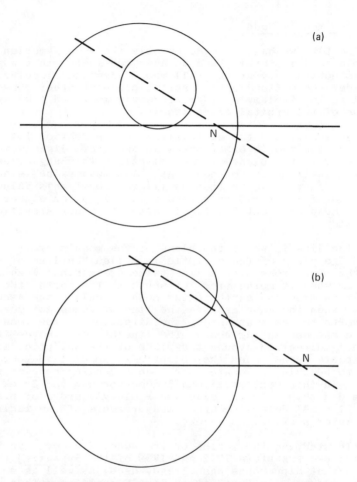

Figure VI.3   The moon and the earth's shadow during lunar eclipses. In both parts of the figure, the horizontal line denotes the ecliptic, the slanting broken line is the line of motion of the moon, the large circle is the earth's shadow, and the small circle is the moon. N is a node of the lunar orbit. The angle between the moon's motion and the ecliptic has been exaggerated. In part (a), the moon passes so close to the center of the shadow that it is totally eclipsed. In part (b), the moon misses the center of the shadow by a greater distance and it is only partially eclipsed.

The horizontal line in both parts of Figure VI.3 represents the plane of the ecliptic, and the large circle in both parts is the outline of the earth's shadow. Under average conditions, the apparent radius of the shadow is slightly more than 41'.† The apparent radius of the moon's disk is about 15' 33", under average conditions. The sum of the two radii is somewhat more than 56' under average conditions.

Clearly no part of the moon can be eclipsed by entering the shadow unless the distance from the center of the moon to the center of the shadow is less than the sum of the radii. For simplicity in discussion, let me use 1° for the sum; in any particular case, we must use the value that actually applies for that case. At most full moons, the moon is several degrees from the center of the shadow and there is no eclipse. On the average, there are about 1.54 lunar eclipses per year,‡ or about 1 eclipse for every 8 full moons. There is an eclipse only if the moon is closer than 1° to the center of the shadow when the moon is full.

In both parts of Figure VI.3, the slanting broken line denotes the line of motion of the moon's center. The correct angle between this line and the ecliptic is the inclination, slightly more than 5°; I have exaggerated the angle in the figure. When the node N is close to the center of the shadow, as it is in part (a) of the figure, the moon comes so close to the center that it is totally eclipsed. When the node is farther away, as it is in part (b), only part of the moon passes within the shadow, and we have a partial eclipse.

The magnitude of the eclipse, when it is not total, means the fraction of the moon's diameter that lies within the shadow at the middle of the eclipse.‡ It is clear from the figure that the magnitude depends upon the distance from the moon to the node at the middle of the eclipse. The magnitude also depends upon the apparent sizes of the earth's shadow and of the moon. As I have said, these depend upon the distance from the earth to the moon; to a lesser extent, the apparent size of the shadow also depends upon the distance from the earth to the sun.

In both parts of Figure VI.3, the moon is north of the ecliptic and moving to the left (east) away from the ascending node. There are three other possible configurations. The moon may be south of the ecliptic and moving toward the ascending node, it may be north and moving toward the descending node, or it may be south and moving away from the descending node. In a partial eclipse, when the moon is

---

†This value is readily derived from the information in Section 9E of the Explanatory Supplement [1961].

‡Oppolzer [1887] lists 5200 lunar eclipses in a span of 3369 years.

‡For simplicity, I am neglecting a number of small effects in this discussion. The interested reader can find an extensive discussion in Section 9E of the Explanatory Supplement [1961].

north of the ecliptic, it is eclipsed on its south side.

Ptolemy uses a quantity that we may call ψ, the mean argument of the latitude, which is the angle in the plane of the moon's orbit from the ascending node to the position of the mean moon.† Clearly, if we know both ψ and $L_{(\!(}$, the mean longitude, we can locate the node.

First, we want ψ', the amount by which ψ changes in a day. Hipparchus [Syntaxis, Chapter IV.2] searched for two eclipses with the same magnitude of eclipse, with the moon on the same side of the ecliptic and near the same node, and with the moon at the same distance from the earth.‡ In this way, Hipparchus found that the mean argument of the latitude went through 5923 full revolutions in 5458 months. With the length of the month already known, this gives

$$\psi' = 13;13,45,39,40,17,19 = 13.229\ 350\ 330\ 8 \text{ degrees/day}$$

$$(\text{VI}.13)$$

Ptolemy checks this value by using two eclipses. The first was observed at Babylon on -490 April 25. The middle of the eclipse came in the middle of the sixth hour of the night, and the moon was eclipsed by 2/12 of its diameter on the southern side. Hence the moon was north of the ecliptic, and it was near the descending node. Ptolemy concludes that the middle of the eclipse was 256 years plus 122 days plus $10\frac{1}{4}$ hours after his standard reference time, in mean time.‡

The second eclipse was observed in Alexandria on 125 April 5, and the middle of the eclipse came at 3 3/5 hours before midnight.* Again the moon was eclipsed by 2/12 of its diameter on the southern side, and it was near the descending node. Ptolemy concludes that the middle of the eclipse came at 871 years plus 256 days plus 8 1/12 hours, mean time, after the reference epoch. Hence the interval between the eclipses was 615 years plus 133 days plus 21 5/6 hours.

---

†Argument of the latitude, without the qualifier "mean", is the angle from the node to the actual moon.

‡Which means that the anomaly γ must have had the same value at both eclipses, to the accuracy needed.

‡Ptolemy takes noon, apparent time at Alexandria, on -746 February 26, as the epoch to which he refers all of his ephemerides. This was the beginning of the Egyptian year in which the Babylonian king Nabonassar began his reign. According to many modern sources [Neugebauer, 1957, p. 98, for example], Ptolemy says that he has almost complete lists of lunar eclipses from the reign of Nabonassar onward. I have not located this statement in the Syntaxis, but I have not tried hard to do so.

*Ptolemy does not say that he observed this eclipse, and its date is before any observation that Ptolemy claims to have made. Ptolemy attributes several observations made in Alexandria just before his own time to an astronomer named Theon. This may be an observation made by Theon that Ptolemy forgot to attribute.

Since the circumstances are parallel, the actual moon
has made an integral number of revolutions with respect to
the node during this interval, but the mean moon has not. At
the first eclipse, the calculated anomaly is 100;19 degrees,
and e_c = -5;0 degrees. At the second, the calculated anomaly
is 251;53 degrees and e_c = +4;53 degrees. Hence the mean
moon has made 9;53 degrees less than a full number of revolutions. If we use Equation VI.13, however, we find that ψ
has changed by 10;2 degrees less than a full number of revolutions. Hence Equation VI.13 must be increased by the
amount that will give 0;9 in the interval between eclipses.
The increase in ψ' is 0;0,0,0,8,39,18 degrees per day, which
makes

$$\psi' = 13;13,45,39,48,56,37 = 13.229\ 350\ 998\ 7\ \text{degrees/day}.$$

$$(VI.14)$$

I have verified all this arithmetic.

Note that Ptolemy has only one significant figure (0;9
degrees or 9') in the amount to add to Hipparchus's value.
However, he adds a number with five significant figures.

Now he wants to find ψ_0, the value of ψ at the reference
epoch. For this, he says that he needs two eclipses with the
same magnitude, in which the moon is at the same distance
from the earth, with the moon on the same side of the ecliptic, but near opposite nodes.[†] For the first of these, he
will use the eclipse of -719 March 8, which he used in finding the radius of the epicycle. The moon was eclipsed by one
quarter of its diameter from the southern side, and it was
near the ascending node. For the second, he will use an
eclipse that Hipparchus used. This was observed at Babylon
on -501 November 19, at 6 1/3 hours after sunset. The moon
was again eclipsed by one quarter of its diameter from its
southern side, but it was near the descending node. Ptolemy
concludes that the eclipse came at $10\frac{1}{4}$ hours, mean time,
after noon at Alexandria. The interval between the eclipses
was 218 Egyptian years plus 309 days plus 23 1/2 hours, mean
time.

When we use Equation VI.14, we find that ψ increased by
17 610 full revolutions plus 160;4 degrees.[‡] As we find from
the anomaly, the real moon was 59' behind the mean moon at
the first eclipse and 13' behind at the second. Hence the
real moon has travelled 17 610 revolutions plus 160;50 degrees with respect to the node. It was ahead of the ascending node by, say, X degrees at the first eclipse and it was
behind the descending node by the same amount at the second
eclipse. Hence, X = $\frac{1}{2}$(180 - 160;50) = 9;35 degrees.

At the first eclipse, then, the real moon was 9;35

[†]I believe that Ptolemy is mistaken. He could equally well
use eclipses with the moon near the same node but on opposite sides of the ecliptic. However, maybe he could not
find a pair of eclipses with this property.

[‡]I verify all the following arithmetic with discrepancies
of less than 1'.

degrees ahead of the ascending node and it was 0;59 behind
the mean moon. Hence the mean moon was 10;34 degrees beyond
the ascending node. That is, $\psi = 10;34$ at the time of the
first eclipse. From this, we readily find

$$\psi_0 = 84;15 \quad \text{degrees} \tag{VI.15}$$

at the reference epoch. We should note that Ptolemy does
not measure ψ from the ascending node, as modern astronomers
do, but from the position of the mean moon at maximum latitude. Hence, Ptolemy finds 354;15 for what he calls the
motion of the mean moon in latitude.

TABLE VI.5

TIMES AND MAGNITUDES OF FOUR LUNAR ECLIPSES

Date	Hour		Magnitude, digits		
	Stated by Ptolemy	Calculated from Ptolemy's tables	Stated by Ptolemy	Calculated from Ptolemy's tables	Calcula from moder theor
-490 Apr 25	22;15	21;59	2	1.8	1.0
125 Apr 5	20;05	20;21	2	1.8	1.7
-719 Mar 8	23;10	23;10	3	2.4	1.5
-501 Nov 19	22;15	22;14	3	2.4	2.2

Now we want to investigate the authenticity of these
eclipses, which we can do with the aid of Table VI.5. This
table lists the four eclipses just discussed, in the order of
their appearance in Ptolemy's text. First we have the hour
at which the middle of the eclipse occurred, in mean time,[†]
both as stated by Ptolemy and as calculated from Ptolemy's
tables. Then we have three values of the magnitude of the
eclipse. The first is the value that Ptolemy states in his
discussion of the eclipse, while the second is the value that
I calculate from his eclipse tables.[‡] The third is the value
calculated from the modern theory of eclipses.

The magnitudes are stated in digits, which is a common
unit for giving the magnitude of an eclipse. One digit, in
this context, is the fraction 1/12. Thus, for example, when
the magnitude is given as 3 digits, the fraction 3/12 = 1/4
of the moon's diameter is eclipsed.

For the first two eclipses, the time calculated from
Ptolemy's tables disagrees by 16 minutes with the time that
he states. This makes it seem unlikely that the eclipses
were fabricated. Against this, however, we have the evidence
of the magnitudes. For each eclipse, the magnitude calculate

[†]Ptolemy states the hour as measured from noon. I have
changed this to the hour as measured from midnight.

[‡]These tables appear in Chapter VI.8 of the Syntaxis.

from Ptolemy's tables is 1.8 digits, which he would have rounded to 2 digits, the stated value. However, the eclipses were far from having the same magnitude; as the last column shows, the magnitudes differed by 0.7 digits. We should remember that it was vital, in Ptolemy's use of these eclipses, for the magnitudes to be equal. We easily find, either from Ptolemy's eclipse tables or from elementary geometry, that a change of one digit corresponds to a change of about 30' in the argument of the latitude. If the magnitudes differed by 0.7 digits, the difference between the values of the argument of the latitudes was about 21', although Ptolemy takes the difference to be zero. Thus, when Ptolemy used the eclipses to make a correction of 9' to Hipparchus's results, he did so in the presence of a measurement error of 21'. That is, his correction was completely illusory, and a competent astronomer should have realized this.

This is a good place to investigate the accuracy with which the magnitude of a lunar eclipse can be measured with the naked eye, before we take up the last two eclipses in Table VI.5. Earlier [Newton, 1970, p. 215], I studied measurements of the magnitudes of three eclipses made by Chinese astronomers and of twelve eclipses made by Islamic astronomers; all the measurements were made by the naked eye between 585 January 21 and 990 April 12. The standard deviation of the error in the measured magnitude came out to be 0.98 digits, which I shall round to 1 digit for simplicity. Thus, in comparing the magnitudes of two eclipses, as Ptolemy did with the eclipses of -490 April 25 and 125 April 5, the standard deviation of the difference is 1.4 digits, which amounts to about 42' in the position of the node. Any correction smaller than this amount was meaningless with the data available to Ptolemy, unless it were based upon many pairs of eclipses.[†]

Now let us turn to the eclipses of -719 March 8 and -501 November 19, which Ptolemy uses to find the value of ψ_0, the mean argument of the latitude at the fundamental epoch. He has already used the eclipse of -719 March 8 in evaluating the lunar eccentricity, and we know that it was fabricated. There is no obvious reason why Ptolemy should prefer one value of ψ_0 to another, and thus there is no obvious reason why he should use fabricated data in finding ψ_0. To be sure, the eclipse of -719 March 8 was fabricated, but for another purpose, and his use of it may have been merely a convenience. The value found for ψ_0 would still depend upon the second eclipse, and there was no apparent reason to fabricate it. Nonetheless, the time measured for it agrees with Ptolemy's tables within 1 minute. This agreement is possible but suspicious, and we must admit the possibility that this eclipse was fabricated for a reason that is not obvious.

There is an oddity about the magnitudes of these eclipses. They had to have the same reported magnitude in order for Ptolemy to use them as he did, and Ptolemy reported the magnitude of 3 digits for both. If they were fabricated, we

---

[†]To reduce the standard deviation from 42' to 9' would require using about 25 pairs of eclipses.

should expect the magnitudes to be calculated from Ptolemy's
tables. However, as the table shows, the magnitude calculated from his tables is 2.4 digits for both, and I can think
of no plausible computing procedure that yields 3 digits.
Further, as the last column shows, the correct magnitudes
were 1.5 and 2.2 digits.

It is possible that Ptolemy had genuine records of these
eclipses and that he altered the times for his own purposes
of fabrication. He had no interest in the magnitudes except
their equality, and thus he would not need to alter the reported magnitudes. Thus it is possible that the magnitudes
were reported as 3 digits and that Ptolemy left them unaltered. This explanation requires the error in the magnitude
for -719 March 8 to be 1.5 digits. This is large but not
impossible; the Islamic record of the eclipse of 856 June 22
[Newton, 1970, p. 215] shows this error.

To summarize this section, we already know that the
eclipse of -719 March 8 was fabricated, at least so far as
its time is concerned. The analysis given in this section
gives no strong evidence either for or against the fabrication of the other eclipses in Table VI.5. It is possible
that some reported circumstances of these eclipses are genuine and that others are fabricated. The evidence given in
Appendix C, which relates to all eclipses that Ptolemy dates
by means of the Babylonian calendar, suggests that the records of -501 November 19 and -490 April 25 are fabricated.
However, since there is no other evidence, I shall put these
eclipses, along with the one of 125 April 5, in the category
called "may be genuine". It is clear that one cannot safely
use any of the data from these records in astronomical research.

10. A Summary

The topics treated in this chapter seem to have a be-
wildering variety even though they are linked by a unifying
theme, that of studying the longitude of the full moon. A
brief summary of the chapter may therefore be useful.

The moon is closer to the earth and moves faster than
any other body studied in Greek astronomy. Even at the level
of accuracy of naked eye observations, the study of the moon
involves the study of parallax and the equation of time.

The longitude of the sun can be described by a simple
epicycle or eccentric model. The motion of the moon is more
complex. If we find an epicycle model that fits the motion
reasonably well at, say, the full moon, it will not fit well
at the quarter moons. Ptolemy's plan is to find first an
epicycle model that will work at the full moon. He will then
find the modifications that are needed at other phases. He
assumes without proof that the model devised for the full
moon will also fit at the new moon. This model has five
parameters that must be found by analysis of observations.

Ptolemy adopts a value of $n_{\mathbb{C}}$, the mean motion in longitude, which he attributes to Hipparchus, who allegedly found

it from the analysis of data covering a time span of about 350 years. However, the value is actually Babylonian in origin. On a tentative basis, Ptolemy also adopts a value of $\gamma_{\mathbb{C}}'$, the daily change in the anomaly, that Hipparchus had found; he later makes a trivial change in $\gamma_{\mathbb{C}}'$ on the basis of his own observations.

In order to find the remaining parameters, Ptolemy says that he will use lunar eclipses, because they are unaffected by parallax and hence give more accurate results. Ptolemy is wrong in this. Parallax is not a source of error because it can be calculated and the observations corrected for its effects. The edge of the earth's shadow is not sharply defined and hence observations of lunar eclipses are not particularly accurate. It should be better to use occultations of stars by the moon.

Ptolemy first uses a triad of eclipses observed in Babylon in the years -720 and -719 (Table VI.1). From this triad, he deduces the radius of the lunar epicycle, or, what is equivalent, the maximum value of the equation of the center. He also deduces the values of the mean longitude and the anomaly at the time of the middle eclipse of the triad. He does the same things for a triad of eclipses which he claims to have observed himself in the years +133, 134, and 136 (Table VI.1). He finds a value for the maximum equation that is almost identical to the one he found from the ancient eclipses. By comparing the anomalies from his triad and the ancient triad, he makes the small correction to $\gamma_{\mathbb{C}}'$ that has been mentioned, but he finds that no change is needed to the value of $n_{\mathbb{C}}$.

All this is based upon fabrication. He fabricated all the eclipses which he claims to have observed himself, and he fabricated the middle eclipse in the ancient triad. We cannot reach a firm conclusion about the other two eclipses in the ancient triad, but the weight of the evidence inclines toward their fabrication.

Ptolemy then says that Hipparchus had found values of the maximum equation by using two triads of eclipses, one from the years -382 and -381, the other from the years -200 and -199. He says that Hipparchus's values of the maximum equation differ from each other and from the value he has found. He says that this happened because Hipparchus did not analyze the data correctly. When the data are analyzed correctly, he says, all triads yield the same value of the maximum equation, and he proceeds to alter the data so that they will yield consistent values.

This may be the blackest thing that Ptolemy has done. The epicycle model in fact cannot agree with the longitude of the full moon to the accuracy of naked eye observations. Thus Hipparchus's results were probably correct. Because Ptolemy does not understand the situation, he falsely accuses Hipparchus of incompetence, and he alters the data to fit his model. In doing so, he destroys the original data.

Finally, Ptolemy takes up a topic that is not connected with the longitude of the full moon. This is the study of the node of the moon's orbit and how it moves. In this study, he uses the eclipse of -719 March 8, which we know to be fabricated. He also uses records of the eclipses of -501 November 19, -490 April 25, and +125 April 5. The last of these is close to Ptolemy's own time, but he does not claim that he observed it. The last three eclipse records may be genuine.

Since the study of the node involves the fabricated record of -719 March 8, all aspects of Ptolemy's theory of the moon that we have studied in this chapter rest upon fabricated data.

CHAPTER VII

THE LONGITUDE OF THE MOON AT ANY PHASE

1.   Measuring the Longitude of the Moon

In the preceding chapter, we gave the Greek theory used
to calculate the latitude and longitude of the moon during
lunar eclipses. It is reasonable to assume that the same
theory describes the position at any full moon and not just
at one that results in an eclipse. The Greek astronomers
went on to assume that the same theory is valid at the new
moon. I have not noticed any place in Greek writing on
astronomy where this assumption is justified, but we may
accept it.

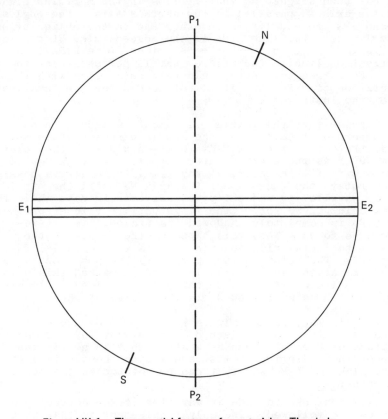

Figure VII.1   The essential feature of an astrolabe. The circle
marked with the symbols $NP_1E_1SP_2E_2$ is the edge of a metal ring
mounted so that it can rotate about the axis NS. The axis NS is
aligned so that it is parallel to the earth's axis. The angles NP_1 and
SP_2 are equal to each other and to the obliquity of the ecliptic, and
line P_1P_2 is perpendicular to the line E_1E_2. Line E_1E_2 is carried
on another metal ring, and is calibrated in degrees. The point at
the center of the figure is 0°, the point E_2 is 90°, and so on.
By rotation about NS, the ring E_1E_2 can always be aligned
parallel to the ecliptic, so that the calibrated circle measures
celestial longitude.

Now we turn to the longitude of the moon at any phase, and not just at the syzygies. The first question to be discussed deals with how the longitude is to be measured. In the opening part of Chapter V.1 of the Syntaxis, Ptolemy says that the theory will be based upon observations that Hipparchus made as well as upon observations that he has made with the aid of the instrument that he is going to describe; he calls this instrument an astrolabe. He does not describe the instrument that Hipparchus used, but it must have been an instrument that could measure the same quantities as the astrolabe.

The most essential feature of the astrolabe is shown in Figure VII.1. The circle marked with the symbols $NP_1E_1SP_2E_2$ represents a metal ring that can rotate about the axis NS. At an appropriate time, suppose that this ring has been aligned so that it lies in the meridian plane, with the axis NS parallel to the earth's axis. The angles NP_1 and SP_2 are equal to each other and to the obliquity of the ecliptic. The line E_1E_2 is the edge-on view of a circle that is inscribed on another metal ring. Line P_1P_2 is a construction line in the figure that is perpendicular to E_1E_2. For strength, there could be a metal ring with the orientation of P_1P_2, but it is not needed for the functioning of the instrument.

Suppose that the initial alignment is made at noon at the summer solstice.† Then the sun is directly above E_2. We calibrate the circle E_1E_2 in degrees and mark the point E_2 as $90°$. As the earth rotates, it will carry the metal circles with it. However, at any time, it is always possible to rotate the rings about the axis NS until the circle E_1E_2 is parallel to the ecliptic. If this is done while the sun is shining, we rotate the rings until the inner part of ring E_1E_2 is completely shadowed, as Ptolemy himself instructs us to do. When this happens, the sun is in the plane of the metal ring and the ring is then parallel to the ecliptic.‡ The point E_2 then points to the position where the sun is at the summer solstice, and we marked this point $90°$. The circle E_1E_2 then directly reads celestial longitude, and the point called $0°$ is the center of the figure.

In this way, we can always read the longitude of the sun whenever the sun is visible. If we want to read the longitude of a point, such as a star, that is not in the plane of the ecliptic, we must add a ring whose axis lies in the plane of E_1E_2 and that can be rotated in the necessary manner.‡

---

†The alignment can be made at any time if the correct procedures are followed. The only requirement is that NS be parallel to the earth's axis.

‡This neglects the parallax of the sun, which is only a few seconds of arc.

‡The axis of the ring must be able to move along E_1E_2, and the ring must be able to rotate about its axis. The position of the plane of this ring then gives the longitude of the point. The ring itself can be calibrated to read latitude.

Using the astrolabe does not require any knowledge of
the solar motion, or any tables. Within the limits of observational error, it gives the longitude of the sun with
respect to the real equinox, no matter where the user thinks
that the equinox is.

The main drawback to this instrument, so far as I can
see, is that the rotation of the earth destroys the alignment rather rapidly, so that the instrument must be read
quickly. I have experimented with a simplified model of an
astrolabe that George Bush of this Laboratory kindly made
for me, and I believe that aligning the ring E_1E_2 with the
ecliptic can be done with an accuracy of one or two minutes
of arc. If the sun is more than a month from either solstice, the resulting error in the longitude of the sun
should be no more than, say, 5', if the longitude circle
could be read at leisure. An error of more than, say, 15'
seems rather unlikely.

If a wrong value of the obliquity is used in constructing the astrolabe, there will be a further error for this
reason. However, using it a few times near each solstice
will quickly show if the obliquity used is wrong. If Ptolemy had really used the astrolabe as he says he did, he
would soon have discovered the correct value of the obliquity.

If there is any object whose latitude and longitude are
known, the astrolabe can be lined up so that it reads the
known coordinates of the known body. The coordinates of any
other object in view can then be read. In this use of the
astrolabe, however, errors in the "known" coordinates will
show up one-for-one in the coordinates read for the second
object.

The meaning of the term astrolabe has changed through
the course of history. Probably by the second half of the
medieval period, it had come to mean an instrument for measuring the elevation angle of a celestial object above the
horizon. By the time this happened, the kind of instrument
just described was often called an armillary sphere, of
which modern telescope mounts are a development.

2. Five Measurements of Solar and Lunar Position

In Chapters V.3 and VII.2 of the Syntaxis, Ptolemy
gives the results of two observations of the sun and moon
that he claims to have made himself using the astrolabe that
was just described. In a parallel fashion in Chapters V.3
and V.5 he gives the results of three observations of the
sun and moon that Hipparchus had made, but he does not say
how Hipparchus made them. I shall now summarize the circumstances of the observations as Ptolemy gives them, in the
order that they appear in his text.

(1) Ptolemy observed the sun and moon just after sunrise on 139 February 9, at $5\frac{1}{4}$ hours before noon, apparent
time. The sun appeared at 318 5/6 degrees and the moon appeared at 219 2/3 degrees in longitude. The parallax of the

moon in longitude was negligible. The time since the basic
epoch was 885 (Egyptian) years plus 203 days plus 18 3/4
hours, in both apparent and mean time. At this epoch, the
position of the sun calculated from his tables was 318 5/6
degrees, just as it appeared on the astrolabe. The position
of the mean moon was 227;20 degrees, so e_c (equation of the
center) was -7 2/3 degrees.† The anomaly was 87;19 degrees,
which, he says, is the value that makes e_c a maximum. This
record is in Chapter V.3.

   (2) Hipparchus observed the sun and moon on the date
that corresponds to August 5, in a year that needs some
discussion. Ptolemy states the year directly, and he states
it indirectly by giving the time since the fundamental epoch.
In both Halma's Greek text and his translation, the year
must be -126 according to one statement and -127 according
to the other. In both Heiberg's Greek text and Manitius's
translation, the years are -128 and -127, respectively. Altogether, three different years are stated. Luckily, the
position of the moon resolves the matter. On -127 August 5
the moon was in the correct place, and on August 5 in both
-128 and -126 the moon was in a different part of the heavens. Thus the correct date is -127 August 5.

   The observation was made when two-thirds of the first
hour of the day‡ at Rhodes had passed. The measured longitude of the sun was 128 7/12 degrees and the measured longitude of the moon was 42 1/3 degrees. Again the parallax
in the longitude of the moon was negligible. Thus the true
angle between the sun and moon was $86\frac{1}{4}$ degrees. When the
time is changed from hours of the day to uniform hours, the
time since the basic epoch was 619 years plus 314 days plus
17 5/6 hours of apparent time, or 17 3/4 hours of mean time;
Rhodes and Alexandria are on the same meridian, according
to Ptolemy. At this time, Ptolemy calculates that the longitude of the sun was 128;20 degrees. That is, Hipparchus's
measurement was in error by 15'.‡ The position of the moon
was therefore 42;05 degrees rather than 42 1/3 degrees as
Hipparchus measured it. The mean longitude of the moon at
this time was 34;25 degrees, so that e_c = +7 2/3 degrees.
The moon's anomaly was 257;47 degrees. This record is in
Chapter V.3 of the Syntaxis.

   (3) Hipparchus observed the sun and moon at the beginning of the second hour of the day on -126 May 2. The longitude of the sun was 37 3/4 degrees and the apparent longitude of the moon was 351 2/3 degrees. When it was corrected

† Ptolemy does not use the minus sign. He expresses this by
  saying that the longitude of the true moon was less than
  the longitude of the mean moon by 7 2/3 degrees.

‡ "Hour of the day" is similar to "hour of the night" that
  was used in the preceding chapter. It is a twelfth of the
  interval from sunrise to sunset. Until perhaps the 14th
  century, an hour in European writing frequently meant either
  an hour of the day or an hour of the night, depending upon
  the circumstances.

‡ Ptolemy does not explicitly write the last sentence, but
  it is clearly implied by what he does write.

for parallax, the longitude of the moon was 351 11/24 degrees.† The time was 5 2/3 hours before noon, and it was thus 620 years plus 219 days plus 18 1/3 hours in apparent time, or 18 hours in mean time, after the basic epoch. Ptolemy calculates that the correct longitude of the sun was 37;45, as Hipparchus observed it. The mean longitude of the moon was 352;13 degrees, and the equation of the center was -0;46 degrees.‡ The anomaly was 185;30 degrees. This record is in Chapter V.5 of the Syntaxis.

(4) Hipparchus observed the sun and moon at 9 1/3 hours of the day on -126 July 7. The longitude of the sun was 110 degrees less 1/10,‡ and the longitude of the moon was 149 degrees. There was no parallax of the moon in longitude.* The time was 4 hours after noon, so that the time after the basic epoch was 620 years plus 286 days plus 4 hours, apparent time, or 3 2/3 hours, mean time. At this time, the calculated longitude of the sun was 100;40 degrees rather than 100;54 as Hipparchus measured it. Hence the longitude of the moon was 148;46. Its anomaly was 333;12 degrees. This record is in Chapter V.5 of the Syntaxis.

(5) Ptolemy observed the moon at sunset on 139 February 23, when the 60th degree of the zodiac was in the meridian plane.⚹ The longitude of the sun was read as 333°, and the apparent longitude of the moon was 92 1/8 degrees east of the sun. Half an hour later, when the point at $67\frac{1}{2}$ degrees was in the meridian, the star Regulus was visible and was 57 1/6 degrees east of the moon in longitude. He calculates that the longitude of the sun was 333;03 degrees at the time of the first observation.# The moon, being 92 1/8 degrees to the east, was at 65;10 degrees (Ptolemy has dropped 30″). In the next half an hour, the moon moved by 15′, but its parallax decreased by 5′. Hence, at the time of the second observation, its apparent longitude was

---

†The fraction is written as 1/8 + 1/3. Ptolemy's correction for parallax is $12\frac{1}{2}$ minutes of arc; I calculate $14\frac{1}{4}$ minutes. The implication of Ptolemy's account is that Hipparchus made the parallax correction, I believe.

‡The data lead to -0;45,30, which Ptolemy rounds to -0;46 rather than to -0;45.

‡It is unusual to see a decimal fraction in Greek astronomy.

*I calculate that the parallax was 15 seconds of arc, which is negligible.

⚹In several places, Ptolemy states the point of the zodiac (the ecliptic circle) that is in the plane of the meridian. This implies a familiarity with using celestial positions for measuring time. He could have read the degree of the zodiac on his astrolabe if he had actually made the observation.

#Ptolemy usually states the interval from the basic epoch, but he does not do so here. By his tables, the interval would be the same in both apparent and mean time, and we have no difficulty in reconstructing the interval that he used in the calculation.

65;20 degrees.† Hence the longitude of Regulus (α Leonis)
was 65;20 + 57;10 = 122;30 degrees. This record is in Chapter VII.2 of the Syntaxis.

Ptolemy uses the first four of these observations in
developing the theory of the moon and he uses the last one
in deriving the rate of precession of the equinoxes. We
shall not take up the use of this observation until Section
IX.3, but it is convenient to study the authenticity of all
these observations together. We can do this with the aid
of Table VII.1.

TABLE VII.1

FIVE OBSERVATIONS OF LUNAR POSITION
WITH RESPECT TO THE SUN

Date	Longitude of the sun, degrees			Longitude of the moon, degrees	
	Measured	Calculated by Ptolemy	Calculated[a] by me	Measured	Calculated[a] by me
139 Feb 9	318;50	318;50	318;44	219;40	219;50
-127 Aug 5	128;35	128;20	128;20	42;05	42;04
-126 May 2	37;45	37;45	37;45	351;27,30[b]	351;24
-126 Jul 7	100;54	100;40	100;43	148;46	148;34
139 Feb 23	333;00	333;03	333;04	65;05	65;04

[a]These are values that I calculate from Ptolemy's tables; they
do not come from modern tables.
[b]Stated as 351 degrees plus ⅓ plus ⅛.

For each observation, the table gives three values of
the longitude of the sun and two values of the longitude of
the moon. Let us consider the solar longitudes first. The
first one is the measured longitude, the second is the value
calculated by Ptolemy, and the third is the value that I
calculate using Ptolemy's tables. The values that he calculates and that I calculate agree as closely as we can expect except for the first observation. Unless he made a
simple slip in calculation, I do not understand how he calculated 318;50 degrees for the longitude of the sun. The
reason cannot be an error in his tables, because I get the
same value when I use Equation IV.6 directly.

In judging the authenticity of the observations, the
important comparison is that between the measured values and
those calculated by Ptolemy. In order to interpret the comparison, we should remember that Ptolemy's solar table is
based upon the equinox and solstice observations made by
Hipparchus (Section V.3). Hence the measurements attributed
to Hipparchus should agree with the values calculated by
Ptolemy, within a reasonable amount of measurement error.
By Ptolemy's time, however, as we saw in Section V.4, the

---

† I calculate that the parallax change was 9′ rather than 5′.

error in the solar table had grown to more than 1° because
it was based upon a wrong value for the length of the year.
Thus the values measured by Ptolemy should disagree with the
calculated values by more than 1° if they are authentic.

Both values measured by Ptolemy agree closely with the
calculated values, exactly in one instance and within 3' in
the other. The probability that this happened by chance and
not by fabrication is [Part II] between 10^{-5} and 10^{-6} for
each observation. We may say without hesitation that the
observations which Ptolemy claims to have made were fabricated.

Of the three observations attributed to Hipparchus,
those of -127 August 5 and -126 July 7 disagree with the
calculated values by about 15', which is reasonable if they
are genuine. The observation of -126 May 2 agrees exactly,
at the level of rounding used, but this could be an accident.
We should defer judgment on this observation until we can
study the matter in more detail, including the use that
Ptolemy makes of the observation. Part of the additional
detail comes from the longitude of the moon.

Since Ptolemy uses the first four observations in developing his tables of the moon and the fifth in developing
his star table, he does not directly state a calculated
longitude of the moon. Hence, for the moon, Table VII.1
gives only two longitudes of the moon for each observation.
The first is the measured value and the second is that which
I calculate by using Ptolemy's tables of the moon.

For Ptolemy's observation of 139 February 9, the measured and calculated values disagree by 10'. For Hipparchus's observation of -127 August 5, they agree within 1'.
This seems to contradict the conclusion just reached: If
Ptolemy's observation is fabricated, we expect close agreement; and if Hipparchus's observation is genuine, we expect
a moderate discrepancy. Before we can resolve this apparent
paradox, we must describe Ptolemy's model of the moon in all
its details.

3.   Ptolemy's Model of the Lunar Motion

Almost three centuries before Ptolemy, Hipparchus had
found that E, the maximum equation of the center for the
moon, is considerably greater at the quarter moons than at
the new or full moons. That is, as an observed phenomenon,
Hipparchus discovered the evection. Ptolemy says [Chapter
V.3] that he finds the same thing as Hipparchus from observations that he has made with the astrolabe. He then presents the record that we numbered (1) in the preceding section, in which he found that the true position of the moon
was 7 2/3 degrees behind the position of the mean moon. This
was at the last quarter, on 139 February 9.

He now does something that stands in strong contrast
with what he did about lunar eclipses and finding E at the
full moon. In that matter, he asserts (Section VI.6 above),
Hipparchus could not analyze the observations correctly and

his results were not reliable. Here, however, he emphasizes
the agreement between his results and what Hipparchus found.
Hipparchus observed the moon at the last quarter on -127
August 5 (record (2) in the preceding section) and also found
that the true position was 7 2/3 degrees from the mean position, but the true position was ahead on this occasion rather
than behind.

According to Equation VI.11, the maximum equation of
the center equals $\sin^{-1} r$ if we take the deferent radius to
be unity. If instead we let the deferent radius have any
value R, we have

$$E = \sin^{-1} (r/R) . \hspace{4cm} (VII.1)$$

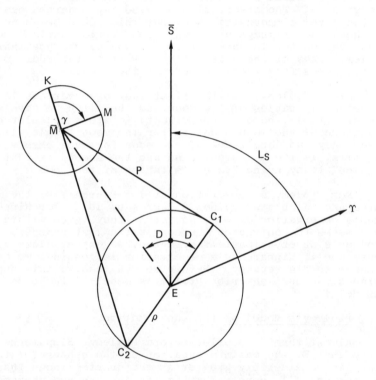

Figure VII.2   The epicycle-deferent system that Ptolemy uses
for the moon. Υ is the direction from the earth E to the vernal
equinox and $\overline{S}$ is the direction to the mean sun. $\overline{M}$ is the mean
position of the moon, and the angle $\Upsilon E\overline{M}$ is the mean longitude
of the moon. Angle $\overline{M}E\overline{S}$ is the mean elongation D, which increases
in the counterclockwise direction. A line $C_1 EC_2$ rotates uniformly
in the clockwise direction with respect to the line $E\overline{S}$, so that
angle $C_1 E\overline{S}$ is numerically equal to D. As the point $\overline{M}$ moves,
the distance $C_1\overline{M}$ remains constant. Distances EC_1 and EC_2 are
equal. The mean anomaly γ is measured in the clockwise
direction from the line $\overline{M}K$, which is the extension of $C_2\overline{M}$.
M is the true position of the moon.

-150-

If E is greater at a quarter moon than it is at a syzygy, the ratio r/R must correspondingly be greater, as Ptolemy points out explicitly in the last sentence of his Chapter V.2. He then makes what turns out to be a grievous error. Since the epicycle radius r is constant, he says, the deferent radius R must be smaller at the quarter moons than it is at the syzygies.†

If we accept Ptolemy's statement, we can readily calculate the ratio between the radius R at the quarters and syzygies. At the syzygies, sin E = 5;15 ÷ 60 = 0.0875. At the quarters,‡ E = 7 2/3 degrees and sin E = 0.133 410. The ratio is 0.0875/0.133 410 = 0.655 873. Thus, if R = 60 at the syzygies, it must be 39.35 at the quarters. This is 39;21 in sexagesimal notation. In his final model, Ptolemy uses the value 39;22. Thus Ptolemy needs to find a mechanism, based upon uniform circular motion, that will change the deferent radius from 60 at the new moon to 39;22 at the first quarter, back to 60 at the full moon, to 39;22 again at the last quarter, and finally back to 60 again at the next new moon. His mechanism for doing this is shown in Figure VII.2.

In the figure, E denotes the center of the earth, ♈ denotes the direction to the vernal equinox, and $\overline{S}$ denotes the direction to the mean position of the sun. The angle ♈E$\overline{S}$ is the mean longitude of the sun and the angle ♈E$\overline{M}$ is the mean longitude of the moon. That is, both of these angles increase uniformly with time.

While the line E$\overline{M}$ rotates uniformly around the point E, its length does not remain constant. At the same time that it rotates counterclockwise around E, another line EC_1 rotates clockwise around E at the same rate, in such a way that the angle $\overline{M}E\overline{S}$ and $C_1E\overline{S}$ are always equal to each other and to the mean elongation D of the moon from the sun. The radius EC_1 must be found from the data, as must the distance $C_1\overline{M}$, both of which are constant. For the moment, let us ignore the other elements of Figure VII.2. Let P denote the distance $C_1\overline{M}$‡ and let ρ denote the distance EC_1.

Following Ptolemy, let us continue to use 60 for the radius of the deferent (that is, the distance E$\overline{M}$) at the syzygies. The configuration of Ptolemy's mechanism at the

---

†When I say that this is an error, I mean that it was wrong in terms of the information known to Ptolemy, not merely that it is wrong in the light of modern knowledge. As I shall show in the sequel, Ptolemy not only should have known better, but he actually did know better. He persisted in advocating a theory that he knew to be wrong, and he used fraudulent data in order to do so.

‡If Ptolemy gives evidence to show that conditions are the same at both first and last quarters, I have not noticed it. However, it is perhaps plausible to assume that there was a large body of well-known evidence that Ptolemy did not cite.

‡The letter P is supposed to be a capital rho, not the Latin letter with the same shape.

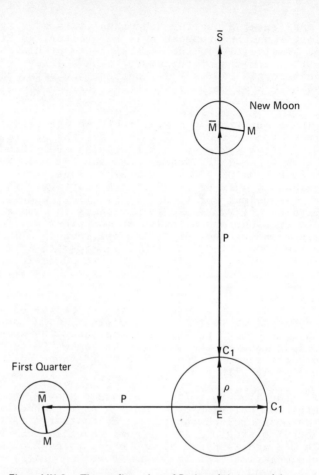

Figure VII.3   The configuration of Ptolemy's lunar model at
the new moon and at the first quarter. At the new moon, the
distance $E\overline{M}$ equals $P + \rho$ and at the first quarter it equals $P - \rho$.
The points that are labelled in this figure have the same meanings
that they have in Figure VII.2. For both positions of $\overline{M}$, the point
M is drawn in the position that puts it farthest in longitude from $\overline{M}$.
The configurations at the full moon and at the last quarter are
symmetrical to those shown. The figure is approximately to
scale.

new moon and at the first quarter is shown in Figure VII.3.
At the new moon, D = 0 and both ρ and P are directed toward
the mean sun $\overline{S}$. At the first quarter, D = 90°. ρ then extends to the right of E in the figure while P extends to
the left of C_1. At the new moon, the deferent radius $P + \rho$
= 60. At the first quarter, the deferent radius $P - \rho$
= 39;22. Hence

$$P = 49;41,$$

(VII.2)

$$\rho = 10;19.$$

The epicycle radius $\overline{MM}$, which we have been denoting by r, is still $5;15$ in this system of units.

The epicycle radius $\overline{MM}$ is drawn in Figure VII.3, for both the new moon and the quarter, in the position that gives the maximum equation of the center. We see that the maximum equation (the angle subtended by the radius) is indeed larger at the quarter moon than at the new moon.

The mechanism that has been described handles the evection at the syzygies and quarters but not at any other phase. We can see this in modern terms by keeping only the two largest terms in Equation VI.4, which are the terms needed to describe the evection, and evaluating them at some other phase, say at $D = 45°$. The two terms in question then give us

$$6°.29 \sin M + 1°.27 \cos M,$$

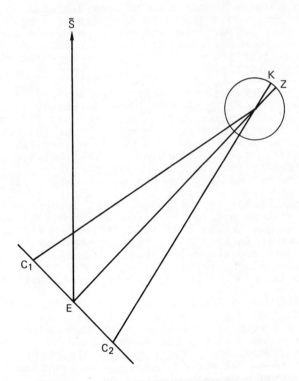

Figure VII.4 The configuration of Ptolemy's lunar model at the time of the observation of −126 May 2. The moon was slightly to the right of the line EZ as seen from the earth E, but the mean anomaly was 185;30 degrees. Hence the mean anomaly must be measured from a point K rather than from Z.

which does not vanish when M = 0 if we measure M from the line joining E and $\overline{M}$. In order to get the term $1°.27 \cos M$, we must measure M from some other direction.

Ptolemy does not word the matter this way, of course, because he had no reason to suspect the correct form of the evection term, so far as we know. He found from measurements that a further mechanism is needed. The measurements that he used [Chapter V.5 of the Syntaxis] are the measurements of -126 May 2 and -126 July 7 in Table VII.1, which Ptolemy attributes to Hipparchus.

The observation of -126 May 2 was made midway between the last quarter and the new moon, when D was 315;32 degrees by Ptolemy's calculation and 315;34 by mine. The situation is shown in Figure VII.4, which is approximately to scale. For clarity, the point $\overline{M}$ is not labelled, but it is the center of the epicycle, which appears to the upper right in the figure. Any letter that appears in both Figure VII.2 and VII.4 has the same meaning in both figures. Since the line $E\overline{M}$ rotates counterclockwise, the point $\overline{M}$ is 315° ahead of the mean sun $\overline{S}$.† The mean anomaly of the moon was 185;30 degrees by Ptolemy's reckoning and 185;28 by mine. If the anomaly were measured from the point Z (which is on the extension of $E\overline{M}$), as it has been in previous uses of the epicycle, the moon would appear to the left of $\overline{M}$ by a small amount; I have not drawn in the position of the moon because there is not room for it in the figure. That is, the moon would be ahead of the mean moon and the equation of the center would be positive.

From Table VII.1, we see that the measured longitude of the moon was 351;27,30 degrees. The longitude $L_{\mathbb{C}}$ of the mean moon was 352;13 degrees by Ptolemy's calculation and 352;10 degrees by mine. Hence e_c, the equation of the center, is -0;45,30 if we use Ptolemy's figures; he immediately replaces this by -0;46, which is reasonable. That is, the moon is to the right rather than to the left of the line $E\overline{M}Z$. From the observation, we can locate where the moon actually is on the epicycle, and from this position we measure around the epicycle by the amount of the anomaly (185;30 degrees) to the point K. Then we extend the lines $K\overline{M}$ and C_1E until they meet in the point C_2, and we calculate the distance EC_2. In the scaling used in Equations VII.2, he finds that $EC_2 = 10;18$. I have not carried the calculations to completion and cannot make an explicit comparison, but I can safely say that Ptolemy's calculations are as accurate as the situation warrants.

In this way, Ptolemy has proved by measurement that the distance EC_1 and EC_2 are equal, to high accuracy. He verifies this by using the observation of -126 July 7. At the time of this observation, the lunar elongation D (angle $\overline{MES}$) was almost exactly 45°. The configuration, shown in Figure VII.5, is almost the mirror image of the previous configuration. In discussing this observation, I shall give only

---

†I am dropping the small excess over 315° for simplicity in the discussion, although it must be taken into account in quantitative calculations.

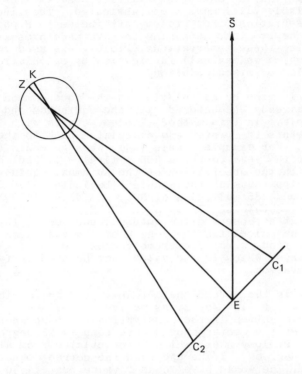

Figure VII.5   The configuration of Ptolemy's lunar model at the time of the observation of −126 July 7. The moon was somewhat to the left of the line EZ as seen from the earth E, but its position was not consistent with the value of the mean anomaly, which was 333; 12 degrees. The mean anomaly must be measured from point K rather than point Z.

The measured longitude of the moon was 148;46 degrees and the longitude of the mean moon was 147;20 degrees. Hence e_c = +1;26 degrees. The mean anomaly was 333;12 degrees. If this value of the mean anomaly were measured clockwise from Z, we would have e_c = +2;33 rather than 1;26 degrees, so that the mean anomaly must be measured from some other point. As before, we locate the position of the epicycle, we measure off the amount of the mean anomaly in order to find point K, and we locate the point C_2. Ptolemy finds that the distance EC_2 is 10;20.

To summarize the situation, Ptolemy finds from the observations of 139 February 9 and −127 August 5 that the distance EC_1 is 10;19 in his system of scaling. From the observation of −126 May 2 he finds that EC_2 equals 10;18 and from the observation of −126 July 7 he finds that EC_2 equals 10;20.

4. Fabrication by Calculation and Fabrication by Miscalculation

We may say at once that all five observations of the moon in Table VII.1 have been fabricated. This includes the three observations that Ptolemy attributes to Hipparchus as well as the two that he claims to have made himself. In fabricating these observations, Ptolemy has used two distinct methods which we may call fabrication by calculation and fabrication by miscalculation.

Fabrication by calculation is the more common sort that we have already encountered with the equinoxes and solstices, for example. In this method, Ptolemy starts with the result that he wants to "prove" and calculates back to the data that he needs. For example (Table V.4), when he wants to establish that the year contains 365 + (1/4) - (1/300) days, he starts with the observation of the autumnal equinox attributed to Hipparchus at the midnight beginning -146 September 27. He adds 285 multiples of 365 + (1/4) - (1/300) days to this time, obtaining 07 hours on 139 September 26, and he claims that he observed the autumnal equinox at this time. Finally, he calculates the length of the year from the two times and finds a value extremely close to 365 + (1/4) - (1/300) days, which is the value that he used in fabricating the observation.

This is the method that Ptolemy used in fabricating the observation of -126 May 2, as we may see by asking what he should have gotten if the observation had been accurate. The equation of the center should have been -1;23 degrees,† but the value Ptolemy uses, which is calculated from his tables, is -0;46 degrees. If he had used the correct equation of the center, he would have found a value near 14;40 for the distance EC_2 instead of the value 10;18 that he did find. Thus, if the data really did lead to the value 10;18, it must have been because of an accidental error of observation.

Let us suppose that the standard error of an observation is 15'. This is enough to change the value found for EC_2 by 1;48, so 1;48 is the standard deviation of a single value of EC_2 under the conditions that existed. However, the value that he finds for EC_2 agrees within 0;1 of the preassigned value 10;19. That is, the value lies in a preassigned zone of total width 0;2, which is 0.0185 times the standard deviation. The probability that this happened by chance is about 0.0074, about 1 chance out of 135.

We have similar results for the observation of -126 July 7. The equation of the center should have been about +2;19 degrees, but Ptolemy uses +1;26 degrees. If he found 10;20 for EC_2, it must have been by chance, because he should have found a quite different value. The probability that the value 10;20 resulted from accidental errors of observation

---

†I calculated this value by using the sum of Equations VI.4 and VI.12. It is doubtful if this value is in error by as much as 10'. It is certainly not in error by enough to affect our conclusions significantly.

is again about 1 out of 135. The joint probability that both values agree by chance with 10;19 within an error of 0;1 is about 5.5×10^{-5}, which is negligible.[†] In other words, the data did not happen by chance. They were fabricated.

In both these cases, then, Ptolemy started with the requirement that the distance EC_2 should be 10;19, equal to EC_1. We cannot know with certainty how he proceeded to fabricate the data, but there is an obvious procedure. When we draw the line $C_2\overline{M}$ in either Figure VII.4 or VII.5, we can locate the point K by a simple calculation. We then calculate the mean anomaly and measure it around the epicycle in the clockwise direction from K. This gives us the desired position of the moon on the circumference of the epicycle, and we can then calculate its longitude by using the radius $\overline{EM}$ of the deferent when the elongation is 45°. We also calculate the position of the sun at the same time, and this gives all the data that we need.

In the observation of -127 August 5, in contrast, we have fabrication by miscalculation. The only quantity that Ptolemy actually uses from the record of the observation is the angle 86;15 degrees between the observed positions of the moon and sun. He calculates that the longitude of the sun was 128;20 degrees, so that the position of the moon appeared to be 42;05 degrees. The mean longitude of the moon was 34;25 degrees, just 7;40 degrees from its apparent position, and this is the value that Ptolemy wants. In order to get this value, Ptolemy states, simply but erroneously, that there was no parallax in longitude. Actually, according to Ptolemy's theory, if the longitude of the moon as seen from Alexandria was 42;05 degrees, its geocentric longitude was 42;14 degrees. This would make the equation of the center be 7;49 rather than 7;40 degrees.

By an oddity, the geocentric longitude calculated from Ptolemy's lunar theory (Table VII.1) is 42;04 degrees, in close agreement with the value that Ptolemy uses. However, the longitudes calculated for the sun and moon from his tables do not agree with those that were originally measured. This makes it difficult to decide exactly what Ptolemy did. He may have fabricated the data by calculation and then disguised the fact by stating that the measured values were slightly different. If he did this, he did not calculate the parallax either when he was fabricating the data or when he was analyzing them. That is, he used miscalculation whether the original data were genuine or fabricated.

I tend toward the conclusion that the data which he

[†]The reader who is versed in the theory of errors will recognize that the width of the zone (0.0185 times the standard deviation) does not by itself determine the probability. Estimating the other factors needed requires making some assumptions. This situation arises several times in the course of this work. Instead of making assumptions, I give instead the maximum probability that is possible for the given width of the zone.

attributes to Hipparchus in this observation are genuine and that he found he could use them for purposes of fabrication by miscalculating the parallax. However, there is no sound basis for this conclusion that I know of, and one should not use the data in astronomical research.

The alleged observation of 139 February 9 involves both calculation and miscalculation. We remember that Ptolemy claims to have measured the longitude of the sun as 318;50, and this is also the value that he calculates. Thus we know that the observation is fabricated; the calculated and measured values should disagree by more than 1°. Ptolemy says that the measured longitude of the moon was 219;40 degrees and that there was no parallax in longitude, so that this was also its correct geocentric longitude. This statement is not correct. If the longitude as seen from Alexandria was 219;40, the geocentric longitude according to his parallax theory was 219;49 degrees. Since the mean longitude was 227;20 degrees, he should have found that the equation of the center was -7;31 degrees rather than -7;40.†

Ptolemy now makes another serious miscalculation. For the observation of 139 February 9, he calculates that the mean anomaly is 87;19 degrees (I calculate 87;18), and he says that this is the anomaly that gives the maximum equation of the center. However, the anomaly that gives the maximum equation E (see just below Equation VI.11) is 90° plus E. That is, if the maximum equation is 7;40 degrees, the maximum occurs when the anomaly is 97;40 degrees, not 87;19. If the equation equalled -7;40 degrees for an anomaly of 87;19 degrees, as Ptolemy says it did, the maximum would have been 7;47. If the equation equalled -7;31 as Ptolemy would have found if he had calculated the parallax correctly, the maximum would have been about 7;38 degrees.

We may now make an educated guess about what Ptolemy did. His fabricated observation must meet two conditions: the moon must be at a quarter (it probably does not matter which one) and the mean anomaly must be near 98°. When he tried the quarter moon that came on 139 February 9, he found that the anomaly was only about 87°. It would have been close to the desired value on the next day, but then the moon would have been too far beyond the quarter. Rather than search for another time when both conditions would have been met, he merely stated that the anomaly was the value that he wanted. In order to fabricate the observed position of the moon, he took the calculated mean longitude of the moon, namely 227;20 degrees, and subtracted the maximum equation that he wanted, obtaining 219;40 for the desired longitude.

This does not explain why he handled the parallax incorrectly. It would have been simple for him to say that the observed longitude was 219;31. If this sounded implausib̶

---

†Note that Ptolemy would have gotten E = 7;31 degrees from this record and 7;49 from the observation that he attributes to Hipparchus. That is, the observations are not consistent in spite of Ptolemy's claim that they are.

-158-

for the result of a measurement, he could have said that the
observed longitude was 219;30 degrees; the geocentric longitude was thus 219;39, and thus the mean longitude was 227;19
degrees. He could simply have stated this value, or he could
have changed the time by 2 minutes. Better, no one would
have quarreled if he took the parallax to be 10', giving
219;40 degrees for the geocentric longitude, which is the
value that he wants. Laziness is the only reason I can think
of for his not calculating the parallax; he certainly did not
forget about it, as we know from the fact that he states its
value erroneously.†

The observation of -126 July 7 still poses a problem. I
think there is no doubt that the observation was fudged, but
the position of the moon calculated from Ptolemy's tables
disagrees by 14' with the longitude that Ptolemy uses. Further, he says that the anomaly was 333;12 degrees, but I
calculated 333;01 from his tables. The answer to the problem
comes from the time, I believe. Ptolemy says that the time
of the observation, omitting the years and days, was 4 hours
after the epoch in apparent time but 3 2/3 hours in mean time.
In my calculations, I used the mean time that he states. However, if I use 4 hours, I get 333;12 degrees for the anomaly,
just as Ptolemy does, and I get 148;45 degrees for the longitude where he uses 148;46. There is little doubt, I think,
that Ptolemy accidentally used the apparent time rather than
the mean time when he fabricated the observation.

As I said earlier, Ptolemy uses the observation of 139
February 23 in deriving the rate of precession of the equinoxes. I shall defer discussion of this observation to the
appropriate place in Section IX.3.

5.   The Author of the Fraud

When the solar observations discussed in Chapter V were
fabricated, the fabricator did not need any specialized
knowledge. The fabrication was done by means of famous older
observations combined with Hipparchus's length of the year.
Thus we could readily envisage that Ptolemy had an assistant
or an apprentice who betrayed him by fabricating data that
he was supposed to observe.

When the lunar eclipses discussed in Chapter VI were
fabricated, the situation was slightly different. The quantities needed to start the fabrication are the daily changes
$n_{\mathbb{C}}$ and $\gamma_{\mathbb{C}}'$ of the moon in mean longitude and anomaly, and
the maximum equation E of the center when the moon is at a
syzygy. The fabricator could have used the values of $n_{\mathbb{C}}$ and
$\gamma_{\mathbb{C}}'$ that Hipparchus had found, and our hypothetical apprentice could well have known these values. However, judging
from what Ptolemy says, Hipparchus did not have a value for
E. Hipparchus, quite correctly, had found different values

---

†For both 139 February 9 and -127 August 5, the correct
parallax is 9'. Ptolemy would not have thought that this
value is negligible. For the observation of 139 February
23, the parallax is only 5', but Ptolemy includes it in his
analysis of the observation.

of E from different sets of eclipses, but the fabrication has been done using the specific value E = 5°. However, the apprentice, when he decided to fabricate the data, could have chosen to use 5° as a convenient round number that was reasonably accurate. Thus, with slightly lower probability, we may still accept fabrication by the apprentice.

The fabrication of the observations in Table VII.1 presents a greatly different situation. This fabrication requires the use of Ptolemy's fully developed lunar model, including the values of all his parameters. If the fabricator were an assistant or an apprentice, we now have a remarkable situation.

To start with, the apprentice must have developed the lunar model himself. If so, not only did the apprentice do Ptolemy's observing and calculating for him, but he also invented the lunar model that we have been crediting to Ptolemy. We may indeed ask what did Ptolemy do himself if this is so. If the apprentice fabricated the data and presented them to Ptolemy, Ptolemy then proceeded to invent the same models. I think we may safely rule out the possibility that Ptolemy invented the identical models in complete independence of the apprentice.

Of course, we can still find ways around this difficulty, but our task gets steadily more and more difficult as we try to uphold the hypothesis of the guilty apprentice. Perhaps the apprentice managed to implant the ideas into Ptolemy's mind so subtly that Ptolemy mistook them for his own. Or, perhaps, Ptolemy had the basic ideas for the models and discussed them thoroughly with the apprentice while telling him to find the kind of data needed in order to test them. The apprentice could then have chosen the parameters, combined them with the basic mechanism of the model, and fabricated the data which Ptolemy proceeded to use.

On this basis, it would have been the apprentice who calculated the parallax incorrectly for the observations of 139 February 9 and -127 August 5. It is too much of a coincidence to assume that Ptolemy independently made the same errors when he analyzed the data, so we must assume that the apprentice also made the analysis and that Ptolemy merely wrote it up. This still does not explain the significant mistake about the anomaly on 139 February 9.

As we have seen, the fabricator of this observation incorrectly had the maximum value of e_c (the equation of the center) come when the anomaly was 87;19 degrees, although the correct value is 97;40. This is not a trivial error; it is vital to the conclusion that is drawn from the data. But Ptolemy actually writes that 87;19 degrees is the value of the anomaly for which e_c has its greatest value; it is not a mere matter of putting this value on paper in the middle of a string of calculated values. If we attribute this sentence to the apprentice, then the apprentice also wrote the Syntaxis and the apprentice becomes the person we have been calling Ptolemy.

There are known cases in history in which a work written by one person has by accident become attributed to another. Now there was a person named Ptolemy to whom a number of works have been attributed, including works on optics, geography, and astrology. It is perhaps conceivable that the Syntaxis was written by someone else and accidentally attributed to the author of these other works. In order to allow for this possibility, I shall adopt a definition: When I write "Ptolemy" in the rest of this work, I mean the author of the Syntaxis. It is beyond the scope of this work to enquire whether the author of the Syntaxis also wrote the other works in question.

The reader may say that the point about the anomaly on 139 February 9 is a small one. It may be small, but to me it seems highly significant. It concerns a vital part of using the observation in establishing the lunar model. I have thought of no explanation for Ptolemy's sentence about the anomaly except the explanation that he is responsible for the fabrication of the data.

6. The Accuracy of the Lunar Model in Longitude

The lunar model of Figure VII.2 provides a way of calculating both the longitude of the moon and its distance from the earth. In Section VIII.5 I shall take up the accuracy with which it gives the distance. Here I shall consider only its accuracy in the lunar longitude.

Dreyer [1905, p. 200] writes, after he has summarized Ptolemy's models of the sun, moon, and planets: "Nearly in every detail (except the variation of distance of the moon)† it represented geometrically these movements almost as closely as the simple instruments then in use enabled observers to follow them, ..." Many writers since Dreyer, and probably many before him, make similar statements. I cannot find any basis for the statement in a detailed study of Ptolemy's models. I think that it is an illustration of the well-known principle that we may call the immortality of error.

We may state this principle in the following manner: Suppose that an error made by a writer A has somehow been published, and suppose further that a later writer B quotes and cites the error, accepting it as correct. The error then becomes immortal and cannot be eradicated from the scholarly literature. I do not maintain seriously that the principle has no exceptions. However there are distressingly many examples for which the principle is valid, and any reader can probably furnish his own examples.

In order to test the accuracy of Ptolemy's lunar model, I first calculated the longitude of the moon from Brown's theory‡ for 51 times spaced 139 days apart. The total span

†The parenthesis is Dreyer's.

‡Brown [1919]. Actually I used the refinements of Brown's theory developed by Eckert, Jones, and Clark [1954]. This is the current basis for the lunar ephemerides found in the major national ephemeris publications.

covered by the data is about 19 years. At the end of the
span, both the sun and moon are reasonably close to the
positions they had at the beginning. I then prepared a
computer program to calculate positions directly from Ptolemy's theory for the same times, and compared the results
from the two theories. Since there are errors in Ptolemy's
values of $n_{\mathbb{C}}$ and $\gamma_{\mathbb{C}}'$, the comparison depends upon the average epoch involved in the comparison. In order to avoid
this difficulty, I first added a constant to all of the longitudes calculated from Brown's theory in order to make the
average longitude the same for both theories. Finally,
after making the initial comparison on this basis, I varied
the parameters in Ptolemy's theory until it gave the best
fit to the value from Brown's theory. In this variation, I
kept the values of $n_{\mathbb{C}}$ and $\gamma_{\mathbb{C}}'$ constant. The parameters that
were varied were the values of L_0 and γ_0 at the initial
epoch, the radius r of the epicycle, and the distances EC_1
and EC_2 in Figure VII.2.† I denote these distances by ρ_1
and ρ_2, respectively.

TABLE VII.2

A COMPARISON OF PTOLEMY'S LUNAR PARAMETERS
WITH THOSE THAT GIVE THE BEST FIT

Quantity	Ptolemy's Value	Value That Gives the Best Fit
L_0	41°.367	41°.376
γ_0	268°.817	268°.775
r	0.087 500	0.092 929
ρ_1	0.171 944	0.143 900
ρ_2	0.171 944	0.163 040
E_1	5°.020	5°.332
E_2	7°.664	7°.497
σ	0°.581	0°.558

The comparison is summarized in Table VII.2. In the
first five lines, the table lists Ptolemy's values and the
values that give the best fit for the five parameters L_0,
γ_0, r, ρ_1, and ρ_2. The distances are given in a scale for
which the radius of the deferent is unity when the moon is
at a syzygy. With this scaling, the distance P in Figure
VII.2 is $1 - \rho_1$. In the next two lines, the table lists

†In discussions of Ptolemy's model, it is usually said that
the distance EC_1 and EC_2 are equal, and they are equal in
Ptolemy's final version of the model. However, he starts
his discussion by assuming that they may be different and
"proves" from measurements that they are equal. Hence it
is appropriate to let the distances be different in explor-
ing the accuracy of his treatment.

the derived quantities E_1 and E_2. E_1 is the maximum equation
of the center when the moon is at a syzygy and E_2 is the maximum when the moon is at a quarter. Finally, the table gives
the value of σ, the standard deviation of the error.

With Ptolemy's parameters, the maximum error found is
$1°.08$ and the standard deviation is $0°.581$, about $35'$. Since
the apparent diameter of the moon at its average distance is
only about $33'$, the error is greater than the apparent size
of the moon much of the time. Thus the theory cannot be
called a great success, it seems to me. Certainly it does
not predict the longitude within the accuracy of the observations that the Greek astronomers could make.

We also see that the limitation is in the model itself
and not in the values of the parameters. When the parameters
are adjusted to give the best fit, the standard deviation
falls only to $0°.558$, about $33\frac{1}{2}$ minutes of arc.

Tannery [1893, p. 211ff] shows that Ptolemy's model includes the effects of the evection and about half of the term
in Equation VI.4 that is called the variation. This is the
term $0°.66 \sin 2D$, in which D is the lunar elongation. On
the basis of this, I wrote in Part I† that Ptolemy's model
describes the evection and also "agrees in part" with the
variation. After examining the situation in detail, I realize that the statement is not correct, as we can see from two
arguments.

First, Tannery derives a formal expression for the equation of the center that is given by Ptolemy's model, and he
expands this expression in a series of powers of the distances
r and ρ. The expansion contains terms that correspond to the
terms $6°.29 \sin M$, $0°.22 \sin 2M$, and $1°.27 \sin (2D - M)$ in
Equation VI.4, but with slightly different numerical coefficients. The remaining term in the expansion equals a numerical coefficient multiplied by

$$\sin 2D[\cos (2D + M) + 2 \cos (2D - M)],$$

in the notation used in this work. The value of the coefficient is about $0°.30$. This value, when multiplied by the
maximum value of the quantity in brackets, amounts to about
$0°.53$, fairly close to the coefficient $0°.66$ in the variation. Since the quantity just exhibited is proportional to
$\sin 2D$, just as the variation is, Tannery says that it corresponds to a part of the variation.

Probably because the matter was not an important one,
Tannery did not examine the situation carefully. If he had,
he would have realized that the quantity in brackets has an
average value of zero when averaged over the anomaly M, for
every value of D. Thus the term that Tannery found does
not correspond to the variation at all; it is simply a

---

†Thus doing my part to maintain the principle of the immortality of error.

measure of how Ptolemy's model fails to agree with the evection.†

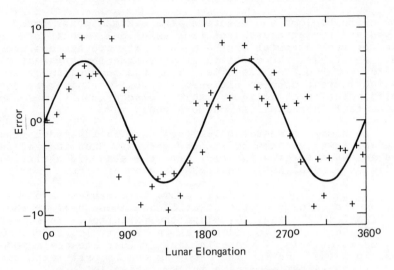

Figure VII.6   The error in Ptolemy's lunar model, plotted as a function of the lunar elongation. The error is taken to to be the value from Brown's theory of the moon minus the value from Ptolemy's theory. The +'s are errors calculated at 51 times spaced 139 days apart. The curve is the function $0^{o}.66 \sin 2D$, in which D is the elongation.

Second, let us look at the errors in Ptolemy's model at individual times, plotted as a function of the elongation D. This is done in Figure VII.6. The +'s in the figure are the 51 individual errors that were calculated, while the curve is the function $0^{\circ}.66 \sin 2D$. The curve passes rather well through the center of the errors. That is, the errors seem to consist of the variation $0^{\circ}.66 \sin 2D$ plus other

---

†Dreyer [1905, p. 256] points out that Tannery's term "obviously ... has nothing in common with the variation, except that" it is proportional to sin 2D. Since the matter is obvious, he does not give an explanation. On the same page, he makes a remark that I do not understand. He says that adding the variation to Ptolemy's theory would "spoil the latter and make its maximum error rise to more than a degree." That is, correcting an error or removing an omission increases the total error, according to his statement. He probably did not mean to say this, but I do not know what he did mean. Actually, as we have seen, the maximum error in Ptolemy's theory is more than a degree. I have tested the effect of adding the variation to Ptolemy's theory by the method already described. The result is to drop the maximum error from $1^{\circ}.08$ to $0^{\circ}.75$, and to drop the standard deviation from $0^{\circ}.58$ to $0^{\circ}.31$. Adding the variation to Ptolemy's theory does not spoil it; it improves it.

miscellaneous smaller effects that depend upon other quantities. Thus the model does not seem to include any part of
the variation.

This finding raises some questions. The device of making the radius of the deferent smaller at the quarters than
it is at the syzygies gives reasonable accuracy in the equation of the center at those phases. Why did Ptolemy decide
to make a further correction at the octants (where the elongation is an odd multiple of 45°)? Having decided this, how
does it happen that he chose the parameter ρ_2 so that it has,
fairly closely, just the value demanded by the evection? The
last question needs some explanation.

In Equation VI.4, let us replace M (which is measured
from perigee) by $\gamma + 180°$, since γ is measured from apogee,
and let us keep only the two largest terms. Then

$$e_c = -6°.29 \sin \gamma - 1°.27 \sin (2D - \gamma).$$

Let us choose D to equal 45°, for example; any other odd
multiple of 45° will give a similar result. Then

$$e_c = -6°.29 \sin \gamma - 1°.27 \cos \gamma.$$

When $\gamma = 0$, this gives $e_c = -1°.27$. This in turn tells us
the point K in Figure VII.5, which must be the origin of
the anomaly if the model is to represent the evection correctly. The result is that the angle from Z to K must be
13°.1 when D = 45°. Ptolemy's tables give the angle as
12°.0.

However, his tables do not work as well at some other
values of D. For D = 60°, for example, we find $e_c = -1°.10$
when $\gamma = 0$. If we use the appropriate radius of the deferent
for D = 60°, we find that the angle from Z to K has fallen to
10°.3, while Ptolemy's tables have it rising to 13;04 = 13.07
degrees.

Leaving the last point aside, Ptolemy presumably had no
reason to suspect that the evection should be described by a
term proportional to sin (2D - γ), and thus he presumably
could not have used the preceding argument. However, if Hipparchus's measurement on -126 May 2 was a valid one before
Ptolemy altered it by fabrication, he would have known that
something was wrong with his model when D = 315°. At this
time, γ was 185;30 degrees by Ptolemy's computation, and
this leads to $e_c = +0°.665$ degrees, almost exactly 0;40. At
this time, the correct equation of the center was about
-1;22 degrees, and a valid observation would have led to a
value fairly close to this. Hence the anomaly needed to be
modified by a fairly large amount; Ptolemy's model leads to
$e_c = -0;46$ for this observation.

For the observation of -126 July 7, matters come out
differently. The value of γ used by Ptolemy was 333;12 degrees, which makes $e_c = 2;32$ degrees. This is fairly close
to the correct value, which is about 2;19 degrees. Thus the
anomaly does not need a large modification. However, the

average modification required by both observations is reasonably close to the amount that Ptolemy adopted.

For his final model, Ptolemy fabricated the data so that
ρ_1 and ρ_2 are equal. I do not believe that Ptolemy based
this equality upon the average of a large body of results.
For one thing, he should have had the modification come close
to $13°.1$ when $D = 45°$ if he had done so, but he makes it come
out $12°.0$. For another, Ptolemy frequently chooses to fabricate his observations so that some important parameter will
come out to an even number, or so that two parameters will
come out equal. We saw what is almost surely an example of
the former in Section VI.7, where he found the maximum equation of the center at the syzygies. We shall see further
examples of the latter in later parts of this work. I think
he simply chose to make ρ_2 equal to ρ_1 because he liked the
equality.

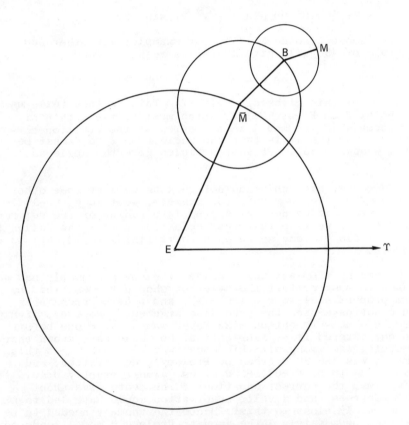

Figure VII.7   The double epicycle model applied to the moon.
E is the earth and the line E ♈ points toward the vernal equinox.
M is the moon and $\overline{M}$ is the mean moon. Thus angle $\overline{M}$E ♈ equals
the mean longitude of the moon. The line from $\overline{M}$ to B points
toward the lunar apogee, and the angle $\overline{M}$BM equals twice the
lunar elongation. Line BM rotates counterclockwise. If distance
E$\overline{M}$ is chosen to be unity, the epicycle radii are 0.109 801 and
0.022 164.

There is a further question that I have deferred until now. How did Ptolemy settle upon 7;40 degrees for the maximum equation of the center when the moon is at a quarter? In Part I, before I discovered his process of fabrication by miscalculation, I took Ptolemy's word for it that Hipparchus's observation (-127 August 5) led to 7;40 degrees. We know now that Hipparchus's observation does not lead to this value, and hence we must ask how Ptolemy chose it. The question is particularly cogent because Ptolemy has fabricated both his observation and the one attributed to Hipparchus in order to give this value.

I have not thought of a satisfactory answer to this question. I do not think of a "numerological" reason why Ptolemy should choose 7;40 degrees, but there may be a reason that I am unaware of. It is of course possible that Hipparchus left a large number of observations whose average was 7;40 degrees. This value is in error by about $0°.11$. Since the variation term with its large coefficient of $0°.66$ does not affect readings made at the quarter moons, a reasonable number of observations of reasonable accuracy would give an average that is this close.

Figure IV.5 shows what I called the double epicycle model. It gives a method of varying r, the distance from $\overline{M}$ to the moon, rather than R, the radius of the deferent. When we apply it specifically to the moon, we get the model shown in Figure VII.7. The angle $\overline{M}E\Upsilon$ is the mean longitude of the moon. The first epicycle, whose center is $\overline{M}$, has a radius $\overline{M}B$ that points toward the position of the lunar apogee. Thus it rotates with a period of about 8.85 years. The angle $\overline{M}BM$ is equal to 2D; M is the position of the moon. The radius of the first epicycle is chosen so that it gives a maximum equation of $6°.29$. The radius of the second epicycle is chosen so that it will change this value by $1°.27$ when both radii lie along the same line.

The standard deviation of the error in this model is $0°.545$. Thus we see from Table VII.2 that the double epicycle is slightly more accurate than Ptolemy's model, whether we use Ptolemy's parameters or those that give the best fit. This is not an important advantage of the double epicycle model. However, it does have two important advantages. The first is that it gives a much better representation of the lunar distance than Ptolemy's model does, as we shall see in the next chapter. The second is that the double epicycle gives essentially the same accuracy in longitude as Ptolemy's model, but it is much simpler. If it had been used, astronomers would have found it easier to discover the systematic discrepancy called the variation. As matters went, the variation was not discovered until about 1600. Further, when the variation was discovered, the model could have been modified to allow for it. The resulting model, though much more accurate, would still have been no more complicated than Ptolemy's model.

As I said in Section IV.6, Copernicus [1543, Chapter IV.3] was the first person in Europe to use a double epicycle for the moon, so far as we know, but the model was used about two centuries earlier by the Islamic astronomer ibn ash-Shatir

[Neugebauer, 1957, p. 197]. We do not know whether Copernicus invented the model independently or not.

I have also calculated the accuracy that would result
from using a single epicycle of radius chosen to give a maximum equation of $6°.29$. The standard deviation is $1°.094$,
about twice that which we get from either Ptolemy's model or
from the double epicycle. Thus Ptolemy's model does achieve
some success in predicting the geocentric longitude of the
moon, since it halves the error that we get from a single
epicycle. Further, the model brings out the systematic difference between the syzygies and the quarters. This improvement is bought at the expense of a severe loss in accuracy of
the lunar distance, as we shall see in the next chapter.

A large error in the lunar distance means a large error
in the parallax, which enters into the topocentric longitude.
This is what astronomers actually observe. Thus it is not
clear that Ptolemy's model gives any improvement in the observed longitude. Overall, it seems to me that Ptolemy's
lunar model must be counted as a failure, by the ordinary
standards of scientific success.

CHAPTER VIII

THE SIZES AND DISTANCES OF THE SUN AND MOON

1. Eclipse Theory

The theory of eclipses follows directly from the theories of motion of the sun and moon, and it requires little information other than that needed to describe the motions. For this reason, the theory will not receive much attention in this work. However, it should be discussed to some extent from two points of view. First, it furnishes impressive testimony about the achievements of Greek astronomy. Second, it requires a few measurements that also come into estimating the sizes of the sun and moon.

Ptolemy devotes Book VI of the <u>Syntaxis</u> to eclipses. Since the moon is obviously at a syzygy when an eclipse occurs, and since Ptolemy's theory of the moon at the syzygies differs but little if at all from earlier theories, it is likely that everything in this book was already known. However, Ptolemy does create one error in the subject that was apparently absent from earlier work.

Ptolemy first gives a table that helps to find the mean syzygies, that is, of the times when the mean longitude of the moon minus the mean longitude of the sun is a multiple of 180°. This table is also especially designed to help find the anomalies of the moon and sun at these times, as well as the latitude of the moon. These quantities can then be used to find the times of the true syzygies, by finding the corrections to the mean syzygies required by the equations of the center for the sun and moon. They can also be used to find the latitude of the moon at the time of the syzygy.

Ptolemy then gives tables for the magnitude of an eclipse as a function of the latitude of the moon at the time of the syzygy,[†] for both solar and lunar eclipses. For lunar eclipses, there is little if any effect of parallax, and these tables are good for any point on earth. For solar eclipses, parallax is important, and these tables are good only for an observer at the center of the earth or, what amounts to the same thing, at the point where it is local noon at the center of the eclipse and for whom the sun is in the zenith. For either kind of eclipse, the calculated magnitude depends upon the apparent diameters of the solar and lunar disks; this is the point of contact between eclipse theory and the (real) sizes of the sun and moon. In his tables, Ptolemy takes the apparent diameter of the sun to be constant. For each kind of eclipse, he then gives a table that applies when the moon has its maximum apparent diameter and another that applies when it has its minimum apparent diameter. For each of the four tables generated,

---

[†]More precisely, he uses the argument of the latitude as the independent variable.

he gives a quantity from which we may calculate the duration of the eclipse.

For lunar eclipses, these tables give us almost all that we need, once the time and the latitude of the full moon have been found. If the moon is not at an extreme of its apparent size, we must interpolate for the size that it does have. Finally, we must decide, for a given place, whether the moon is above or below the horizon.

For the sun, matters are more difficult, since the circumstances depend markedly upon the point of observation. Ptolemy describes how to calculate the circumstances for any place, say for point A. The first task is to calculate the parallax of the moon for point A, using the lunar coordinates calculated for the time of the geocentric conjunction. If there is a parallax in longitude, as there usually is, the time of the apparent conjunction at A will be different from the time calculated. Thus a new time must be calculated, which means in principle that a new parallax must be calculated, and so on. The problem cannot be solved exactly, but the successive approximations just outlined quickly bring us to an acceptably accurate solution. When the time is found, the apparent latitude of the moon as seen from point A can be calculated. Now the earlier tables can be used to give the circumstances.

Ptolemy refers to Hipparchus a few times in his discussion of eclipses, but his references do not say explicitly whether Hipparchus was in possession of the full theory or not. Since the theory does not involve the evection, eclipse theory can be based upon the simple epicycle. Ptolemy says that Hipparchus had developed both the epicycle and eccentric models for the moon, although we cannot be sure what value he used for the maximum equation of the center. Ptolemy also tells us that Hipparchus made several studies dealing with the parallaxes of the sun and moon, and that he used eclipses in doing so [Syntaxis, Chapter V.11]. If this is so, Hipparchus was in possession of all the elements needed to calculate the circumstances of a solar eclipse at any known point.

Thus, probably by the time of Hipparchus, Greek astronomers were able to calculate the circumstances of solar eclipses, and this ability necessarily includes the ability to predict eclipses and their visibility at specified places. It is interesting to estimate the accuracy with which they could do so.

The accuracy is limited by the accuracy of the solar and lunar ephemerides used, and by the accuracy of geographical coordinates. In both Hipparchus's and Ptolemy's

†I believe that I have read, in a source that I cannot put my hand on at the moment of writing, a statement that Hipparchus calculated all eclipses over a span of 600 years. This seems somewhat implausible, because of the enormous amount of computation involved. Still, the statement, if indeed it exists, supports the viewpoint that eclipse theory was known to Hipparchus.

theories, there are significant errors in the mean rates of
motion of the sun and moon, but there is only a small error
in their relative rate. We can neglect this source of error
for times near the year 0. In addition, we found a standard
deviation of $0°.581$ in Ptolemy's theory of the moon, but this
includes the effects of the variation, which does not matter
at the syzygies where eclipses occur. When the variation is
eliminated, the standard deviation in Ptolemy's lunar theory
(Section VII.6) is about $0°.31$, and this would also have been
about the error in Hipparchus's theory at the syzygies. The
maximum error in the solar ephemeris is about $0°.37$, and the
standard deviation is about $0°.26$. The standard deviation in
the relative position of the moon and sun comes out to be
about $0°.40$. Hence the standard deviation in the time of an
eclipse is about 0.80 hours, during which time the earth rotates by $12°$.

We saw in Section III.5 that Ptolemy had an error of
about $21°$ in the longitude of the eastern Mediterranean with
respect to its western end. In effect, this is the error in,
say, the longitude of Gibraltar with respect to Alexandria.
For simplicity, let us take $10°$ as the standard deviation of
a longitude in the classical world, and let us neglect errors
in latitude, even though we found some significant ones. If
there is an error of $12°$ in the orientation of the earth at
the time of an eclipse, and if there is a further error of
$10°$ in positioning points on the earth, the overall error
(standard deviation) in longitude is about $16°$.

That is, if a Greek astronomer near the year 0 calculated that a certain point would be exactly at the center of
a solar eclipse, the standard deviation in his result would
be $16°$ in longitude. This means that the center was $16°$
away in longitude.

By plotting the paths of a number of solar eclipses, I
have estimated that a solar eclipse is total over a band with
a width of about $8°$ in longitude, in a typical case [Newton,
1972a].† The error in the calculation of a Greek astronomer
is thus about twice the width of the band of totality. That
is, if he calculated that an eclipse would be total at a specific point, he was probably wrong. However, it would be
large there. In other words, a Greek astronomer could reliably predict that an eclipse would be impressive at a particular point, but he could not predict reliably that it
would be total there.

2.   Early Studies of the Solar and Lunar Distances

Several centuries before the time of Ptolemy, Greek
astronomers had fairly good estimates of the size and distance of the moon, at least in relation to the radius of

---

†The reader who is accustomed to thinking that the band of
totality is narrow may be surprised by this figure. This
is the width of the east-west direction, but most eclipse
bands run in a nearly east-west direction. The typical
band is quite narrow in its cross section.

the earth.† They also knew that the sun was much larger than the earth and much farther away than the moon, although their estimates fell far short of the actual situation.

We do not know all of the methods by which the Greek astronomers attempted to estimate these quantities, but we do know two of the most important. One method gives the ratio of the solar and lunar distances directly. The other gives the sum of the horizontal parallaxes of the sun and moon. From these relations, we can find both distances. Then, by measuring the apparent diameters of the solar and lunar disks, we can calculate the physical dimensions of the sun and moon, assuming that they are spheres like the earth.

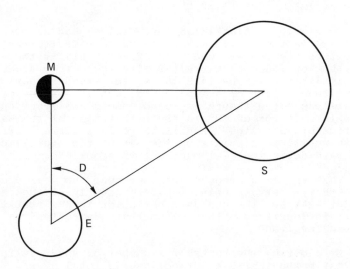

Figure VIII.1   A method of finding the ratio between the solar and lunar distances. When the moon is exactly at the quarter, the lines from it to the earth and sun form a right angle. The ratio of the lunar to the solar distance is the cosine of the elongation angle D. This figure is reproduced from Part I by permission of the Quarterly Journal of the Royal Astronomical Society.

The first method was used by Aristarchus of Samos, the astronomer who proposed the heliocentric theory of the solar system. The work in which he used this method [Aristarchus, ca. -280] is his only work that has survived. Heath [1913], who has edited and translated this work, also gives a large amount of related information obtained from other Greek astronomical writing.

---

†In all the discussion of the sizes and distances of the sun and moon, I shall use the radius of the earth as the unit of length. In this way, we do not need to worry about the accuracy with which Greek astronomers knew the size of the earth itself.

The method is shown in Figure VIII.1. When the moon is exactly half full, the line from it to the earth is perpendicular to the line from it to the sun. More accurately, I should speak of the line to the observer, because the result is clearly affected by parallax. If the moon is exactly halved to an observer for whom it is on the meridian (an observer on the line proceeding from the moon to the earth in the figure), it is more than half full for an observer near the letter E in the figure, and it is less than half full for an observer on the opposite side of the earth.

If we neglect the problem of parallax, we may say that the ratio of the solar to the lunar distance is the secant of the angle D, which is the elongation of the moon at the time of the observation. Aristarchus takes the value of D to be 87°,† so that the ratio is

$$\sec 87° = 19.1073. \tag{VIII.1}$$

Aristarchus does not word the matter in terms of the secant, of course. Nor does he word the matter in the way that we first expect, which is to say that the ratio is 19. Instead, he says that the ratio is between 18 and 20. This does not seem to be a way of expressing an observational uncertainty; it is apparently a limitation on his knowledge of the trigonometric functions.‡ For simplicity in discussing Aristarchus's results, I shall use the value from Equation VIII.1, but the reader should remember that this is not what Aristarchus does.

Dreyer [1905, p. 182] lists several values of the ratio of the solar to the lunar distance that were used by various Greek astronomers and that were obtained by unknown methods; they range from 9 to 30. The correct ratio is close to 400, so that all their estimates were seriously low. Even so, however, they all correctly concluded that the sun is much larger than the earth.

To illustrate this point, let us assume that the average value of the lunar distance is 60, which is close to being correct. Then the solar distance is 60 × 19.1073 = 1146. The apparent radius of the solar disk is about one quarter of a degree. This means that the physical radius of the sun is almost exactly 5 earth radii, and the volume of the sun is about 125 times that of the earth. This is lower than Aristarchus's estimate. He said that the ratio of the volumes lies between 6859/27 (= 254) and 79 507/216 (= 368).

---

†At least one writer has maintained that earlier astronomers used Aristarchus's method, but with other values of D. See Heath [1913, pp. 329-332].

‡Heath [1913, pp. 333-336] shows how Aristarchus probably obtained the limits of 18 and 20 for sec 87°, as well as the limits that he needed for other trigonometric functions that arise in his work.

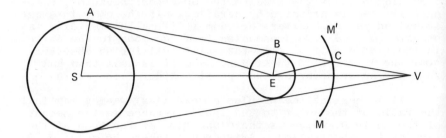

Figure VIII.2   A method of finding the sum of the solar
and lunar parallaxes. Points S and E are the centers of the sun
and earth. The curve MM' is part of a circle whose radius is
the lunar distance. The method is based upon the result from
elementary geometry that the sum of the angles AES and
CEV equals the sum of the angles BAE and BCE. This figure
is reproduced from Part I by permission of the Quarterly
Journal of the Royal Astronomical Society.

The method of finding the sum of the solar and lunar
parallaxes is shown in Figure VIII.2. The circles with
centers S and E represent the sun and earth, and the line
ABV represents the edge of the earth's shadow.[†] The arc MM'
is part of a circle whose radius is the lunar distance.
Suppose that we had a large screen placed at the distance
EC in the figure. The shadow would be projected onto the
screen, and the angle CEV is the apparent radius of the
shadow, at the distance of the moon. I shall denote this
radius (angle) by ρ_U.

Other angles in Figure VIII.2 have important interpretations. The angle AES is the apparent radius $\rho_\odot$ of the
sun's disk. The angle EAB is the parallax of the sun. That
is, it is the difference in direction to the point A on the
sun as seen from E and B.[‡] Similarly, the angle ECB is the
parallax of the moon. From elementary geometry, we easily
find that

$$\angle \text{ AES } + \angle \text{ CEV } = \angle \text{ EAB } + \angle \text{ ECB}.$$

If we give these angles their astronomical interpretations,

[†]More precisely, this is the edge of the umbra. To the
naked eye, the part of the moon that appears to be eclipsed
is only the part that lies within the umbra. At least,
this is so for most observers. Some practiced observers
can observe a darkening of the part that is in the penumbra.

[‡]Specifically, this is the horizontal parallax (Section VI.1)
Until further notice, the unmodified term parallax will refer to the horizontal parallax.

we have

$$\rho_\odot + \rho_U = \Pi_\odot + \Pi_{\mathbb{C}}. \tag{VIII.2}$$

The apparent radius $\rho_\odot$ of the sun can be measured by direct observation and the shadow radius ρ_U can be measured during a partial lunar eclipse; I shall explain the method of doing this later. Substitution of these values into Equation VIII.2 then gives the sum of parallaxes while Equation VIII.1 gives the ratio of the parallaxes.[†] By the simultaneous use of both equations, we can find both parallaxes and hence we can find both the solar and lunar distances. These distances, of course, are expressed in the system of units in which the radius of the earth is 1.

The discussions of Aristarchus's work that I have read leave some confusion about what he does, partly because of the manner in which he introduces the subject. He starts by stating six hypotheses:[‡]

1. That the moon receives its light from the sun.

2. That the earth is in the relation of a point and centre to the sphere in which the moon moves.

3. That, when the moon appears to us to be halved, the great circle which divides the dark and the bright portions of the moon is in the direction of our eye.

4. That, when the moon appears to us to be halved, its distance[‡] from the sun is then less than a quadrant by one-thirtieth of a quadrant.

5. That the breadth of the (earth's) shadow is (that) of two moons.

6. That the moon subtends one-fifteenth part of a sign of the zodiac.

The editor has supplied the words in parentheses to clarify the text. The first three hypotheses are needed to justify the geometrical constructions that Aristarchus will use, and we may doubt that Aristarchus intends for the second hypothesis to be taken literally. Hypothesis 4 says that the elongation when the moon is halved is $87°$, hypothesis 5 says that ρ_U is twice $\rho_{\mathbb{C}}$, and hypothesis 6 says that the apparent diameter of the moon is $2°$, so that $\rho_{\mathbb{C}} = 1°$.

Aristarchus then says that he will prove three propositions from the hypotheses:

[†]More accurately, Equation VIII.1 gives the ratio of the sines of the parallaxes. This fact merely increases slightly the difficulty of solving the equations without affecting the general argument.

[‡]These are quoted from the translation cited in the references.

[‡]We would say the elongation.

1. The distance of the sun from the earth is greater than eighteen times, but less than twenty times, the distance of the moon.

2. The diameter of the sun has the same ratio (as aforesaid) to the diameter of the moon.†

3. The diameter of the sun has to the diameter of the earth a ratio greater than that which 19 has to 3, but less than that which 43 has to 6.

There may be discussions of Aristarchus's work which make it clear what he does, but the ones which I have seen say that Aristarchus proved his propositions from his hypotheses, and they fail to point out that this is impossible. The hypotheses and propositions between them contain five unknown distances. They are the diameters and distances of both the sun and moon, and the breadth of the earth's shadow (at the distance of the moon). However, the hypotheses contain only three relations between these five quantities, and the geometry of the shadow furnishes one more. It is not possible to find five unknown quantities from four relations The answer to this paradox is that Aristarchus tacitly introduces two more relations which take the form of inequalities and these added relations supply the additional information needed. Because of the space that it takes, I shall put the treatment of this subject in Appendix B.

The data that Aristarchus uses contain two serious errors. The amount by which the elongation differs from a quadrant (90°) is far too large, making the ratio of the solar to the lunar distance far too small. The correct amount is closer to 10′ than it is to 3°. However, in view of the difficulty of making the measurement, the error is not surprising.

The other serious error concerns the angle subtended by the moon (hypothesis 6). The angle that Aristarchus gives is too large by a factor of nearly 4. I have seen three proposed explanations of this fact.

The first explanation is that there has been a scribal error. However, the angle is written out in words in at least two places in the Greek text, and Aristarchus clearly uses 2° in his calculations. Hence a scribal error does not seem possible.

The second explanation is that 2° was actually a measured value known to Aristarchus. Heath [1913, p. 311] describes a measurement of the time that the sun takes to move its own diameter, taken from a source that I have not consulted. This measurement assigns 1/216 of a day as the time, which makes the apparent diameter equal to 1;40 degree

---

†That is, the physical diameters of the sun and moon also have a ratio that lies between 18 and 20.

‡However, a Babylonian measurement, also cited by Heath, gives 1/30 hours for the same time, which gives 0;30 degrees for the apparent diameter.

The idea is then either that a measurement now lost gave 2°,
or that Aristarchus chose to round a measured value to 2°.

The third explanation is that Aristarchus deliberately
chose a false value because he wanted to give a specimen
calculation to illustrate the method. We have encountered
this idea before in modern discussions of Greek astronomy,[†]
and it has usually seemed to have little to recommend it.
Here, however, two things seem to support it. First, Archimedes [ca. -225] writes: " .. Aristarchus discovered[‡] that
the sun appeared to be about one 720th part of the circle
of the zodiac." This says that 30′ is the value that Aristarchus thought to be correct[#] and, further, it implies
that Aristarchus was the first to measure this value. Second, the propositions quoted above, which are the ones that
Aristarchus takes as primary, do not depend upon the apparent
size used for the moon. That is, the results that Aristarchus
emphasizes are the same whether he uses a reasonable or an
unreasonable value for the apparent size of the moon. Since
this is so, he might have felt justified in using a value
that, though wrong, does not affect his main results.

In spite of these considerations, I suggest that the
correct explanation is none of these. When we follow Aristarchus's work in detail, we find that he does not want an
accurate value for the apparent diameter of the moon; what
he wants is a safe upper limit. I suggest that he chose 2°
because it was certainly an upper limit. When he uses this
value, he is careful not to state any results that are invalid because it is wrong. The reasons for this suggestion
are given in Appendix B.

The limits that Aristarchus assigns to the diameter of
the sun are 6 1/3 and 7 1/6. The average is 6 3/4, whose
cube is about 308. Thus, as I said earlier, Aristarchus
found that the sun is much larger than the earth. This
makes it physically unreasonable that the sun should revolve
around the earth; doing so is like making a 1 pound tail wag
a 308 pound dog.[*]

[†] See Section III.4. About this point, we must continually
ask: How can a false number have more pedagogical value
than an accurate one?

[‡] The underlining is mine. Heath [1913, p. 312] says that
Archimedes uses the word ευρηκοτοs; I have not seen the
Greek text.

[#] Archimedes refers to the apparent size of the sun while
hypothesis 6 refers to the apparent size of the moon. One
of Aristarchus's intermediate results is that both apparent
diameters are the same.

[*] Aristarchus would have no reason to suspect the large difference in density between the sun and earth, even if the
concept of density had been formulated in his time. Presumably Archimedes, about half a century after Aristarchus,
was the first to formulate the concept of density in a
clear fashion, although there was a vague realization of
it at an earlier date.

After Archimedes gives the apparent size of the sun
that Aristarchus discovered, he says that he (Archimedes)
made an instrument to measure the apparent diameter of the
sun. He finds that it is between 1/164 and 1/200 of a right
angle. The average of these values is almost exactly 30',
which is slightly too small.

Hipparchus was also interested in the sizes and distances
of the sun and moon, and there are many references to his results, although his own writings on the subject have been
lost. Unfortunately, ancient references to his work are not
consistent, and modern references to the same ancient writings
are not consistent. Ideally, I should consult the relevant
and extant ancient writings myself in order to decide what
they contain, but there is a limit to the research that can
be done in the course of this work. Since Heath seems to be
more thorough than the other modern authorities that I have
read on this particular matter, I shall follow him. Doing
so involves some risk, but the risk does not seem to be
large.

Pappus, a mathematician who flourished around +300, has
apparently left us most of the information on this subject
that has survived, and Heath [1913, pp. 341-343] gives a
long quotation from his commentary on the Syntaxis. According to Pappus, Hipparchus wrote two books about the size and
distance of the sun and moon. In the first book, he used an
eclipse of the sun that was total near the Hellespont but
that was only about 4/5 total at Alexandria.† From this, and
presumably from other evidence that Pappus does not mention,
Hipparchus deduced that the lunar distance varies from 71 to
83 earth radii; Pappus does not say what he found for the
solar distance.

In his second book, Hipparchus modified his results considerably. From many considerations, Pappus says, Hipparchus
concluded that the lunar distance varies from 62 to 72 2/3
earth radii and that the solar distance is 490 earth radii.
However, there is convincing evidence that an initial "2"
has been dropped from the last number, and that Hipparchus
actually found 2490 earth radii for the distance. Writers
other than Pappus say that Hipparchus found that the sun has
1880 times the volume of the earth; this makes the radius
of the sun equal 12.342 earth radii, and makes the apparent
size of the sun equal to 17' 2", about 1' too large. The
figures also give 1' 23" for the parallax of the sun and 51'
3" for the parallax of the moon at its average distance.
Hence the sum in the right member of Equation VIII.2 is 52'
26".

This is not consistent with what Ptolemy says about
Hipparchus. In Chapter IV.9 of the Syntaxis, Ptolemy writes

---

†Many writers claim to have proved that this was the eclipse
of -128 November 20, but all such proofs involve a logical
fallacy. All of the dates [Newton, 1970, p. 262] -309
August 15, -281 August 6, -189 March 14, and -128 November
20 are possible, and there may be other possibilities that
I have overlooked.

that Hipparchus took the apparent diameter of the moon to be 1/650 of a circle,[†] which is equivalent to taking $\rho_{\mathbb{C}} = 16'37''$. In the same place, Ptolemy says that Hipparchus took ρ_U to be $2\frac{1}{2}$ times as large, or $41'32''$, but he does not say what Hipparchus took for $\rho_{\odot}$. If he took it to be the average value of $\rho_{\mathbb{C}}$, the sum on the left of Equation VIII.2 is $58'9''$. If he took it to be the value given in the preceding paragraph, namely $17'2''$, the sum is $58'34''$. In either case, the sum on the left is far bigger than the sum on the right, if both Pappus and Ptolemy have reported accurately. It does not matter much which sum we use for the left side in the rest of the discussion. For the sake of definiteness, I shall use $58'9''$, assuming that Hipparchus took the apparent diameter of the sun to be the same as the average value for the moon.

In Chapter V.11 of the <u>Syntaxis</u>, Ptolemy says that Hipparchus used solar eclipses, presumably in connection with lunar eclipses, in attempts to find a greatest possible value of the solar parallax as well as a least possible value. Ptolemy does not state either limit that Hipparchus found, but we can make some estimates. If we use the ratio of distances that Pappus attributes to him, which is 37, the solar parallax is $1'32''$, corresponding to a distance of 2242 earth radii, and the lunar parallax is $56'37''$, corresponding to a distance of 60.7 earth radii. If he neglected the solar parallax, the lunar parallax is $58'9''$, corresponding to a distance of 59.1 earth radii. Both values are lower than those which Pappus attributes to Hipparchus.

We should be suspicious of anything Ptolemy says, especially in view of what he says about the eclipses in Table VI.2. If he is telling the truth here,[‡] Hipparchus made at least three studies of the solar and lunar distances and sizes, with steadily improving results. I am tempted to accept what Ptolemy says here, but for a somewhat illogical reason: The values that Ptolemy attributes to Hipparchus are rather accurate, but the other values are seriously in error. It seems unlikely to me that Hipparchus would have stopped with a result as inaccurate as taking 67 1/3 earth radii for the average distance to the moon,[‡] when methods that give more accurate results were available to him.

In any case, we may conclude that Hipparchus thought that 20 is too small for the ratio of solar and lunar distances. <u>Dreyer</u> [1905, p. 185] says that Hipparchus was not alone in <u>this</u>, that Posidonius "perceived that the sun is very much more than twenty times as far off as the moon," and that he made a "remarkable attempt to determine the actual size of the sun .." Posidonius, according to the account, knew that no shadows are cast at Syene at the summer

---

[†]This is presumably the average value.

[‡]Pappus, as quoted by <u>Heath</u> [1913, p. 413], repeats the values that Ptolemy gives, but this does not confirm Ptolemy. Pappus may have copied him instead of using an independent source.

[‡]The correct value is about 60.3.

solstice within a circle whose diameter is 300 stades. This says, among other things, that the latitude of Syene is equal to the obliquity; we met this peculiar error first in Section III.4.

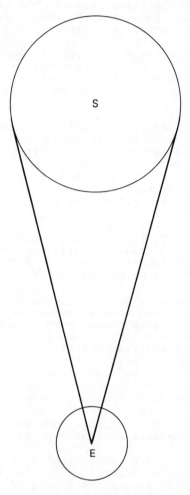

Figure VIII.3   The region on the earth within which there is no shadow. The angle subtended by this region is the same as the apparent diameter of the sun.

This also says something about the apparent size of the sun, as we can see with the aid of Figure VIII.3. In the figure, the circle marked S represents the sun and the circle marked E represents the earth. The angle drawn from the point E is the apparent diameter of the sun. Within the section that this angle cuts from the surface of the earth, some sunlight always falls vertically, and some light falls

on all sides of a vertical object. Thus there are no shadows
within the section, but the edge of the section is not sharply defined.

The ratio of the radius of the unshadowed circle to the
actual radius of the sun is the same as the ratio of the
earth's radius to the sun's distance. The latter ratio is
$\sin \pi_\odot$, by definition. Posidonius says that the radius of
the sun is 1 500 000 stades if the ratio of the sun's distance to the earth's radius is 10 000. This is equivalent
to assuming that $\sin \pi_\odot = 0.0001$.

I am afraid that I do not see anything remarkable in
this result. It gives the physical size of the sun if the
solar parallax is known, but so do the other methods described. So far as I can see, it is only an interesting but
inaccurate way of finding the apparent (angular) radius of
the sun. We do not know the value of the stade used, but if
we assume that it is the same one in which the circumference
of the earth is 252 000 stades, the apparent radius of the
sun comes out to be about 13'. The correct value is about
16'. We do not expect the result to be accurate because the
edge of the unshadowed circle is not sharply defined, as I
have already remarked.

Our word myriad is derived from a Greek word with two
meanings. One meaning was that of an indefinite but large
number. The other was that of the specific number 10 000,
which is the ratio that Posidonius used. It is also the
largest number that can be written conveniently in Greek
numerals. When Posidonius said that the distance to the
sun was a myriad of earth radii, I wonder if he was merely
implying that the distance was indefinite. However, his
usage does seem to imply a belief that the number was larger
than 1200, which is the value implied by Equation VIII.1.

Heath [1913, p. 348] suggests a different origin of
Posidonius's value of 10 000. In the work called The Sand
Reckoner, [Archimedes, ca. -225], Archimedes wants to generate a very large number based upon the distance to the
sun, which he is going to estimate from other data. In order to be safe, he deliberately chooses data that will, he
thinks, exaggerate the distance to the sun by a large factor.
Even with his deliberately exaggerated data, he proves that
the distance to the sun is less than 10 000 times the radius
of the earth.† Posidonius, in Heath's suggestion, ignored
the "less than" and took 10 000 for the distance.

In order to leave nothing uncertain about the matter of
the solar and lunar parallaxes, as Ptolemy says at the beginning of Chapter V.12 of the Syntaxis, he will use a method
different from those that have been described. He will measure the lunar parallax directly. Then he will use lunar
eclipses to find the parallax of the sun. I fear that Ptolemy is mistaken about the value of his method. As we shall

---

†Archimedes would probably be astonished to learn that he
  had not exaggerated after all. The correct value is about
  23 500.

see in what follows, it is distinctly inferior to the earlier
methods. They at least gave values of the solar distance
that are sensible even if they are too small. Ptolemy's
method, in contrast, can give results that are actually nonsense. We shall in fact see it do so in Section VIII.8,
where we use some of Ptolemy's own data.

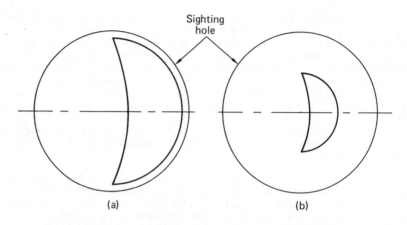

Figure VIII.4   Sighting at the moon when it is near a quarter
(about half full). If the sighting is to be accurate when the moon
is asymmetrical, the apparent sizes of the moon and the sighting
hole must be about equal, as shown in part (a). The situation
shown in part (b) would lead to inaccurate results.

3.   Ptolemy's Parallax Instrument

In Chapter V.12, after making the remark just mentioned,
Ptolemy describes the instrument that he claims to have used
in measuring the parallax of the moon. It was really a device for measuring the meridian elevation angle of the moon.
First he erected a vertical column, and he describes the pre-
cautions taken to ensure that it was vertical. Then he connected two arms to the vertical column in such a way that
they could swing in the plane of the meridian, and he describes the method of orienting the instrument so that the
arms can move in that plane. The column and both arms were
about 2 meters in length.

One arm carries two sights which are used in aiming it
at the moon; I shall call this the sighting arm. The sight
nearer the eye has a small hole; it may be called the peep
sight. The sight away from the eye has a larger hole. Although Ptolemy does not mention the matter explicitly, the
fact that he intends to use it at the quarter moon imposes a
restriction upon the design which is described in Figure
VIII.4. When the moon is about half full, we cannot aim

accurately at the moon unless the apparent size of the moon is almost equal to the apparent size of the sighting circle. This situation is shown in part (a) of the figure. With this arrangement, it is possible to align the center of the moon accurately with the center of the sight in spite of the lack of symmetry of the moon. The situation shown in part (b) would not be admissible in a method with pretensions to accuracy.

As the arm carrying the sights is moved in order to point it at the moon, a pin at the end of the sighting arm slides in a groove in the other arm. After the moon is aligned in the sights, the arms are apparently clamped in some way, yielding a triangle. The angle between the vertical and the line of sight is then found by measuring the sides of the triangle. Specifically, matters are so arranged that the vertical side and the side pointed toward the moon are each constant and equal to 60 in some set of units. The length of the third side is then the chord (Section II.3) of the zenith angle of the moon.

4. The Inclination of the Lunar Orbit

In order to find the parallax angle of the moon,† Ptolemy must compare a measured position with the position of the moon that would be seen from the center of the earth. The latter, of course, must be found by calculation from the lunar theory. The measured position should not be a longitude or a right ascension, because these vary rapidly with time and are difficult to measure accurately. The measured coordinate should be either a declination or a latitude, which vary slowly with time.

From the preceding work, Ptolemy knows all the quantities needed to calculate the geocentric position except the inclination of the lunar orbit. This means the angle between the plane of the lunar orbit and the plane of the ecliptic (the solar orbit). In principle, then, Ptolemy needs to make two observations of position, both of which involve the inclination and the parallax in their analysis. For convenience, he takes advantage of the latitude of Alexandria in order to separate the variables. By doing so, he does not enhance the accuracy of his results. He merely avoids the difficulty of solving two simultaneous equations.

In order to find the inclination, Ptolemy says [Chapter V.12] that he will measure the zenith angle of the moon, using the instrument just described, when two conditions are satisfied simultaneously: The moon must be at the summer solstice and it must be at its northernmost latitude. This is the same as saying that both the longitude of the moon and its argument of the latitude must equal 90°. In turn, this says that the ascending node of the lunar orbit must be at the vernal equinox.

---

†In this section, parallax without modifiers does not mean the horizontal parallax. It means the parallax, in a particular coordinate, that exists in a specific observational situation.

A third condition is that the moon must be in the meridian, but this is not a critical condition, since it is satisfied about once per day. However, the moon must be far enough from the sun that it can be seen clearly, which probably means that the observation should be made between sunset and sunrise. If so, the phase of the moon must be between the first and last quarters.

When the required conditions are met, the declination of the moon equals the obliquity of the ecliptic plus the inclination of the orbit. Since the obliquity is about 24° and the inclination, from approximate readings, is about 5°, the declination is about 29°. That is, the moon is about 29° north of the Equator. Since the latitude of Alexandria is about 31°, the moon is only about 2° from the zenith. Under these conditions, its parallax is negligible.

When he makes the observation under these conditions, Ptolemy says that he always (αει) finds that the angle from the zenith is close to 2⅛ degrees. He claims that he has measured the latitude of Alexandria to be (Section V.6 above) 30° 58'. The lunar inclination is this value minus the measured zenith distance and minus the obliquity of the ecliptic. For the latter, he has "verified" Eratosthenes's value (Section III.3) of 23° 51' 20". He takes the zenith distance of 2⅛ degrees to be the same as 2° 7', and he takes the obliquity to be 23° 51'. This makes the inclination equal 5° exactly.

The correct latitude of Alexandria (Section V.6) is about 31° 13', the correct inclination of the lunar orbit is about 5° 9', and the correct obliquity in Ptolemy's time was about 23° 41'. Thus the zenith distance that Ptolemy always measured should have been about 2° 23' rather than 2° 7'. Hence, in each measurement that Ptolemy claims to have made, his error was about 16', always in the same direction. With the method that Ptolemy describes, a plausible value for the standard deviation is 5'.

Ptolemy does more than find the same value every time. In Chapter V.7 of the Syntaxis, almost at the end of the chapter, Ptolemy says that he and Hipparchus by their measurements have both shown that the inclination is 5°. From the way he states the result, it seems to be his intention to show agreement with Hipparchus to a minute of arc. However, let us assume that he meant to show agreement only after rounding to nearest multiple of 5'. Then each of his measurements lies within a pre-assigned zone whose width is 1 standard deviation and whose center is at 3.2 standard deviations.

He does not tell us how many times is "always". I think it certainly implies at least three times, and we would be justified in assuming it to be even more. To be conservative, let us assume that he made the measurement only three times, and that each measurement gave a result lying in the zone just described. There is less than 1 chance in 10 000 000 that this result happened because of errors in the process of measurement. In other words,

-184-

Ptolemy never made the alleged measurements at all.†

Ptolemy slips badly when he implies that he made the measurement many times. He overlooked the restrictions on date that are required by the conditions. The ascending node of the lunar orbit, as we have said, moves slowly westward around the ecliptic and makes a complete circuit in about 18 2/3 years. It coincided with the vernal equinox on about 126 July 24 and it did not do so again until about 145 March 4 [Part II]. Both dates are outside those usually assigned as the limits to Ptolemy's astronomical activity. That is, the first date is earlier than any other observation that Ptolemy claims to have made, and the second date is later.

We must also ask when the longitude of the moon was 90°. The only times when the node was in about the right place and when the longitude was also 90° were 126 July 7, 126 August 3, 145 February 20, and 145 March 19 [Part II]. On these dates, the difference between the declination of the moon and its maximum value was much less than 1'. However, on 126 June 9, when the longitude of the moon was also 90°, the error caused by the fact that the node was in the wrong place was more than 1', and a month before that it was about 4', an unacceptable amount.

If we assume that Ptolemy was willing to accept an error of about 1' because of departure from ideal conditions, but not an error of 4', there could have been four observations in the summer of 126, or four in the winter and spring of 145, or there could have been a series of observations divided between the two years.

TABLE VIII.1

THE INCLINATION OF THE LUNAR ORBIT
ON VARIOUS DATES

Date	Inclination, degrees
126 Jun 9	5.03
126 Jul 7	5.02
126 Aug 3	5.13
126 Aug 30	5.25
145 Jan 23	5.08
145 Feb 20	5.22
145 Mar 19	5.29
145 Apr 15	5.23

†Various perturbations make the inclination change slightly from one time to another, so that he should not have always gotten the same value if his measurements had been accurate. Regardless of the details of the perturbations, the probability just calculated is correct in its general size. See Table VIII.1 above for values he should have gotten on various dates.

As I mentioned above, various perturbations cause the lunar inclination to change, so that Ptolemy should not have gotten the same result every time. Table VIII.1 shows the values that he should have gotten on the four possible dates in 126 and the four possible dates in 145. With any plausible set of observations, the value ranges over at least 0°.25, or 15'. Observing this change was well within the capability of the method that Ptolemy describes.† The claim that he always got the same value is probably better evidence of fabrication than the probability calculated above, strong as the latter is.

The dates when the observations could have been made bear on the question of Ptolemy's guilt or innocence of the fraud. If Ptolemy were innocent, he must have instructed his hypothetical assistant to make the measurements at the appropriate times, and the assistant then deceived Ptolemy by fabricating the data. However, for reasons that I shall explain in the next section, it is highly unlikely that Ptolemy wanted measurements of the lunar inclination in either year. If this is correct, he did not order the measurements at all. Thus when he said that the measurements always gave the same result, he knew that the measurements had never been made. In other words, his statement was a deliberate falsehood.

The dates are important for another reason. Let us assume, in spite of what has just been said, that the measurements were made in 145. We have seen that a fabricated measurement of the autumnal equinox (Table V.3) was made in 132. Thus the fraud continued for at least 13 years. Alternatively, let us assume that the measurements were made in 126. A vernal equinox and a summer solstice were fabricated in 140, so that the fraud continued for at least 14 years on this basis. Either way, the hypothetical assistant maintained his fraud over a period of at least 13 years.

By considering the circumstances under which the assistant and Ptolemy would have had to work, I estimated in Part II that the assistant, if he really existed, would have had to produce at least 100 observations, all of them fraudulent, over this period of 13 years or more. It is highly unlikely that the assistant could have maintained a continuous and successful fraud for so long and on such a scale.

5. The Lunar Parallax at the Quarter Moon

Figure VII.3 shows the configuration of Ptolemy's lunar model at a syzygy and at a quarter. If the radius of the deferent is 60 at a syzygy, it is 39;22 at a quarter (Section VII.3). The radius of the epicycle is 5;15 in the same units. The greatest possible distance of the moon comes

---

†Even if there were a systematic source of error that kept Ptolemy from getting the correct value, the error should have been nearly constant from one observation to another. That is, he should have seen a variation in the value even if he could not get the correct value.

when the moon is at a syzygy at the same time it is at
apogee; this distance is 65;15. The least possible distance comes when the moon is at a quarter at the same time
it is at perigee; this distance is 34;7. The ratio of the
greatest to the least distance is 1.913.

The apparent size of the moon varies inversely with its
distance, and geometers as good as the Greek astronomers
knew this. Hence, if Ptolemy's model of the lunar motion is
correct, the apparent size of the moon must vary over an extreme ratio of 1.913. Anyone who has watched the moon knows
that the variation of its apparent size is much less than
this. Even if we take the average distances at the syzygies
and at the quarters in Ptolemy's model, the ratio is 60/39;22
= 1.53. Elementary observation shows that the quarter moon
is not 1.53 times as large as the full moon on the average.
Elementary observation shows that there is little if any difference between the average diameters of the quarter and the
full moon. Hence it is clear that Ptolemy's model cannot
represent the distance to the moon correctly.

Almost every writer on the subject points out these considerations about the apparent size and the distance to the
moon, and remarks that Ptolemy surely knew of this defect in
his theory. I concur with this; if Ptolemy made as many observations of the moon as he claims to have made, he must
have known the narrow limits on its apparent size. All these
writers then go on to say that Ptolemy was silent about the
variation of the lunar distance, being content if his model
gave the longitude correctly. This is manifestly not so; it
supplies another vivid illustration of the immortality of
error.

Instead, Ptolemy claims to have measured the parallax†
of the moon at sunset on 135 October 1, when the moon was
simultaneously at the first quarter and on the meridian. He
chose this time because the declination of the moon was near
an extreme, so that it was changing slowly. Hence small
errors in timing and in calculating the longitude of the
moon would have little effect upon the accuracy of the result. Specifically, he measured the distance of the moon
from the zenith, using the instrument that was described in
Section VIII.3 above. He found that the variable side of
his instrument measured 51;35 of the marks that he used,
while both fixed sides measured 60. Hence the ratio of
51;35 to 60 is the chord of the zenith distance, which thus
comes out to be 50;55,03 degrees. Ptolemy gives the result
simply as 50;55 degrees.

I have calculated the circumstances of the observation
in Part I. The longitude of the moon was about 274° and
its latitude was 5°.30 at the epoch given for the observation. The declination of the moon comes out to be -18°.30
(the minus sign means that the moon was south of the Equator),

---

†In the discussion that immediately follows, parallax still
means the value that exists under the particular circumstances of an observation. It does not mean the horizontal parallax.

and Alexandria was at latitude 31° 13'. The correct zenith distance of the moon, after allowing for the effects of parallax and refraction, was about 50°.23, or about 50;14 degrees. Thus the error in Ptolemy's measurement was 41' of arc.

As I showed with the aid of Figure VIII.4, the apparent size of the moon and the sighting hole had to be nearly equal At the time of the measurement, the lunar distance was about 60.4 earth radii, slightly more than its average value, and the apparent diameter of the moon was 31' 1". That is, the error in the measurement was nearly 4/3 times the lunar diameter. No part of the moon could have been seen in Ptolemy's parallax instrument if he had really been pointing it in the direction that he claims.† For this reason, this observation strikes me as the most grossly fraudulent of all the measurements that Ptolemy claims to have made, even though it does not have the largest error.

Before we study Ptolemy's treatment of this observation, let us consider some more implications of the dates. Ptolemy says that he built his observing instrument for the purpose of measuring the parallax of the moon. In order to find the parallax, he must make two measurements of the zenith distance of the moon when it is in the meridian and, in order for him to have good accuracy, the zenith distances must be widely separated. He made the measurement of 135 October 1 when the moon was near the winter solstitial point. He wants the other measurement to be made when the moon is near the point of the summer solstice.

As we saw in the preceding section, he made the other measurement either in 126 or in 145. He asks us to believe that he made this other measurement at a time that is more than 9 years from 135 October 1. This means that he made one measurement of the pair that he needed in order to find an important result, one that is vital to his lunar theory, and that he waited 9 years to make the other one. If this is required by astronomical phenomena, he would have no choice but to wait. The question is: What wait is required by the circumstances of astronomy?

The answer is two weeks. The moon was near the winter solstice on 135 October 1. Either two weeks before or two weeks after, the moon was near the summer solstice and its zenith distance was greatly different. Thus, if he indeed did want a measurement of the moon on 135 October 1 and told his hypothetical assistant to make one, it is almost ridiculous to assume that he had wanted the parallel observation in either 126 or 145. There are only two possibilities that I can see. If he asked for the measurement in 135, he asked for the other one 2 weeks earlier or later, not more than 9 years later, in 145. On the other hand, if he had asked for the measurements in 126, it seems ridiculous to me to assume that he waited for more than 9 years, until 135, to ask for the other.

----

†Of course, some of it might have been visible around the edge of the plate that carried the front sight.

Let us try the following justification: Ptolemy measured the inclination in 126 because he needed it for some purposes, but at that time he had not realized the need for the second measurement to find the parallax. When he did realize the need, say in 130, he waited until 135 for the definitive measurement because the circumstances would be best in that year. This does not work either. I showed in Part II that there are about six favorable opportunities per year for making the second measurement. Of these, as I have shown, the opportunity on 135 October 1 is actually the worst rather than the best. The only explanation for choosing 135 October 1 is that it was the first possible date after Ptolemy realized the need. This is highly unlikely.

In other words, it makes no sense that I can see to assume that Ptolemy would have wanted observations made at the times that he claims. That is, Ptolemy did not specify the conditions of certain measurements which a deceitful assistant falsified. Only Ptolemy can be responsible for the fraud, in my opinion.

Let us now analyze the observation of 135 October 1 as Ptolemy did. Ptolemy chose the conditions of the observations so that the ascending node of the moon was at $180°$ while the longitude of the moon was about $270°$. Hence the moon was at its farthest position north of the ecliptic, and it was in fact just north of the winter solstice, which lies at longitude $270°$. If the desired conditions had been met exactly, the zenith distance of the moon would have equalled the latitude of Alexandria plus the obliquity minus the lunar inclination. That is, if we use Z_g for the zenith distance that would have been seen from the center of the earth, we would have

$$Z_g = 30;58 + 23;51,20 - 5;0 = 49;49,20 \text{ degrees}$$

if we use Ptolemy's values for the various quantities. However, the moon was not exactly at longitude $270°$ and the node was not exactly at $180°$. After Ptolemy makes the necessary small corrections, he finds $Z_g = 49;48$ degrees.

Since the measured zenith distance was 50;55 degrees, the parallax under the conditions of the measurement was 1;07 degrees. In the notation of Equation VI.1, the difference $\lambda_T - \lambda$ is the parallax, namely 1;07 degrees, and the difference $\lambda - \lambda_o$ is the geocentric angle 49;48 degrees. Equation VI.1 then tells us that R = 39;50 in sexagesimal notation while Ptolemy gets 39;45. This difference is not important. The correct distance, as I remarked a moment ago, was 60.4, or 60;24 in sexagesimal notation.

Now we must ask what the distance was in the scale in which the deferent has a radius of 60 at the syzygies. At the time of the observation, the elongation of the moon from the sun was about 78;13 degrees and the mean anomaly γ was about 262;20 degrees. Hence the distance was close to the average distance that the moon has at the quarter points. Ptolemy, in a detailed calculation that I have not checked but that is certainly close to correct, finds that the distance was 40;25 if the deferent radius is 60. Since the actual distance was 39;45 earth radii, according to the observation, the deferent radius at the syzygies is 59 earth radii.

When we use the conventional value of 60 for the deferent radius at the syzygies, we find 5;15 for the epicycle radius, as we saw in Section VI.6. When we use 59 earth radii, we must change the epicycle radius to 5;10 earth radii. Hence the greatest possible distance to the moon in Ptolemy's model is 64;10 earth radii. This is the distance that I used in discussing Ptolemy's model of the universe in Section IV.8.

The lunar distance at the syzygies had already been studied extensively by the methods described in Section VIII.2, but we do not know whether there was a value that had been accepted as being the most accurate. The data which Pappus attributes to Hipparchus lead to a wide range of values, but the data mentioned in Section VIII.2 which Ptolemy attributes to Hipparchus lead to average values between 59.1 and 60.7, and these are close to Ptolemy's result.

Thus the alleged observation of 135 October 1 was clearly fabricated to show that Ptolemy's lunar model gives a lunar distance at the quarter moons that is consistent with known results at the syzygies. As I mentioned above, because of its implications for the apparent size of the moon, Ptolemy almost surely knew that his model gives seriously wrong values for this distance, and he would have known this independently of any fraudulent practices of a hypothetical assistant. If he did know this, then he knew that the observation of 135 October 1 was fabricated; we cannot charge this fabrication to an assistant.

When the observation was fabricated, it had to start from an assumed value for the mean lunar distance at the syzygies. Because of knowledge that was already well established, the assumed value had to be close to 60 earth radii. The lunar model then gave the distance that had to result from the alleged parallax observation and hence the distance itself. The only thing we do not know about this process is why Ptolemy chose a syzygy distance of exactly 59 earth radii rather than some neighboring value. I shall discuss this question in detail in Section VIII.7.

Let us return to the question of the accuracy of the alleged observation. Under the conditions described, the standard deviation of the measurement should have been about 5′ of arc. However, the actual error is 41′ of arc, about 8 standard deviations. Further, the measurement agrees, within a minute or so of arc, with a value that we can assign in advance.† The probability that this could happen by chance is less than 1 part in 10^{14}.

---

†The parallax was slightly more than 60′ of arc. The distance to be found was uncertain by only about 1 part in 60. Hence the parallax had to agree with a pre-assigned value within about 1′.

Let me close this section by quoting from Part I:
".. if we believe that Ptolemy really observed the moon on
135 October 1, we must believe that he found the moon to be
in a position where he could not even have seen it through
his instrument, that he combined the erroneous lunar position with measurements of latitude,† of the obliquity, and
of the lunar inclination that are all in error by amounts
of around 10′, and that these errors by accident happened
to agree, to high accuracy, with his seriously wrong theory
of the moon. This agreement requires more than an error of
eight standard deviations. It requires the error to lie
within a pre-assigned interval whose width is about 1/5 of
a standard deviation and whose center is at about eight
standard deviations.

"These requirements still do not exhaust the things
that we must believe if we accept Ptolemy's measurement of
the lunar parallax. Ptolemy goes on to use his erroneous
lunar distance in a determination of the solar parallax,
which I shall describe below. His treatment of the solar
parallax requires that the measurement of 135 October 1
agree with his lunar model and with all his erroneous parameters within a tolerance of about 1/2 minute of arc or
better."

Ptolemy finds the solar parallax by using Figure VIII.2
and the associated mathematics summarized by Equation VIII.2.
In order to use the figure, he must have values of $\rho_\odot$ and ρ_U.
I shall discuss how he found these values before I take up
the solar parallax directly.

6. The Apparent Diameters of the Sun and Moon

At the beginning of Chapter V.14 of the Syntaxis, Ptolemy says that he will reject the usual ways of finding the
apparent sizes of the sun and moon, such as those based
upon water clocks or measuring the times required for rising
at the equinoxes, because they cannot give accurate results.
I do not understand the "or". The idea back of using rising
times is to find the time between the first appearance of
the disk above the horizon and its complete separation from
the horizon. The water clock is not an alternative, it is a
means for measuring this time.‡

Let "clearing time" denote the interval between the
appearance of the upper and lower limbs of the sun. The remark about the equinoxes may refer to the way that the clearing time varies with the seasons. From p. 26 of the Explanatory Supplement [1961] we find that the hour angle h, the
elevation angle a, the declination δ, and the latitude ϕ of
the observer are connected by

$$\sin a = \sin \delta \sin \phi + \cos \delta \cos h \cos \phi.$$

If we measure $\rho_\odot$ in minutes of arc, we easily find that the

---

†Of Alexandria.

‡There could, however, be timing methods used at a configuration other than sunrise or sunset.

clearing time is approximately $2\rho_\odot/15 \cos \delta \cos \phi$ minutes of time. Thus the clearing time comes closest to representing $2\rho_\odot$ at the equinoxes when $\delta = 0$. However, this is not an important point because we can calculate $\rho_\odot$ from the clearing time at any season, and Ptolemy may have had something quite different in mind.

Ptolemy then says that he built an instrument of the kind described by Hipparchus. He does not tell us much about it, but it apparently consists of an arm about 2 meters long on which a sight of some sort slides. The idea seems to be that the sight is moved back and forth until it has just the same apparent size as the sun or moon. In fact, the sighting arm of Ptolemy's parallax instrument by itself sounds very much like Hipparchus's instrument.

Ptolemy says that this instrument gives poor results for the apparent diameters of the sun and moon. He states a reason that I do not quote because I do not understand it. He says that the instrument, however, can be used to give an accurate comparison of the diameters.† In this way, he has found that the apparent diameter of the sun does not change appreciably; actually it changes from about 31' 31" to about 32' 35". He also says that the apparent diameter of the full moon equals the diameter of the sun when the moon is at its greatest distance from earth; we saw in the preceding section that this distance is 64 1/6 earth radii. He points out that this disagrees with the results obtained by earlier astronomers. They found that the diameters are equal when the moon is at its average distance.‡

The earlier astronomers were considerably more accurate than Ptolemy. At the average distances, the apparent diameter of the sun is about 16' 1" while that of the moon is somewhat smaller, about 15' 33". The apparent diameter of the moon at its greatest distance is about 14' 42", much smaller than that of the sun.

Ptolemy's result about the diameters has a consequence that is immediately contrary to observation. An annular eclipse is one that occurs when the apparent diameter of the moon is less than that of the sun. Thus, at a place where the centers are in line during the eclipse, the moon cannot cover the sun completely, and a small ring of the sun remains visible around the edge of the moon. However, if the smallest diameter of the moon is equal to the diameter of the sun, the eclipse is just total. Under any other conditions, the diameter of the moon would be bigger than that of the sun. Thus annular eclipses could not occur if Ptolemy's result were correct. Actually, there are more annular eclipses than

---

†I believe that this is impossible. So far as I can see, if the instrument can compare the diameters accurately, it can also measure them accurately.

‡Here he is dealing only with the full moon. Hence this means the average distance that can occur when the moon is at a syzygy. This is 59 earth radii in Ptolemy's model.

total ones.

In some earlier writing,† I deduced from this that
Greek astronomers might not have learned to discriminate
between annular and total eclipses. Though it is plausible,
this deduction seems to be wrong. <u>Dreyer</u> [1905, p. 142]
quotes Simplicius as saying that eclipses are sometimes annular and sometimes total. Since Simplicius is one of the
seven philosophers who breathed the dying breath of Greek
philosophy,‡ this might seem to reflect knowledge later than
Ptolemy. However, in the passage in question, Simplicius is
using the difference between annular and total eclipses as
evidence that the distance to the moon is variable. He is
using this evidence to show why early philosophers rejected
a lunar theory of Aristotle which required the lunar distance to remain constant. Thus, if Simplicius has understood and reported the situation correctly, a matter that
is far from certain, Greek astronomers and philosophers long
before Ptolemy knew that there is such a thing as an annular
eclipse.‡

Now if a person is going to err in measuring the com-
parative sizes of the sun and moon, he is going to err in
the direction of making the sun too big. The reason for
this is the great brilliance of the sun, which always tends
to increase the apparent size of an object. However, Ptolemy erred in the direction of making the sun too small, and
I see no ready explanation of this error. Hence I believe
that Ptolemy fabricated his result. I shall explain later
why Ptolemy may have been forced to fabricate a result that
was in contradiction to ordinary observation and to the
established results of earlier astronomers.

Because none of the other methods work, Ptolemy says,
he will turn to the use of lunar eclipses as the way for
determining the apparent sizes of the sun and moon. Further,
since the apparent size of the sun is the same as that of
the moon when the latter is at its greatest distance, he will
use eclipses that take place when the moon is at apogee. Thus
he will get the apparent sizes of the sun and moon simultaneously.

Before I take up Ptolemy's method, let me point out
what Ptolemy has accomplished by his remarks. We need to
recognize first that the use of lunar eclipses is a poor
way, not a good way, of finding the lunar diameter. It is
certainly poorer than the use of a sight at a variable distance. The reason is that the disk of the bright moon is
sharply defined against the dark sky and its size can be
compared easily and accurately with the size of a circle
<u>that is made to</u> surround it. In using lunar eclipses to

†<u>Newton</u> [1970, p. 114], for example.

‡Simplicius was one of the seven members of the academy at
Athens who were still there when Justinian ordered it
closed in the year 529.

‡<u>Heath</u> [1913, p. 313] says that Simplicius is actually quoting a writer of the 2nd century, who was thus nearly contemporary with Ptolemy.

find the apparent size of the moon, we are dependent upon
judging when the edge of the earth's shadow crosses the
moon. The shadow is not well defined, and the resulting
measurement is not accurate. Hence a method using eclipses
cannot be recommended on the grounds of its accuracy.

The method of eclipses does have one property that dis-
tinguishes it from the other methods that Ptolemy discusses
and rejects. It can be used only at the full moon, which is
the only phase at which the moon can be eclipsed. The other
methods can be used at any phase. Now we remember that Ptolemy's lunar theory predicts a grievously wrong size for the
moon at any phase other than a syzygy. What Ptolemy has
done is to direct our attention away from methods that would
reveal his error and to focus our attention upon the only
method that cannot reveal it. I find it hard to believe
that Ptolemy did this by accident, and I find this to be
strong evidence that Ptolemy is perpetrating a deliberate
fraud.

I believe that what Ptolemy says about the defects of
Hipparchus's instrument is part of his fraud. I can think
of no legitimate reason why this instrument should not give
reliable results, and in fact earlier results are considerably more accurate than those which Ptolemy gives, as we
shall see. However, Ptolemy was on the spot here. He had
to avoid using the instrument and to avoid quoting results
obtained with it, but he could not give a sound reason for
this avoidance because there is none. Hence he was reduced,
I believe, to writing something that he hoped would pass for
sense although it is not.

In order to find the apparent diameters of the sun and
moon, and to find the quantity ρ_U at the same time (see
Equation VIII.2), Ptolemy uses two lunar eclipses that occurred when the moon was at its greatest distance from the
earth. He does this in the middle of Chapter V.14 of the
Syntaxis.

The first eclipse was observed in Babylon on the date
that we call -620 April 22. The eclipse began at the end
of the 11th hour of the night, and, when the eclipse was
greatest, one quarter of the moon's diameter was eclipsed
from the southern side. Ptolemy calculates that the interval from the beginning to the middle of the eclipse was 1
hour of the night,† so that the middle of the eclipse came
at 6 hours of the night after midnight, which at the time
amounted to 5 5/6 ordinary hours. Since the time difference
between Babylon and Alexandria is 50 minutes in Ptolemy's
reckoning, the middle of the eclipse was 5 hours after midnight, Alexandria time. The time from Ptolemy's fundamental
epoch was 126 years plus 86 days plus 17 hours in apparent
time or 16 3/4 hours in mean time. At this time, Ptolemy
calculates that the center of the moon was 9 1/3 degrees
from a node and hence that its latitude was 48′ 30″.

---

† In Chapter VI.5 of the Syntaxis, however, he takes the
interval to be 0.5 ordinary hours for an eclipse in which
1/4 of the diameter was eclipsed.

The other eclipse was also observed in Babylon, on the
date we call -522 July 16. The northern half of the moon's
diameter was eclipsed at 1 hour before midnight. At Alexandria, then, the middle of the eclipse was 1 5/6 hours before midnight, or at 22;10 hours. This was 224 years plus
196 days plus 10 1/6 hours in apparent time, or 9 5/6 hours
in mean time, after the fundamental epoch. At this time,
Ptolemy calculates that the moon was 7 4/5 degrees from a
node and hence that its latitude was 40' 40".

Since half of the moon's diameter was eclipsed in the
second eclipse, the edge of the earth's shadow passed through
the center of the moon, which was 40' 40" from the ecliptic.
Since the center of the shadow necessarily lies in the ecliptic, the radius ρ_U of the shadow is therefore 40' 40". This
applies only to the radius at the greatest lunar distance.

In the first eclipse, when one quarter of the moon's
diameter was eclipsed, the center of the moon was 48' 30"
from the ecliptic. The difference between 48' 30" and 40'
40", namely 7' 50", is thus half of the moon's apparent
radius when it is at its greatest distance. Hence the moon's
radius at its greatest distance is 15' 40". By Ptolemy's
results, this is also $\rho_{\odot}$, the apparent radius of the sun.

The correct value for the lunar radius at greatest distance is about 14' 42", so Ptolemy's value is considerably
too large for the moon. However, it is considerably too
small for the sun, whose average apparent radius is about
16' 1".

TABLE VIII.2

TIMES AND MAGNITUDES OF FOUR LUNAR ECLIPSES

Date	Hour		Magnitude, digits		
	Stated by Ptolemy	Calculated from Ptolemy's tables	Stated by Ptolemy	Calculated from Ptolemy's tables	Calculated from modern theory
20 Apr 22	4;45	4;37	3	2.9	1.7
22 Jul 16	21;50	22;00	6	6.0	6.3
73 Apr 30[a]	2;00	1;50	7	6.9	7.5
40 Jan 27	22;10	22;14	3	2.8	3.2

[a]The eclipse was on -173 May 1 in Alexandria time.

We can study the authenticity of these eclipses with
the aid of Table VIII.2, which has the same format as Table
VI.5. The table also includes two eclipses that will be
discussed in Section VIII.8. The eclipse dated -173 April
30 in the table came after the midnight that began -173 May
1 in Alexandria, but it came slightly before midnight, and
hence on -173 April 30, in terms of Greenwich time, according to modern calculations. Hence it is customary to call
this the eclipse of -173 April 30 in modern writing.

The times of the four eclipses as stated by Ptolemy disagree with those calculated from his tables by amounts ranging from 4 to 10 minutes. Hence a test based upon the times does not reveal any fabrication. The magnitudes as stated by Ptolemy also do not agree particularly well with those calculated from modern theory. They do agree well with those calculated from Ptolemy's eclipse tables, but that is to be expected because he uses all four eclipses to determine the parameters that form the basis of his eclipse tables. The fact that the stated magnitudes do not agree exactly with those calculated from his tables shows that he made some minor adjustments in the parameters before he constructed his tables.

Let us assess the accuracy of this method, which Ptolemy takes as the only reliable way of finding the apparent size of the moon. We saw in Section VI.9 that the standard deviation in measuring the magnitude of a lunar eclipse was almost exactly 1 digit. We want to find the effect of errors of this amount on $\rho_{\mathbb{C}}$, ρ_U, and the sum of $\rho_{\mathbb{C}} + \rho_U$. Let $\Delta\rho_{\mathbb{C}}$ denote the absolute value of the change produced in $\rho_{\mathbb{C}}$, and so on.

First suppose that the magnitude for -620 April 22 had been reported as 2 digits rather than 3 digits, but keep the magnitude reported for -522 July 16 as 6 digits. The value of $\rho_{\mathbb{C}}$ calculated from the eclipses changes from 15' 40" to 11' 45", a change that is close to 4', while the value of ρ_U is not changed. Hence, to an accuracy that is sufficient in this approximate discussion,

$$\Delta\rho_{\mathbb{C}} = 4', \qquad \Delta\rho_U = 0, \qquad \Delta(\rho_{\mathbb{C}} + \rho_U) = 4', \qquad \text{(VIII.3a)}$$

Now keep the magnitude for -620 April 22 at 3 digits but suppose that the magnitude for -522 July 16 had been reported as 7 digits rather than 6 digits. The value calculated for $\rho_{\mathbb{C}}$ again becomes 11' 45". The value calculated for ρ_U becomes 42' 37$\frac{1}{2}$" and the value of the sum becomes 54' 22$\frac{1}{2}$". If we round the changes slightly, we have

$$\Delta\rho_{\mathbb{C}} = 4', \qquad \Delta\rho_U = 2', \qquad \Delta(\rho_{\mathbb{C}} + \rho_U) = 2'. \qquad \text{(VIII.3b)}$$

In the actual case, errors in the magnitudes of the two eclipses occur independently of each other. The actual standard deviation of each parameter is then the square root of the sum of the squares of the individual errors in Equations VIII.3a and VIII.3b. Hence the standard deviations $\Delta\rho_{\mathbb{C}}$, and so on, are

$$\Delta\rho_{\mathbb{C}} = \sqrt{32} = 5' \, 40", \qquad \Delta\rho_U = 2',$$

$$\Delta(\rho_{\mathbb{C}} + \rho_U) = \sqrt{20} = 4' \, 28". \qquad \text{(VIII.4)}$$

The error in the apparent radius of the moon is about a sixth of its value. A person could hold a measuring stick at arm's length, compare it with the moon, and do as well as this.

In other words, of all the methods of finding $\rho_{\langle}$ that could enter into a serious discussion, the method of using eclipses looks like the worst one rather than the best one. While the error analysis just given was beyond the capacity of the science of Ptolemy's time, we can still reasonably expect Ptolemy to realize the basic conclusion. It would not take a sophisticated analysis to show that the errors are something like those in Equations VIII.4. Further, Ptolemy claims that he has studied the accuracy of the various methods and has chosen the eclipse method in preference to all others. Either his study was incompetent or he has misled us about its conclusions.

While the test in Table VIII.2 does not reveal any fabrication, there has almost surely been some fabrication in connection with the eclipses. The reasons for this conclusion cannot be explained conveniently until we come to Section VIII.8. If the basic observations are genuine, there has been fabrication by miscalculation.

7. The Size and Distance of the Sun

Ptolemy now has all the quantities in Equation VIII.2 except $\Pi_{\odot}$, the (horizontal) parallax of the sun. The radius $\rho_{\odot}$ of the sun is constant at $15' \, 40''$. When the moon is at its greatest distance of 64 1/6 earth radii, its parallax $\Pi_{\langle}$ is $\sin^{-1}(1/64.167) = 53' \, 35''$. At the same distance, the radius ρ_U of the earth's shadow is $40' \, 40''$. Hence

$$\Pi_{\odot} = 2' \, 45''. \tag{VIII.5}$$

The corresponding solar distance is 1250 earth radii.

Ptolemy does not solve the problem in this simple way. Instead he draws a line from point C in Figure VIII.2 perpendicular to the line SEV. Then, assuming that the radius BE of the earth is unity, he proceeds to solve for the sides of all the important triangles in the figure. This process finally gives him the length of the line SE, which is the distance to the sun. In this way, he gets [Chapter V.15] that SE is 1210 earth radii. This corresponds to a parallax of $2' \, 50''$. Since the value is quite sensitive to rounding and other details of the calculations, the difference between this and the value in Equation VIII.5 is not surprising. In discussion, it does not matter appreciably which value we use, and I shall use Ptolemy's value of 1210 for the mean solar distance.

Let me digress for a moment in order to establish some results that were used in Section IV.8. We saw in Section V.1 that the eccentric distance for the sun is 1/24 of its deferent radius, which is the same as its mean distance. If the mean distance is 1210, the deferent radius is almost exactly 50. Hence the greatest and least solar distances are 1260 and 1160. Ptolemy does not use the greatest and least distances in the Syntaxis, but he does use them in another work in establishing his model of the universe, as we discussed in Section IV.8.

Ptolemy's method of finding the solar distance, once the lunar distance is known, is equivalent to saying that the sum of their parallaxes is a known constant. If he should increase his estimate of the lunar distance, this would decrease the lunar parallax and therefore increase the solar parallax. This in turn would mean a smaller solar distance. Ptolemy refers to this property of the analysis in discussing his model of the universe (Section IV.8), but without explaining it. Since the lunar parallax is much larger than the solar parallax, the solar parallax and hence the solar distance is highly sensitive to a small change in the lunar distance. Goldstein [1967, p. 10] demonstrates the same result by using Ptolemy's complicated method of solving for the solar distance.

Now let us return to the main question. Since the angle AES in Figure VIII.2 is known, the physical radius AS of the sun can be calculated in units of earth radii. Likewise, the radius of the moon can be calculated from its distance and apparent size. In this way, Ptolemy finds [Syntaxis, Chapter V.16] the following values relative to the earth:

radius of moon = 0;17,33,   volume of moon = $1/39\frac{1}{4}$,

radius of sun = $5\frac{1}{2}$,   volume of sun = 170.

$$(VIII.6)$$

In Equations VIII.4, we saw that the standard deviation of a measurement of $\rho_{\mathbb{C}} + \rho_U$ is $4' \, 28''$. If the value of $\Pi_{\mathbb{C}}$ is exact, this is also the standard deviation in $\Pi_{\odot}$.[†] Since $\Pi_{\odot}$ comes out to be only $2' \, 45''$, this means that negative values of the parallax are within less than one standard deviation if we use Ptolemy's method. A negative parallax is, of course, physical nonsense.

On the other hand, if we use Aristarchus's method, we take the sum $\Pi_{\mathbb{C}} + \Pi_{\odot}$ to be, say, $20.1 \, \Pi_{\odot}$; see Equation VIII.1 Now if the standard deviation of $\rho_{\odot} + \rho_U$ is $4' \, 28''$, the standard deviation of $\Pi_{\odot}$ is only $13''$. Because of errors in the coefficient 20.1, the actual standard deviation is of course much larger than this. However, Aristarchus's method can never yield a negative value and thus it is superior to Ptolemy's method.

In fact, as we shall soon see, Ptolemy's method and data do in fact yield a negative solar parallax, or at least so small a distance that the sun comes out to be smaller than the earth. I say this in spite of Equations VIII.6; when Ptolemy finds these values, he uses only selected data. Other data that he uses in other contexts yield violently divergent results, which he ignores.

The lunar distance that Ptolemy uses is 64 1/6. If we multiply this by Aristarchus's ratio from Equation VIII.1, we get a solar distance of 1225, almost exactly what Ptolemy gets. This is indeed remarkable. All of the quantities

---

[†]Since Ptolemy takes $\rho_{\odot} = \rho_{\mathbb{C}}$, the error in $\rho_{\odot} + \rho_U$ (Equation VIII.2) is the same as the error in $\rho_{\mathbb{C}} + \rho_U$.

involved in finding this value, beginning with the latitude
of Alexandria, are in error by amounts ranging up to about
40'. Yet the final result for the parallax of the sun agrees
with a pre-assigned value, namely the value demanded by
Aristarchus's ratio, within a few seconds of arc. However,
there are some obstacles in the way of deciding that Ptolemy
fabricated his observations in order to yield this result.

The first obstacle is that it involves multiplying
Aristarchus's ratio by the greatest lunar distance. As we
saw in Section VIII.2, Aristarchus does not say which lunar
distance is involved and, indeed, the variation of the lunar
distance may not have been appreciated in his time. If we
are going to use his ratio, it would be natural to apply it
to the mean distance rather than to the greatest distance.
Still, there is nothing that would keep Ptolemy from applying
it to the greatest distance if he chooses.

The second obstacle is that there is a striking coincidence involved. In Section IV.8, we studied Ptolemy's scheme
for finding the size of the universe. In his scheme, he sets
the greatest distance of one body equal to the least distance
of the next body, and so on. Thus the least distance of Mercury equals the greatest distance of the moon, and so on.
According to Table IV.1, this means that the greatest distance
of Venus, which equals the smallest distance of the sun, must
be 18.541 298 times the greatest distance of the moon. Since
the solar eccentric is 1/24 of the mean solar distance, the
ratio of the mean to the least solar distance is 24/23, so
the ratio of the mean solar distance to the greatest lunar
distance is 19.347 442.

This is fairly close to Aristarchus's ratio, but there
is one important distinction. This ratio must be that of
the mean solar distance to the greatest lunar distance. However, Aristarchus's ratio would most naturally apply to the
mean lunar distance, and we must invoke an arbitrary decision
in order to explain why Ptolemy applied it to the greatest
lunar distance.

Because we do not know for sure what quantities in the
Syntaxis are genuine, if any, we cannot be sure exactly how
Ptolemy went about his fabrication. Further, we often cannot
duplicate his calculations with assurance even when we know
his procedure, because of uncertainties about how he rounded
numbers, how he interpolated in tables, and other computational matters. Thus trying to reconstruct the fabrication
cannot tell us whether Ptolemy used the ratio of Aristarchus
(19.1) or the ratio 19.347 442 which we just derived. The
reason is that we get almost identical results either way.

Let us suppose that Ptolemy started from the values of
$\rho_{\mathbb{C}}$ and ρ_U given by the eclipses of -620 April 22 and -522
July 16 in Table VIII.2. In accordance with what he says
about the lunar and solar diameters in Chapter V.14 of the
Syntaxis (see Section VIII.6 above), he equates $\rho_{\odot}$ and $\rho_{\mathbb{C}}$.
This gives him 56' 20" for the right member of Equation
VIII.2. If he now chooses 19.1 for the ratio of distances,
this gives him 2' 48" for the solar parallax and 53' 32" for

-199-

the lunar parallax. If he uses 19.347 442 for the ratio, he gets 2' 46" for the solar parallax and 53' 34" for the lunar parallax. The differences are trivial.

We can reconstruct the rest of the fabrication with high assurance. For illustration, let us use 53' 34" for the lunar parallax when the moon is at its greatest distance. More specifically, this is the horizontal parallax at this distance. The distance must then be 64.180 earth radii; this is almost the value that Ptolemy gets. When Ptolemy "made" his parallax observation on 135 October 1, however, the moon was not at this distance according to Ptolemy's model, and we must next find the distance that his model demands on 135 October 1.

When Ptolemy used the conventional value 60 for the greatest radius of the lunar deferent curve, he found $65\frac{1}{4}$ for the greatest lunar distance and 40;25 on 135 October 1. Hence, if the greatest lunar distance is 64.180 earth radii, the distance had to be 39.754 earth radii on 135 October 1. The corresponding horizontal parallax is $1°.4414$, but this is not the parallax that he needs. Instead, he needs the specific parallax at this distance when the geocentric angle $\lambda - \lambda_0$ in Figure VI.2 is 49° 48'. By Equation VI.1, this is $1° 7'$, and this is exactly the parallax that Ptolemy claims to have measured on 135 October 1. To this accuracy, we get exactly the same result if we take the horizontal parallax at the greatest distance to be 53' 32".

We might try other possibilities. Instead of starting with the eclipses of -620 April 22 and -522 July 16, Ptolemy might have decided to use the round number 59 for the average lunar distance at a syzygy. If he did this, he would still get 1° 7' for the parallax on 135 October 1. However, he would now have to fabricate the eclipse data of -620 April 22 and -522 July 16.

As a matter of fact, we shall conclude in the next section that Ptolemy did fabricate the eclipse data of -620 April 22 and -522 July 16. At first sight, this seems to confirm the idea of the preceding paragraph, but a closer look shows that it does not. We shall see that Ptolemy fabricated the data to be consistent with values of the apparent radius $\rho_{\mathbb{C}}$ of the moon that were already accepted as correct. That is, instead of starting with the eclipses, Ptolemy started with values of $\rho_{\mathbb{C}}$, and he first fabricated the eclipse data to correspond. Then he proceeded in the manner outlined earlier.

There are undoubtedly still other possibilities. However, on the basis of what we now know, I think the most likely possibility is that Ptolemy started with certain measurements that gave him 56' 20" for the right member of Equation VIII.2 (the sum of the lunar and solar parallaxes, with the moon at its greatest distance), and that he combined this with a selected ratio of the solar to the lunar distance. If this is correct, the remaining question is whether he started with the ratio of Aristarchus or the ratio demanded by Ptolemy's model of the universe. Since we get the same fabricated results either way, the question is unimportant in some sense. However, the question has an interesting facet, and I shall pursue it for a moment.

We remember from Section IV.8 that Ptolemy's scheme for the size of the universe left a gap between the orbits of Venus and the sun. This gap is a result of Ptolemy's errors in arithmetic, but Ptolemy does not seem to understand this point. Actually, if we use the ratio 19.347 442 between the mean solar distance and the greatest lunar distance, the orbit of Venus comes out slightly too large, but we can easily correct this by minor changes in the numbers, changes that are well within Ptolemy's practices with regard to rounding or truncating numbers.

If Ptolemy fabricated his data in order to make his universal model come out right, it is hard to see why he made it come out wrong when he actually came to the description of his model. However, since his model would come out right if he did the arithmetic carefully, whether the data are fabricated for this purpose or not, we have the same problem in either case. That is, the fact that his description comes out wrong does not bear on the question of how he fabricated the data.

If we assume that Ptolemy fabricated the data to make his model come out right, we can explain one problem that I have not been able to answer otherwise. This is the problem of why Ptolemy chose to say that the solar radius $\rho_\odot$ equals the least value of the lunar radius $\rho_{\mathbb{C}}$ when well-established results said that $\rho_\odot$ equals the average value which $\rho_{\mathbb{C}}$ has at the syzygies. Usually Ptolemy tries to keep his results consistent with accepted values.

As we shall see in the next section, $\rho_{\mathbb{C}}$ varied from 15' 40" to 17' 40" at the syzygies, according to accepted results, and Ptolemy felt constrained to keep these limits because he needed them for eclipse theory. If he had chosen to make $\rho_\odot$ equal the average $\rho_{\mathbb{C}}$, he would have had 57' 20" rather than 56' 20" for the right member of Equation VIII.2. This change is enough to change the lunar distance by slightly more than 1 and the solar distance by more than 20. He must have thought at some point in his work that the distances would be consistent with his universal model if he used 56' 20" and that they would be inconsistent if he used 57' 20". If he concluded this, he would naturally think that it was more important to make his model of the universe work than to keep a specific value of $\rho_\odot$. Then, for some reason that we cannot explain on any basis, he later made a mistake in developing the details of his model, without understanding the origin of his error.

In other words, the universal model demanded certain lunar and solar distances, and these distances in turn demanded a certain value of $\rho_\odot$. The exigency is still there even though Ptolemy made mistakes in arithmetic and ended up with a model that is internally inconsistent. Only by admitting this exigency, it seems, can we explain Ptolemy's need to override the accepted value of $\rho_\odot$.

-201-

8. A Negative Solar Parallax, and Related Problems

Ptolemy now has all the information that he needs to calculate the circumstances of eclipses at any lunar distance. Since he knows the apparent radius $\rho_{\mathbb{C}}$ of the moon at its greatest distance, he can calculate its apparent radius at any other distance. Since he knows the size of the sun and its distance, he can calculate the angle AVS in Figure VIII.2; this angle tells how rapidly the earth's shadow converges as we move away from the earth. Hence he can calculate the radius ρ_U of the earth's shadow at any distance.

However, when he wants to construct tables for eclipses that occur when the moon is at its least (syzygy) distance, he does not use the information that results from his model. Instead, he uses another pair of eclipses. He does not point out that he could have used his model instead of using another pair of eclipses. He merely says near the beginning of Chapter VI.5 of the Syntaxis that there is more surety if one relies upon the observed phenomena.

First he uses an observation of the eclipse of -173 April 30 made at Alexandria. The moon was eclipsed by 7 digits on its northern side, and the middle of the eclipse came at $2\frac{1}{2}$ hours of the night after midnight. By Ptolemy's calculation, this was at 2 1/3 ordinary hours after midnight. The interval since the fundamental epoch was 573 years plus 206 days plus 14 1/3 hours in apparent time, but only 14 hours in mean time. For this time, Ptolemy calculates that the moon was 8;20 degrees beyond its descending node, and that its center was 43' 3" south of the ecliptic.

Next he uses an observation of the eclipse of -140 January 27 made on Rhodes. Since this is the time and place of Hipparchus's greatest activity, we would expect that the observation was made by him, but Ptolemy does not mention Hipparchus in this connection. The eclipse began at the beginning of the 5th hour of the night, and, at maximum eclipse the moon was eclipsed by 3 digits on its southern side. Ptolemy assumes that Rhodes and Alexandria are on the same meridian. The beginning of the eclipse was at 2 hours of the night before midnight, which Ptolemy takes to be 2;20 ordinary hours, or 9;40 hours after noon. Half of the duration was 30 minutes, so the middle was at 10;10 hours after noon, he says. This is a serious underestimate. According to Oppolzer [1887], the interval from beginning to middle was 58 minutes, and the interval was about an hour according to Ptolemy's own tables. I shall return to this point later.

Continuing with Ptolemy's treatment of the eclipse, the time from the fundamental epoch to the middle of the eclipse was 606 years plus 121 days plus 10 1/6 hours, in both apparent and mean time. For this time, Ptolemy calculates that the moon was 10;36 degrees beyond its ascending node, and that its center was 54 5/6 minutes of arc north of the ecliptic.

In the second eclipse, the center of the moon was 11' 47" farther from the ecliptic than it was in the first eclipse. This represents one third (4 digits) of the moon's diameter. Hence the apparent diameter of the moon when it is closest to the earth is 35' 21", which Ptolemy rounds to 35' 20". He also calculates that the radius ρ_U of the shadow at the moon's least distance is 46'.

Ptolemy could have calculated the solar parallax from the eclipses of -173 April 30 and -140 January 27 just as readily as from the eclipses of -620 April 22 and -522 July 16. However, he does not mention the solar parallax in connection with the eclipses of -173 and -140, and in fact he puts these eclipses in another part of the Syntaxis altogether. In Part II, I have supplied the missing calculation. The apparent radius $\rho_\odot$ of the sun is still 15' 40", according to Ptolemy. From Ptolemy's lunar model, we readily calculate that $\Pi_{\mathbb{C}}$ is $\sin^{-1}$ (1/53 5/6) = 63' 52". The eclipses yield the result that ρ_U is 46'. Hence, by Equation VIII.2,

$$\Pi_\odot = 46' + 15' \; 40" - 63' \; 52" = -2' \; 12". \qquad \text{(VIII.7)}$$

A negative parallax is physical nonsense.

Finding a negative parallax by Ptolemy's method is not merely a possibility revealed by the error analysis of the preceding section; it has actually occurred. We must ask: Is it only by accident that Ptolemy chooses one set of eclipses to find the solar parallax? Is it only by accident that Ptolemy separates the other set of eclipses by a wide interval from his treatment of the solar parallax?†

Although Ptolemy's method of finding the lunar radius is quite inaccurate, it yields surprisingly accurate results in his hands. He finds that the largest and smallest radii are 17' 40" and 15' 40", while the correct values are about 16' 45" and 14' 42". His values are each too large by about 1', although the standard deviation of his method is (Equation VIII.4) 5' 40". The probability that two successive values would lie within 1' of the correct value, in the face of this standard deviation, is about 0.025, about 1 chance in 40.

We can see this result in another way. The ratio of Ptolemy's radii is 1.125 while the correct ratio is 1.140. If the standard deviation of a single value is 5' 40", the standard deviation of the ratio is 0.48. The probability that Ptolemy's ratio would agree this closely by chance is again about 0.025, or 1 in 40.

With high but not overwhelming confidence we conclude that all four eclipses used in finding the lunar radius were fabricated. Testing the times of the eclipses in Table VIII.2 did not reveal this, because Ptolemy was not really concerned

---

†If Ptolemy had recognized that his method is sensitive to observational error, and if he had accordingly used the average of his results, he would have found 19" for the solar parallax. This is amazingly accurate.

with the times. He was concerned with the radius of the moon, and perhaps the radius of the earth's shadow, at the least and greatest distances, and he fabricated the data to yield specific values of the radii. That is, he calculated the times when the latitude of the moon would have certain values, rather than calculating when the sun and moon would be in opposition.

For the eclipse of -140 January 27, we saw that Ptolemy took the time interval from the beginning of the eclipse to the middle to be 30 minutes although his tables require a value of about an hour. The quantity that he actually tabulates is not the time interval but the angle in minutes through which the moon travels between the beginning and the middle. Since the moon travels 1' in about 2 minutes, he must double the tabular quantity in order to find the time interval. In fabricating the data for this eclipse, he apparently forgot to do the doubling.

There is confirmation from an unlikely source that the eclipse of -522 July 16 was fabricated. A Babylonian record of this eclipse has actually been found [Kugler, 1907, pp. 70-71]. According to the Babylonian record, the northern half of the moon was eclipsed just as Ptolemy says, but the time given disagrees by about 40 minutes with what Ptolemy says. Various people including me have tried to explain or to remove this discrepancy by postulating various meanings of the way the times are stated in the Babylonian record and in the Syntaxis, and I have summarized [Newton, 1970, pp. 136-139] much of the discussion. These discussions were all based upon the assumption that both records are genuine. If Ptolemy's form of the record has been fabricated, there is no longer a problem. The fact that the Babylonian record gives the same magnitude of the eclipse as Ptolemy while giving a different time agrees exactly with what we have just concluded, namely that Ptolemy fabricated the time to give him a specific latitude of the moon when half of the moon was eclipsed.

9.   The Need for Elliptic Orbits

We see that Ptolemy had available some rather accurate values of the apparent radius of the moon. It is not likely that these values were obtained by means of lunar eclipses. In order to obtain values as accurate as those which Ptolemy used, one would need to analyze 30 to 40 pairs of eclipses occurring at the greatest lunar distance and another 30 to 40 pairs occurring at the least distance. Altogether it would take, say, 150 eclipses. It is doubtful that Ptolemy, Hipparchus, or any other Greek astronomer undertook such a labor. In fact, since the eclipses had to occur within a rather narrow range of the moon's anomaly, it is not sure that enough eclipse records even existed.

Thus it seems that there were values of the lunar radius, measured without the use of eclipses, that were already accepted by astronomers and that Ptolemy had to duplicate by his fabricated data. I believe that the most accurate way of measuring the lunar radius is to compare the

moon with a circle that can be moved until the apparent sizes
match. Further, the apparent size of the moon as measured by
the naked eye should be slightly too large because of its
brightness, and the values that Ptolemy uses are indeed
slightly too large.

We conclude that Ptolemy felt impelled to fabricate
eclipse data in order to agree with other data obtained by
the very method he decries. Apparently he had to use the
other data because they were accepted but he had to down-
play the method because it would reveal the gross error in
his lunar theory. In turn, he had to conceal his error further by fabricating his parallax data.

Ptolemy's values for the radius of the earth's shadow
are also rather accurate, but not as accurate as the lunar
radii. He finds that the shadow radius is 40' 40" when the
moon is at its greatest distance and 46' when the moon is at
its least distance. The correct values are about 38' 02"
and 45' 36", respectively.†

Greek astronomers could probably not find the shadow
radius except by using eclipses. However, Ptolemy's method
of using eclipses is clearly not the method to use. What
one should do is to take the radius of the moon from the
sighting or comparison method and then use it in calculating
the shadow radius. Thus, for example, in the eclipse of
-173 April 30, the magnitude was 7 digits and the moon was
43' 3" from the ecliptic. The radius of the shadow is then
43' 3" plus 1/12 of the radius of the moon. In this way,
every eclipse can be used to find the shadow radius, and the
results are much more accurate.

In representing the position of the moon at a syzygy,
Ptolemy uses only the simple epicycle model of Figure IV.1.
The values of the moon's radius and of the shadow's radius
both show that this simple model is seriously in error. When
the epicycle is chosen to give the correct equation of the
center, we see from Section IV.2 that the epicycle gives
twice the correct variation of distance. The values of the
radii show just this effect.

First let us look at the radius of the moon. The ratio
of the greatest to the least radius, as we saw in the preceding section, is 1.125 if we use Ptolemy's values. In
contrast, the ratio of the greatest to the least distance is
the ratio of 64 1/6 to 53 5/6, which is 1.191. That is, the
variation of distance needed by the model is 0.191‡ while

†The radius of the shadow depends slightly upon the sun's
 distance as well as upon the moon's. In calculating these
 values, I have used the average distance of the sun.
‡The variation is this small because Ptolemy did not use the
 epicycle radius that best fits the longitude of the moon.
 He based his epicycle only upon the longitude of the moon
 at syzygy. If he had used the epicycle that best fits the
 entire lunar orbit, the epicycle radius would have been
 about $6\frac{1}{2}$ rather than 5 1/6. The variation of the distance
 would then be close to 0.25.

the variation demanded by the radii is only 0.125.

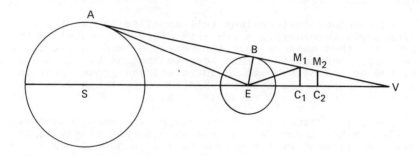

Figure VIII.5   The radii of the earth's shadow at the moon's
greatest and least distances. Line M_1C_1 is the radius at the least
distance and line M_2C_2 is the radius at the greatest distance.

The radius of the shadow casts a similar problem, as
we can see with the aid of Figure VIII.5. S and E denote
the centers of the sun and earth, and the line SEV is the
axis of the earth's shadow. Line ABM_1M_2V is the edge of
the shadow. Lines EM_1 and EM_2 are the least and greatest
distances to the moon respectively (EM_2 is omitted in order
not to confuse the drawing), and M_1C_1 and M_2C_2 are the physical radii of the shadow at these distances. Angles M_1EC_1
and M_2EC_2 are the apparent or angular radii. Since the distance EM_1 is 53 5/6 earth radii and angle M_1EC_1 is 46', the
distance M_1C_1 is 0.720 earth radii, according to Ptolemy's
model. Similarly, since EM_2 is 64 1/6 earth radii and angle
M_2EC_2 is 40' 40", the distance M_2C_2 is 0.759 earth radii.
This is greater than M_1C_1.

That is, according to Ptolemy's model and his data,
the edge of the shadow diverges from the axis instead of
meeting it at the point V. This requires the sun to be
smaller than the earth, in contradiction to other results.

The problems with the lunar radii and with the shadow
radii can both be avoided if the variation of distance is
half that required by the epicycle model.† There are two
models that do give just this variation of distance. One is
the equant model of Section IV.5 and Figure IV.4.‡ The
other is Kepler's model of an elliptical orbit with the
central body (the earth, in this case) at one focus. Furthe
the planets also require a variation in distance just half
that given by the epicycle or eccentric models, as we shall

---

†Or by the eccentric model. The epicycle and eccentric give
identical observed results, as we saw in Section IV.3.

‡I shall prove this in Section XI.1.

-206-

see in later chapters.

In other words, the studies of the moon and the planets made in Greek astronomy show that the epicycle and eccentric models give twice the correct variation of distance. Only the sun seemed to be compatible with these models, because its variation of distance was too small for the Greeks to measure. The equant model was applied to the planets but not to the moon. If Ptolemy had not used fraudulent data to "prove" that his lunar model gave the correct variation of distance, astronomers might have realized much earlier that the moon and the planets share this common problem. Then, perhaps, the world would not have had to wait for the elliptical model until the time of Kepler.

Whether the world would have had to wait this long or not, I think we are entitled to say that Ptolemy's model of the lunar motion did not advance the cause of astronomy.

In order to test whether the equant can be applied successfully to the moon, I have programmed the model of Figure IV.4 for the moon, with the epicycle being used to represent the evection term. The standard deviation of the model for the longitude of the moon is $0°.544$, slightly less than with either Ptolemy's model or the double epicycle model of Figure IV.6. It is better in its representation of the distance, giving a ratio of greatest to least distance of 1.165 where Ptolemy's model gives 1.90; the correct ratio is about 1.140.† If we confine our attention to the syzygies, the equant gives a ratio of 1.068, somewhat less than the correct ratio; Ptolemy's model gives 1.191. The trouble is that the change in the epicycle, which is now used only to represent the evection and not the main eccentricity, is still causing twice the change in distance that it should. These results are based upon the first set of parameters that I tried. Varying the parameters might improve the results somewhat.

Greek astronomers knew that the sun was much larger than the earth and much farther away than the moon, but they greatly underestimated both the sun's size and distance. There are four ways by which they could study the solar distance, once they knew the lunar distance. One way is that used by Aristarchus, which was described in Section VIII.2. A second is the method used by Ptolemy, which is certainly inferior. A third is implicit in Figure VIII.5, which is to find the distances M_1C_1 and M_2C_2. From these distances, and the distances EM_1 and EM_2, we can calculate the distance EV and the angle AEV. Since the angle AES is the apparent solar radius, which has been measured, the angle AEV is known. Hence the triangle AEV can be solved and the solar distance found. This is probably the poorest way, although it gives a useful check on the consistency of other results.

A fourth way is to use both solar and lunar eclipses. Chapter 9 of the Explanatory Supplement [1961] shows that

---

†The double epicycle model of Figure VII.7 gives a ratio of 1.303. This is about twice the correct variation, but it is much better than the variation given by Ptolemy's model.

solar eclipses depend upon the difference $\Pi_{\mathbb{C}} - \Pi_{\odot}$ between the lunar and solar parallaxes. Lunar eclipses, as Equation VIII.2 shows, depend upon the sum of the two parallaxes. Hence, by studying both kinds of eclipse, it is possible in principle to find both parallaxes. On the basis of the comments in Chapter V.11 of the Syntaxis, it is likely that Hipparchus tried to use this basic method.

With the methods available to the Greeks, none of these methods is likely to give accurate results. The basic problem is that the solar parallax is only about 8″.8. This is much less than the errors present in any quantity from which it could be derived.

10. A Summary of Greek Knowledge of the Sun and Moon

We have concluded the main study of the sun and moon in Greek astronomy, although there are many interesting details that we have omitted. The reader who wants to know more about the subject will find a valuable collection of Greek ideas concerning the moon in Plutarch [ca. 90]. We can now usefully summarize the situation.

The Greeks knew that the sun is quite large and far away in terrestrial terms, but they underestimated its diameter and distance by a factor of about 20. They could have estimated the variation of its distance by measuring the change in its apparent radius. However, the change is small and hard to measure, because of the sun's brightness, and it seems that they did not know the variation in its distance by observation; they could only calculate it from theory. They could calculate the longitude of the sun with a standard deviation of about 15′.

Greek knowledge of the sun was complete by the time of Hipparchus (about -130), so far as we know. Certainly the Syntaxis tells us nothing about the sun that was not known to Hipparchus.

The Greek astronomers had rather accurate knowledge of the size of the moon and of its average distance during syzygies. They also had rather accurate knowledge of the variation in the apparent size of the moon; by implication, this gives the variation in the distance. However, they did not use any model of the lunar motion that gives the correct variation of distance. Ptolemy's model gives about twice the correct variation when the moon is at syzygy, and it gives about eight times the correct variation overall. Ptolemy gives no indication that these problems are present. We do not know whether other Greek astronomers knew of these difficulties or not.

In fact, the Greek knowledge of the apparent size of the moon was compatible with an elliptical orbit but not with an epicycle/eccentric model of motion. It was compatible with the equant model, but this model was never applied to the moon, although it was applied to the planets.

With regard to the geocentric longitude of the moon, Ptolemy's model has a standard deviation of about 35′. This is about half of the error that was present in pre-Ptolemaic models of the lunar motion that we know of. Much of this improvement is negated by the error in the lunar parallax.

In the development of his models of the solar and lunar motions, Ptolemy uses many alleged observations. He claims to have made many of them himself; all of these are demonstrably fraudulent. Distressingly, many of the observations which he attributes to other and earlier astronomers are also fraudulent. The best we can say at this point is that some of the observations attributed to others may be genuine. I shall survey this situation in Chapter XIII, Table XIII.2 particularly, after I have completed the review of the Syntaxis.

Note added in proof: A footnote on page 170 refers to a statement that Hipparchus calculated all eclipses over a period of 600 years. I have since come across the source, which is Pliny's Natural History, II.53. Pliny is not always reliable in such matters; he says in II.57 that a lunar eclipse sometimes starts at the eastern edge of the moon and sometimes at the western edge.

CHAPTER IX

THE STARS

1. Introduction

Two sources bring us most of the Greek knowledge of the stars that has survived. One is the Commentary on Aratus and Eudoxus, as it is frequently called, by Hipparchus [ca. -135]. The other is the Syntaxis, of which Books VII and VIII are devoted to the stars.

Eudoxus was a Greek astronomer who flourished around -375. Dreyer [1905, p. 87] describes his theory of homocentric spheres as "the first attempt to account for the more conspicuous irregularities" of the motions of the planets.† He also wrote a book about the appearance of the heavens. About a century later, the poet Aratus either rendered Eudoxus' book into a poem or wrote a poem based upon it. Still later, Hipparchus wrote his commentary on this book and poem; this commentary is the only surviving work of Hipparchus.

In it, he gives a large amount of qualitative information about stellar configurations and almost 900 items of quantitative stellar data. Some of the items are coordinates and some are data from which we may calculate coordinates. He also gives [Hipparchus, ca. -135, pp. 271-281] the positions of "clock stars" which cross the meridian at intervals of about an hour. They were intended for use in time keeping, and, by using them, an astronomer should have been able to tell the time at night with an accuracy of a few minutes. Finally, he gives the declinations of about 40 stars that I have used [Newton, 1974] in finding the obliquity of the ecliptic in the time of Hipparchus.

Hipparchus is also known to have prepared a star catalogue that has been lost.

Ptolemy starts Book VII of the Syntaxis by quoting some star configurations described by Hipparchus and stating that the configurations are still the same. In his first example, Hipparchus had written that the star which we call ℬ Cancri lies 1½ digits to the northeast of the center of the line joining α Cancri and α Canis Minoris (Procyon), and Ptolemy finds no change in this configuration. I do not know what was meant by a digit in this context.

---

†Specifically, this theory was designed to account for the retrograde segments of the planetary motions. Each planet was carried on the equator of a rotating sphere whose axis was carried by another concentric sphere, and so on. Although Aristotle advocated it, the theory had been abandoned by most astronomers by, say, -250. Its greatest defect was that it made a planet stay always at the same distance from the earth, so that it always had the same brightness.

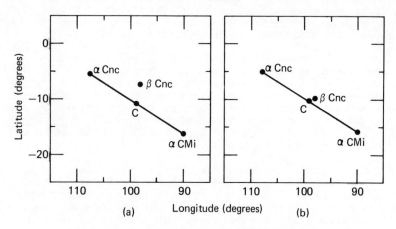

Figure IX.1 The configuration of the stars α Cancri, β Cancri, and α Canis Minoris. Part (a) is based upon the coordinates that Ptolemy gives while part (b) shows the configuration in Ptolemy's time as calculated from modern results. The point marked C is the center of the line that joins α Cancri and α Canis Minoris. Ptolemy says that β Cancri is northeast of C.

Again it looks as if Ptolemy did not make an independent observation. Figure IX.1 shows the situation. Part (a) of the figure shows the configuration according to the coordinates that Ptolemy gives. The star at the left (eastern) end of the straight line is α Cancri and the star at the other end is α Canis Minoris. The star that is not on the line is β Cancri. C is not a star; it is the center of the line. In ordinary parlance, β Cancri is north and slightly west of C. I can see only one way in which the direction can be called northeast, and that is to think of the line as lying in the east-west direction. I believe that this is what Ptolemy means.

However, this interpretation does not yield the correct direction, as part (b) of the figure shows. Part (b) shows the configuration that existed in Ptolemy's time according to calculations based upon modern results. We see that β Cancri is closer to the line than it appears in part (a), but it is northwest, not northeast, of C by any interpretation of the term. Even if Hipparchus made a mistake in locating β Cancri, it is not likely that Ptolemy would have made the same mistake. It is more likely that Ptolemy claimed to have verified the observation when all he did was to use the coordinates from Hipparchus's star catalogue.†

---

†The main difference between parts (a) and (b) of the figure comes in the latitude of β Cancri. Perhaps in order to make it agree with the modern value, Peters and Knobel [1915] give $10\frac{1}{2}$ degrees south as the latitude of β Cancri, although all of the Syntaxis that I know of give $7\frac{1}{2}$ degrees south as the

By this and several other examples, Ptolemy claims to show that the stars always maintain the same relative positions. It would be interesting to investigate the other examples, but I have not taken the time to do so.

After he studies this point, Ptolemy shows in three different ways that the equinoctial points are moving westward at the rate of 1° per century. He also shows that the latitudes of stars remain constant as this happens. As we know from modern astronomy, this is not quite so. The plane of the ecliptic, from which latitudes are measured, is rotating slowly at the rate of about 47″ per century. During the time that separated Ptolemy from Hipparchus, the total rotation was only about 2′, and we do not expect that Ptolemy could find this rotation.

At this point in the Syntaxis [Chapter VII.4], Ptolemy summarizes the situation. The fixed stars keep the same relative positions, but the sphere of the fixed stars rotates slowly about the axis perpendicular to the ecliptic. Thus the latitudes of the stars remain constant while their longitudes increase steadily at the rate of 1° per century. He has shown these facts by the observations he has described (which I shall discuss in succeeding sections) and by similar observations made on the other most brilliant stars.

This makes it appropriate for Ptolemy to construct a catalogue of the fixed stars. The catalogue will give the longitudes which the stars had at the beginning of the reign of Antoninus Pius†; these longitudes will increase by 1° per century. It will also give the latitudes and the magnitudes, which will remain constant. In his catalogue, he will include all the stars that it has been possible for him to observe, down to stars of the sixth magnitude. He has measured the longitudes and latitudes of the stars by using the astrolabe, which he has already described.‡ In using it, he claims to measure the longitude by comparing the star with the moon. At the same time, he sets one circle of the astrolabe into coincidence with the ecliptic so that he can read the latitude.

Thus Ptolemy's star catalogue has four columns. The first column identifies a star, using a description of the sort that I illustrated in Section I.2. The second, third, and fourth columns give, respectively, the longitudes, the latitudes, and the magnitudes. The catalogue is divided into four main parts. The first part contains the stars in the constellations that lie to the north of the zodiacal constellations, and the second part contains the stars of the zodiacal constellations from Aries to Virgo. Ptolemy

---

value. Since Ptolemy's description seems to agree with the value of $7\frac{1}{2}$ degrees, the interpretation of Peters and Knobel requires two errors in the text, which seems unlikely.

†More accurately, he means the first day of the Egyptian year in which Antoninus became the Roman emperor. This day is 137 July 20.

‡And which I have described in Section VII.1.

ends Book VII of the <u>Syntaxis</u> at this point. Book VIII begins with the stars in the remaining zodiacal constellations (Libra through Pisces), and the fourth part of the catalogue contains the stars in the constellations that are south of the zodiacal ones.

Following the catalogue, Ptolemy takes up a few miscellaneous topics in the rest of Book VIII. First is a description of the Milky Way. This is followed by a discussion of various aspects of the rising, culmination, and setting of the stars, with special attention to the circumstances in which a star sets just as the sun rises, or vice versa. Book VIII concludes with the conditions that determine how close a star can get to the sun before it can no longer be seen. I did not notice any information in this part of Book VIII that is relevant to this study and I shall not discuss it.

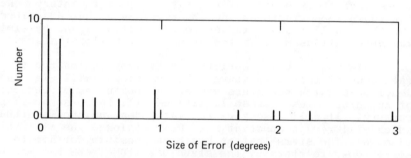

Figure IX.2   A histogram of the errors in the declinations that Hipparchus gives in his <u>Commentary on Aratus and Eudoxus</u>. The most plausible explanation is that the errors out to about $0°.3$ are random errors of measurement and that the larger errors are errors in writing the original manuscript or in copying it.

2.   The Accuracy of Stellar Coordinates

The errors in every source of ancient astronomical data that I have analyzed share a common characteristic. Suppose we make the kind of plot called a histogram; a histogram is a plot that shows the frequency with which errors of various sizes occur. Figure IX.2 is a histogram for the values of declination that <u>Hipparchus</u> [ca. -135] gives in his commentary on Aratus and Eudoxus. I have assumed that the error in a value of declination is the difference between it and the value calculated from modern data. Sometimes we are interested in the symmetry of the errors about zero. In this case, we are not, so the reader should think "absolute value of the error" whenever I write "error" in the immediate discussion.

In Figure IX.2 I consider the horizontal axis to be divided into ranges whose length is $0°.1$. Whenever an error lies in a particular range, I make a vertical mark at the center of the range. If there is already a vertical mark there, I make a new mark on top of the old one, being careful to make all marks with the same length. In this case, there were 9 errors lying between $0°$ and $0°.1$, 8 errors lying between $0°.1$ and $0°.2$, and so on.†

For errors out to about $0°.3$, the number of errors falls off as the error increases, in about the way that we would expect if the errors come from accidents of observation. Beyond about $0°.3$, this is no longer so. The number of errors shows little tendency to fall off with increasing size, and the number of large errors is much greater than we can explain by any reasonable process of measurement. I think that any statistician would say without hesitation that the errors in Figure IX.2 belong to two different populations with different sources and properties.

The most plausible explanation of Figure IX.2 is that the errors less than about $0°.3$ arise from accidents of observation and that the errors larger than about $0°.3$ arise from errors in writing. In these, the original observer may have made a mistake in writing down the result of his observation. Even if he wrote the result correctly, we must remember that Hipparchus's work was transmitted for nearly two millenia by the process of having a scribe read one handwritten copy and write out a new handwritten copy as he did so. If a letter in a word was wrong or hard to make out in the manuscript that was being copied, the copier had a chance of correcting it. If a numeral was wrong or hard to make out, he had no chance of correcting it because, from his point of view, the numerical values were arbitrary. For simplicity, I refer to errors in either the original or in the surviving copies as recording errors. Rarely if ever can we decide whether an error was made in the original record or in a subsequent copy from which all known copies have descended.

Figure IX.2 includes still a third type of error that does not occur often in astronomical writing. In four cases, Hipparchus did not state an accurate value. He merely said that the declination was about such and such a value; it is clear from the context that he rounded to a convenient integral number of degrees. However, this does not explain the largest error, which is $3°$. This occurs for the star μ Draconis. Hipparchus says that the star is 34 3/5 degrees from the pole, but the correct value is close to 31 3/5 degrees. Since it was easy to confuse 1 and 4 in ancient Greek numerals, I feel sure that this is a recording error.‡

---

†Obviously, the maker of a histogram must decide upon a convention to handle the case in which an error lands exactly at the boundary between two ranges. In Figure IX.2, I include such errors in the smaller range.

‡In the time of Hipparchus, 1 was probably written as **A** (an early form of alpha) while 4 was written as Δ. See Peters and Knobel [1915].

By a careful comparison of various manuscripts, it is often possible to detect errors that have occurred in the course of successive copyings and to correct them. Peters and Knobel [1915] have made an extensive study of the star catalogue in the Syntaxis, and theirs is probably the most accurate version of it that exists. Occasionally, by comparing different portions of the same work, it is possible to correct an error that the author made in the original autograph copy; we shall see an example of this in Section IX.5.

TABLE IX.1

THE ACCURACY OF MEASURING STELLAR COORDINATES

Observer	Source	Quantity observed	Standard deviation,
Hipparchus	Ptolemy[a]	Longitude	22.3
Hipparchus	Ptolemy[a]	Latitude	20.8
Hipparchus	Hipparchus[b]	Declination	12.3
Timocharis and Aristyllus	Ptolemy[a]	Declination	8.8
Hipparchus	Ptolemy[a]	Declination	6.6
Ptolemy	Ptolemy[a]	Declination	7.2

[a]Ptolemy [ca. 142].
[b]Hipparchus [ca. -135].

Since we have no way to tell rigorously whether some errors are observing or recording errors, we cannot make a rigorous estimate of the size of the observing errors. However, we get about the same estimate for any reasonable division of the errors between observing and recording. By making a reasonable division, I have estimated the standard deviations of the observing errors in six samples of star data, which are given in Table IX.1. The first column in the table names the person who made the observations, the second gives the written source through which the data reach us, the third gives the quantity observed, and the last gives the estimated standard deviation of the errors of observation in minutes of arc.

The first two lines refer to the longitudes and latitudes of the stars in the star catalogue in the Syntaxis. Although this catalogue occurs in the Syntaxis, and although Ptolemy claims to have made the observations himself, it will be made clear in what follows that he did not do so. It is almost certain that Hipparchus made the observations that Ptolemy

appropriated as his own in the Syntaxis,† and consequently
I have listed Hipparchus as the observer in Table IX.1.

The last four lines refer to four samples of measured
declinations, all of which will be discussed later in this
chapter. We see that the declinations have about half the
error that the latitudes and longitudes have. I believe
that this is inherent in the measurement process. One can
measure declinations by setting up a graduated circle in the
plane of the meridian, with the zero mark being offset from
the vertical by the geographic latitude. Once the instrument is properly set up, declinations can be read on it with
no further adjustment. Latitudes and longitudes, however,
must be read on an instrument like the astrolabe, which was
described in Section VII.1. This is the instrument which
Ptolemy claims to have used in studying the motion of the
moon, and which he claims to have used in measuring the coordinates in his star catalogue. With it, the basic circles
must be aligned parallel and perpendicular to the ecliptic
before it is used, and then the latitude and longitude must
be read quickly before the rotation of the earth destroys the
alignment. It is not surprising that the errors in longitude
and latitude are twice those in declination.

The estimates of the longitudes and latitudes in the
table are not based upon all the stars in the star catalogue,
for two reasons. First, a small error of observation makes
a large error in the longitude of a star at high latitude.
Second, there are more than 1000 stars in the catalogue, and
using all of them in estimating the errors would be highly
laborious. For these reasons, I have based the estimates in
Table IX.1 only upon the stars in the twelve zodiacal constellations. This sample contains about 320 stars, after I
eliminate what seem to be recording errors. I have used the
version of Peters and Knobel [1915] in preparing Table IX.1.

The observations by Timocharis and Aristyllus were made
in Alexandria in about -290 while those made by Ptolemy were
also made in Alexandria, in about +140. It is interesting
that there was no significant improvement in accuracy in the
intervening four centuries. We have found this same situation with other kinds of observation in earlier chapters.
The accuracy of Greek observations of all kinds seems to be
about the same from the earliest ones known to us to the end
of Greek astronomy.

3.   The Longitudes of Regulus and Spica

In Chapter VII.2 of the Syntaxis, Ptolemy says that
Timocharis had measured the longitude of Spica (α Virginis)

---

†There is a bias of about 1°.1 in the longitudes as they are
listed in the Syntaxis. This bias came from the process by
which Ptolemy appropriated the observations from Hipparchus.
The standard deviation of the longitudes listed in Table
IX.1 refers to the errors that are left when this bias is
removed. In the original star catalogue of Hipparchus, it
is probable that the bias was only about 4'.6 [Peters and
Knobel, 1915, p. 17].

to be 8° from the autumnal equinox, and hence at longitude
172°. Hipparchus, however, found it to be only 6° from the
autumnal equinox, at longitude 174°. From this and other
comparisons, Hipparchus concluded, according to Ptolemy,
that the longitudes of stars increase steadily at a rate
that "is not less than 1° per century."† Ptolemy says that
he has found the same thing and the same rate by observing
longitudes with the aid of the astrolabe and comparing the
results with those of Hipparchus, and he gives us an example.

The example is the observation that he made on 139 February 23 which I discussed in Section VII.2 and which I have
analyzed earlier in Part II. Just before sunset on that date,
when the moon was at the first quarter, Ptolemy aligned the
astrolabe with the sun and measured the elongation of the
moon. Half an hour later, after sunset, he measured the difference in longitude between the moon and the star Regulus
(α Leonis). From these measurements, after he corrected for
the parallax of the moon and for its motion during the half
hour between the observations, he found that the longitude of
Regulus was 122° 30'. This is the longitude of Regulus that
we find in Ptolemy's star catalogue. Hipparchus had measured
its longitude to be 119° 50'. Since there had been 2 2/3
centuries between the observations, and since the longitude
had changed by 2 2/3 degrees, the rate of precession is 1°
per century.

As Table VII.1 shows us, the positions that Ptolemy
claims to have measured for the sun and moon agree almost
exactly with those that we calculate from his tables of the
sun and moon. Now, in Ptolemy's time, there was a systematic error of more than 1° in his tables of the sun and moon.
Further, there were oscillating errors with a standard deviation of more than 0°.5 in his tables of the moon. In the
presence of these errors, it is almost incredible that accidental errors of observation would cancel the errors in his
tables. The probability that the agreement shown in Table
VII.1 happened by chance is [Part II] less than 1 in
1 000 000.

In other words, with odds that are greater than
1 000 000 to 1, the observation of Regulus was fabricated.
It was fabricated for the purpose of making the precession
rate come out to be 1° per century, exactly. Probably because it was a round number, Ptolemy decided to fabricate
observations to yield exactly the value which Hipparchus
had given as a lower limit.

As I have already said, Hipparchus found that the longitude of Spica (α Virginis) was 174°, according to Ptolemy.

---

†There were about 160 years between the measurements by
Timocharis and Hipparchus, and a change of 2° in this
time is 1°.25 per century, or 45" per year. If Ptolemy
has quoted Hipparchus correctly, it seems that Hipparchus
was aware of the possibility of error but concluded that
error could not make the rate fall below 1° per century.
According to the theory of Newcomb [1895], the rate at the
time in question was about 49".8 per year.

Ptolemy has also "proved" from the observation of Regulus
that the total precession between Hipparchus's and Ptolemy's
star table is 2° 40'. If these statements are so, the longitude of Spica at the epoch of Ptolemy's table should be
176° 40'.

Ptolemy does not call attention to the fact, but this
is exactly the longitude that he gives for Spica in his star
catalogue. Further, according to his general claim [Syntaxis,
Chapter VII.4] that I have already quoted, he measured this
longitude with the astrolabe. Thus the longitude which he
claims to have measured agrees within rounding error with the
preassigned value obtained from Hipparchus's table and Ptolemy's precession. The rounding error is 5', so the measured
value lies in a preassigned range whose total width is 10'.
According to Table IX.1, this is about 0.5 standard deviations.
However, the correct longitude at Ptolemy's epoch was about
177°.95. The error in the measured longitude is about 1°.28,
or about 3.5 standard deviations. The probability that a
value would lie within a range of 0.5 standard deviations
centered at 3.5 standard deviations is about 1 in 2000.

Thus, with odds of about 2000 to 1, the longitude of
Spica given in Ptolemy's table was fabricated instead of being
observed. The odds that the longitude of Regulus was fabricated have just been estimated at more than 1 000 000 to 1.
With enormously high probability, the only two longitudes
that we can test directly were both fabricated, contrary to
Ptolemy's claim that he observed them.

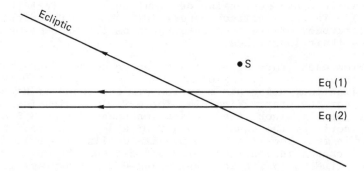

Figure IX.3   Conditions near the vernal equinox at two epochs.
The ecliptic remains almost fixed with respect to the stars; the arrow
indicates the easterly direction along the ecliptic. As the equator
moves from Eq (1) to Eq (2), the distance of the star S from the
equator increases; this distance is the declination of S. In general,
the declination increases with time for a star near the vernal
equinox and decreases for one near the autumnal equinox.

4.   The Declinations of Eighteen Stars

In Chapter VII.3 of the Syntaxis, Ptolemy gives the declinations of eighteen stars as measured either by Aristyllus or Timocharis, by Hipparchus, and by himself. The measurements by Timocharis and Aristyllus were made in Alexandria about -290, those by Hipparchus were made probably in Rhodes about -128, and those by Ptolemy were allegedly made in Alexandria about 137. These three sets of measurements form three of the data samples in Table IX.1.†

Now look at Figure IX.3, which shows circumstances near the vernal equinox at two different times. The slanting line is the ecliptic, which we can take as fixed with respect to the stars.‡ The line marked Eq (1) is the position of the equator at some time. Since east is to the left if north is up, the intersection of the ecliptic with Eq (1) is the vernal equinox at this time. At a later time, the equinox has moved westward and the equator is now the line marked Eq (2). The star S is farther from the equator at the second time than at the first time; in other words, its declination has increased.

With the aid of a similar figure drawn for the autumnal equinox, the reader should be able to convince himself that the following is true: If the longitude of a star is between -90° (270°) and 90° at both times, the declination increases. If the longitude is between 90° and 270° at both times, the declination decreases. If the longitude is close to 90° or 270°, the declination changes little if at all.

Ptolemy has chosen the 18 stars so that 9 had increasing declinations between Timocharis or Aristyllus and himself, while 9 had decreasing declinations. The fact that these lie in the longitude ranges just described indicates that latitudes of stars do not change as the precession increases their longitudes.

From each group of 9, Ptolemy now chooses 3 and calculates the rate of precession that is indicated by the change of declination, using his measurements and those of Hipparchus. For the star α Virginis, for example, which is near the autumnal equinox, the declination changed from 3/5 of a degree north of the equator to 1/2 of a degree south. For a star in this location to change its declination by 66′, it takes a change in longitude of 2 2/3 degrees, he says. Since the longitude of α Virginis has changed 2 2/3 degrees in 2 2/3 centuries, the rate must be exactly 1° per century. Ptolemy finds the same rate from each of the other 5 stars that he considers. He does not use 12 of the 18 stars at all, nor does he use the measurements attributed to Timocharis or Aristyllus. Let us repair these omissions.

---

†For reasons that will be explained in a moment, I used only 12 of the 18 measurements that Ptolemy claims to have made when I prepared Table IX.1.

‡The slow rotation of the ecliptic can be neglected in this approximate discussion.

I studied the declinations measured by Timocharis or Aristyllus in the process of deducing the obliquity of the ecliptic [Newton, 1974]. I concluded that they had a bias of 0°.068 (about 4′) in the setting of the circle with which they measured declinations. As Table IX.1 shows us, they had a standard deviation of 8′.8 in their measurements. These are quite reasonable values. By using the values attributed to them in conjunction with those attributed to Hipparchus, I find a precession rate p given by

$$p = 46.6 \pm 4.5 \quad ''/\text{year}. \tag{IX.1}$$

The value given by the theory of Newcomb [1895] is 49.8″ per year.

TABLE IX.2

PRECESSION RATE CALCULATED FROM
PTOLEMY'S DECLINATIONS

| Star | Declination | | Change | Precession rate |
| | Hipparchus | Ptolemy | | |
	° ′	° ′	′	″/year
η Tauri	15 10	16 15	65	41.1
α Aurigae	40 24	41 10	46	35.1
γ Orionis	1 48	2 30	42	40.1
α Virginis	0 36	− 0 30	−66	36.9
η Ursae Majoris	60 45	59 40	−65	36.0
α Boötis	31 0	29 50	−70	39.5
α Tauri	9 45	11 0	75	54.6
α Orionis	4 20	5 15	55	64.4
α Leonis	20 40	19 50	−50	50.9
ζ Ursae Majoris	66 30	65 0	−90	50.4
ε Ursae Majoris	67 36	66 15	−81	45.1
α Librae	− 5 36	− 7 10	−94	54.2
β Librae	0 24	− 1 0	−84	50.5
α Scorpii	−19 0	−20 15	−75	52.2

Thus the data of Timocharis, Aristyllus, and Hipparchus lead to a value that agrees well with modern results but which disagrees vigorously with the result that Ptolemy claims to have found. This, together with the estimates of accuracy in Table IX.1, indicates strongly that the measurements which Ptolemy attributes to Hipparchus are genuine. We have some confirmation of this with regard to Hipparchus's data. As I have said, Hipparchus [ca. -135] gives the declinations of 40 stars, and 3 of these are in the list that Ptolemy gives. Of the three, the declinations agree exactly for two, namely α Boötis and β Geminorum. For α Geminorum,

Hipparchus states $33\frac{1}{2}$ degrees while Ptolemy says that Hipparchus found 33 1/6 degrees. Since it is easy to confuse $\frac{1}{2}$ and 1/6 in certain forms of Greek numerals, this is probably a copying error and not a genuine disagreement.† Thus, in the instances that we can check, the values which Ptolemy attributes to Hipparchus agree with those that Hipparchus states himself.

Now let us use all 18 stars in calculating the precession rate. The results are shown in Table IX.2, which is copied from Part III by permission of the Royal Astronomical Society. Four of the stars have longitudes so near 90° or 270° that we cannot infer a reliable rate from them. Hence there are only 14 stars in Table IX.2. The table gives the designations of the stars, the declinations that Hipparchus measured, the declinations that Ptolemy claims to have measured, the change, and the precession rate calculated from the change in declination.

The order of the stars in Table IX.2 is not the order in which Ptolemy lists them. I have given first the six stars that Ptolemy uses; within this group, the order is the same as Ptolemy's. Also, within the group that Ptolemy does not use, I have preserved Ptolemy's order. In choosing the 3 that he uses from each group of nine, Ptolemy does not use the first three. He has chosen the three apparently at random. In the first group of 9, he uses the 2nd, 4th, and 5th, while in the second group of 9 he uses the 2nd, 3rd, and 6th.

The values of precession from the group of 6 that Ptolemy uses range from 35.1 to 41.1, although Ptolemy says that each one gives 36 ″/year (1° per century). The estimate of p that we would form from this group is

$$p = 38.1 \pm 1.1 \quad ''/\text{year}. \tag{IX.2}$$

The values that he does not use range from 45.1 to 64.4; note that there is no overlap between the groups. The estimate that we form from the unused group is

$$p = 52.8 \pm 2.0 \quad ''/\text{year}. \tag{IX.3}$$

This should be compared with the value 46.6 ± 4.5 ″/year from the data of Timocharis or Aristyllus (Equation IX.1) and with the value 49.8 ″/year derived from Newcomb's theory.

The group that Ptolemy does not use leads to a reasonably accurate value of the precession, and the declination errors in this group have a standard deviation (Table IX.1) of 7′.2. The bias in reading declinations indicated by this group [Newton, 1974] is 0°.007. Thus this group passes all the tests for genuineness, and I have little doubt that some one did observe these declinations. However, Ptolemy did not use them.

We reach a different conclusion for the group that Ptolemy does use. The individual values of p from this group

---

†Since we do not have either original, we do not know which value Hipparchus really measured.

have nothing in common with the unused group, nor [Part III] do the errors in the values of the declination. Thus there can be little doubt that the group of 6 declinations used by Ptolemy were fabricated.

Pannekoek [1955] also noted that the set of the used stars gives a quite different result from the set of the unused stars, but he draws a different conclusion. He concludes that the entire set of declinations is genuine and that "Ptolemy selected these six stars because† they were favourable to his assumed value of the precession ... Yet we cannot speak of an attempt to deceive his readers; .."

We can see in many ways that the used group is fabricated rather than selected from a body of genuine data. For one thing, it is unlikely that 6 genuine observations even existed which could give rise to the value that Ptolemy found. For the genuine observations, the standard deviation of an individual value of p from the mean is 5.5 "/year. The values of p that Ptolemy uses lie about 2.7 standard deviations from the mean. The chance of finding a value this far from the mean in a preassigned direction is about 0.0034. That is, we expect to find one value of this sort in a sample of about 290. In order to find six values, we would need to search through about 6 × 290 = 1740 values. We may be sure that neither Hipparchus nor Ptolemy measured this many declinations.‡

Further, the search is not merely a matter of looking for specific values. In order to know whether a star would give a favorable result or not, Ptolemy would have had to calculate the precession rate from each star examined. This calculation would have been fairly difficult for him, with the calculating tools available at the time. Since we already know that Ptolemy uses fabricated data in establishing the precession, it is far more likely that he fabricated six favorable cases than that he searched through 1740 cases (if they existed), making a fairly laborious calculation for each one, in the probably vain hope that he would find six favorable ones.

For another thing, the estimates of p in Equations IX.2 and IX.3 differ by about 4.7 times the sum of their standard deviations. I have not calculated the probability that this could happen by chance if the observations are genuine, but it must be of the order of 1 chance in 100 000.‡

For still another thing, the standard deviation of the group of six about its mean is 2.4 "/year while the standard deviation of the group of eight is 5.5 "/year. The smaller scatter of the group of six is to be expected if the group was fabricated. It is almost impossible if the group

---

†The emphasis is Pannekoek's.

‡There are only about 1030 stars in the star catalogue in the Syntaxis.

‡This is about the chance of finding a single value at 4.7 standard deviations from the mean.

of six were the extreme "tail" of a large collection of measured values; the tail of a distribution always shows a large scatter, in my experience.

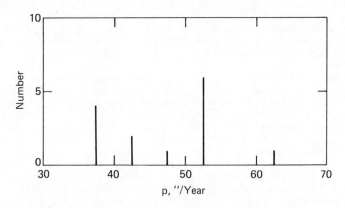

Figure IX.4   The histogram of individual values of the precession. Ptolemy uses the six smallest values of p, but does not present or mention the eight largest values. The histogram shows that the sample of six and the sample of eight have different origins.

Finally, let us look at the histogram of the individual values of p in Table IX.2. The histogram, which is shown in Figure IX.4, is drawn for increments of $5''$/year in p. The six values between 50 and 55 $''$/year, and the two values on either side of it, are the eight values that Ptolemy does not use. They form a plausible population of measured values affected by normal observational error. If they do form such a population, it is highly unlikely that the values less than 45 $''$/year belong to the same population. There is no plausible process that would cause the frequency distribution to have two maxima when the basic measurement is as simple as the measurement of declination. In other words, if the eight values to the right of the histogram are genuine, it is almost certain that the six values to the left are not.

From the standpoint of Ptolemy's actions, it does not seem to me to matter much whether the six values were fabricated or selected. In either case, it seems clear what Ptolemy is trying to do. He is trying to convince the reader that the data which lead to a rate of $36''$ per year are typical, although he knows quite well that they are not. His writing is dishonest whether the data are fabricated or selected. Only if we are trying to assess the quality of ancient observations does it matter whether the data that he uses are selected or fabricated. All the tests we have found say that they are fabricated.

Pannekoek makes an analogy between Ptolemy's selection
of data and the practice of "stripping" widely discordant
values from a collection of data. It seems to me that there
is no resemblance. Stripping is properly applied in situations where there are sound physical reasons for believing
that some values in a string of recorded numbers are not actually data. For example, if a sequence of measurements is
being recorded by a piece of electronic equipment, a short
burst of electronic noise from a nearby source can produce a
reading that is not a measurement of the quantity being
studied. Instead, it is a characteristic of the noise source,
and we should try to eliminate such readings.

The way a situation like this is handled differs in two
important ways from Ptolemy's practice, even if we assume
that the precession data were selected and not fabricated.
In the electronic case, or in analogous cases, what the honest experimenter does in effect is to make a histogram. From
the properties of the histogram, he decides where the set of
actual data lies and he omits the values that do not belong
to the set. However, he never omits more than a small fraction of the data; he does not omit two thirds of the entire
collection, as Ptolemy does. Also, he makes the selection
before he knows what the answer will be. He does not select
data for the purpose of obtaining a specific result, as
Ptolemy does.

In the case of the precession rate that is derived from
measurements of declination, there can be no question of
Ptolemy's being the innocent victim of a dishonest colleague.
Here Ptolemy presents 18 pairs of declinations; exactly six
pairs will lead to the value of precession that he has already established by a fraudulent observation of Regulus,
and none of the remaining pairs will do so. It is not credible that Ptolemy, unknowingly and by accident, chose to use
exactly six pairs or, having decided to do so, that he should
choose precisely the particular six without including a single one of the others.

We must conclude that the writer of the Syntaxis knew
just what he was doing. He deliberately decided to "prove"
a false value of the precession by the use of spurious data.
In order to conceal what he was doing, he mixed the spurious
data with some genuine data, so that he could pretend that
he was using typical data.

5.   Seven Conjunctions or Occultations

After he derives a precession rate from the star declinations, Ptolemy uses seven occultations of stars by the moon
or close conjunctions of stars with the moon. From these
conjunctions or occultations, Ptolemy derives both the latitude and the longitude of the star. With the aid of these
values, he shows that latitudes remain constant while longitudes increase steadily at the rate of 1° per century. All
of the conjunctions or occultations were allegedly observed
by astronomers earlier than Ptolemy.

The interpretation of these observations has raised

several problems. Newcomb [1875], in fact, after studying
the situation, decided not to use these observations in his
research on the motion of the moon, from a fear that they
form a biased sample which Ptolemy chose to support his incorrect precession rate for the equinoxes.† However, as I
have remarked [Newton, APO, p. 150ff], Ptolemy has rather
accurate values for the rates of the sun, moon, and planets
with respect to each other and to the stars. Because he
starts from the wrong length of the year, and builds his
other theories from the motion of the sun, he has an error
in rate that is common to the sun, the moon, the planets,
and the equinoxes. Thus Ptolemy did not need a biased sample.
Unbiased observations could have served his purposes better
than a biased sample.

When I wrote this, it had not occurred to me that Ptolemy would fabricate observations which he attributed to other
astronomers. Now we know that he can do so and that he has
done so, and we see another possibility. His purposes would
be served still better by fabricated observations than by an
unbiased sample. As we shall soon see, all of these conjunctions and occultations have been fabricated. I shall start
the study of them by summarizing the facts of the observations
as Ptolemy gives them, and the deductions that Ptolemy makes
from them, in the order that he uses them. The records are
found in Chapter VII.3 of the Syntaxis.

(1) Timocharis observed in Alexandria on -282 January
29 that the southern half of the moon had covered the eastern
half or third part of the Pleiades at the end of the third
hour of the night. This was the same as 3 1/3 ordinary hours
before midnight, in both mean and apparent time. At this
time, the true (that is, the geocentric) position of the moon
calculated from Ptolemy's tables of the moon was 30° 20' in
longitude and +3° 45' in latitude. After he calculates the
parallax, Ptolemy finds that the apparent position as seen
from Alexandria was 29° 20' in longitude and +3° 35' in latitude. Since the eastern part of the Pleiades was 10' east
of the center of the moon and 5' north of it, the eastern
part was at 29° 30' in longitude and +3° 40' in latitude.

In the final step, note that Ptolemy has assumed that
the northern part of the moon had just touched the eastern
part of the Pleiades. Since Timocharis knows what part of
the moon is covering the eastern Pleiades, he must have seen
the beginning of the occultation, and the recorded time may
be the beginning, as Ptolemy assumes it is. When Ptolemy
wrote (or copied) the "southern part" of the moon, this may
have been a simple mistake for the "northern part". In any
case, the northern part is what he uses in his calculations.

(2) Agrippa observed in Bithynia on 92 November 29, at
the beginning of the third hour of the night, that the southern cusp of the moon was covering the southeastern part of
the Pleiades. The time was 4 hours of the night or 5 ordinary hours before midnight. This is 7 hours after noon in

---

†Newcomb does not seem to suspect that the observations
might be fabricated, however.

Bithynia. Ptolemy takes this to be 6 2/3 hours after noon
in Alexandria, in apparent time, or 6 1/4 hours, mean time.
At this time, by Ptolemy's theory, the true longitude of the
moon was 33° 7' and its true latitude was 4° 50'. As seen
in Bithynia, its apparent longitude was 33° 15' and its apparent latitude was 4°. Hence the longitude of the southeastern Pleiades was 33° 15' and the latitude was +3° 40'.

In this case, contrary to what he did with the first
observation even though the language is similar, Ptolemy assumes that the longitude of the star is the same as the (apparent) longitude of the center of the moon. He takes the
star to be 20' south of the center of the moon.

Ptolemy's treatment of the Pleiades is confusing. According to orthodox Greek mythology, the Pleiades were seven
nymphs who could not run fast enough to escape from Orion.

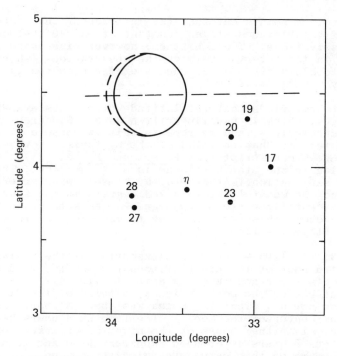

Figure IX.5   The configuration of the moon and of the Pleiades
on the night of 92 November 29 as calculated from modern theory.
The points show 7 Pleiades, namely the stars η Tauri, 17 Tauri,
and so on. The moon was nearly full. The illuminated part of it
is enclosed by the solid curve while the curve needed to complete
its full disc is indicated by the broken semicircle. The broken
line passing through the center of the moon shows its path on
the night in question.

To save them from a fate worse than death, the gods turned them into a constellation of stars. Later, one of them, distressed by the Trojan War, turned into a comet, leaving six stars. The story may be unreliable, but it shows that there should be six Pleiades in Greek writing. However, Ptolemy lists only four. He calls one of them the "eastern and narrowest part of the Pleiades" in the star catalogue in the Syntaxis, and this must be the one involved in the preceding occultations.

There is no agreement about what star, if any, this tabular entry represents. The configuration of the Pleiades (using their pre-war number) is shown in Figure IX.5. Various writers have taken the eastern and narrowest part of the Pleiades to be η Tauri, 27 Tauri, or 28 Tauri. However, 27 and 28 Tauri are so close together that Ptolemy may not have distinguished between them. The most plausible indentification of Ptolemy's star, it seems to me, is both 27 and 28 Tauri, not resolved into separate stars. However, I shall not base any part of the discussion upon this tentative identification.

Ptolemy points out that the latitude of the eastern part of the Pleiades stayed constant at +3° 40′ between the two observations. The longitude, however, increased by 3° 45′ in the interval between the observations, which was close to 375 years. Hence the rate of precession is 1° per century.

Ptolemy finds that the latitude of the eastern Pleiades is 3° 40′, which is usually written as 3 2/3 degrees in the Greek manuscripts, and he repeats this value several times in the text of Chapter VII.3 of the Syntaxis. However, in every Greek manuscript that Peters and Knobel [1915] examined the value in the star catalogue is 3 1/3 degrees. The most likely explanation is that Ptolemy himself made an error in writing: He meant to write 3 2/3 degrees in the star table but accidentally wrote 3 1/3 degrees. As Peters and Knobel [p. 12] show, this error is easy to make in the Greek numerals of Ptolemy's time.

(3) On -293 March 9 in Alexandria, at the beginning of the third hour of the night, Timocharis saw the middle of the moon's rim reach the star Spica (α Virginis). Spica in fact cut behind the moon's disc at a third of its diameter from the north. (That is, Spica was about 5′ north of the apparent center of the moon.) The time is 4 hours before midnight, in either hours of the night or ordinary hours. It is thus 8 hours after noon; apparent time and mean time are the same on this date. By calculation, the true position of the moon was 171° 21′ in longitude and -1° 50′ in latitude. After the correction for parallax is made, the apparent position was 172° 05′ in longitude and -2° in latitude. Spica was 15′ farther east, at 172° 20′ in longitude and -2° in latitude. Ptolemy obviously ignores the explicit statement that Spica was north of the center of the moon.

(4) On -282 November 9 in Alexandria, when half of the 10th hour of the night had passed, the moon being risen from

the horizon, Timocharis saw Spica exactly touch the northern limb of the moon. The time was 3½ hours of the night, or 3 1/8 ordinary hours, after midnight. Hence (!),[†] Ptolemy says, the time was 2½ hours after midnight because this is when the moon and Spica rose. This is apparent time at Alexandria; the mean time was 2 hours after midnight. By calculation, the true longitude of the moon was 171° 30′ and its true latitude was -2° 10′. The apparent longitude was 172° 30′ and the apparent latitude was -2° 15′. Therefore Spica was at 172° 30′ of longitude and -2° of latitude.

(5) On 98 January 11 in Rome, when the 10th hour of the night had ended, "the geometer Menelaos" saw that Spica was hidden by the moon. When the 11th hour of the night had ended, Spica was visible to the west of the center of the moon. It was less than the moon's diameter from its center, and it was equidistant from the two cusps.

Ptolemy ignores the second observation and uses only the first. He is certainly wrong to do so. The uncertainty in position allowed by the second observation is less than the radius of the moon, whereas the uncertainty allowed by the first observation is the full diameter. With no visible basis, Ptolemy assumes that Spica coincided exactly with the center of the moon at the time of the first observation.

This time is 4 hours of the night after midnight at Rome, which is 5 ordinary hours after midnight. The corresponding time at Alexandria is 6 1/3 hours after midnight.[‡] This is apparent time; the mean time is 6 1/4 hours. The calculated position of the moon is 175° 45′ in longitude and -1° 20′ in latitude. The apparent position, after correcting for parallax, is 176° 15′ in longitude and -2° in latitude. Hence this is also the position of Spica.

All of these observations show that the latitude of Spica is -2°. Its longitude increased by 10′ in the interval of about 12 years between the first two observations of Spica. The interval is too short to give an accurate value of the rate, but it is interesting to find that the precession can be detected in an interval of only 12 years, as Ptolemy analyzes the data. The longitude increased by 3° 45′ in the 379 years between the second and third observations. Hence the rate is almost exactly 1° per century.

(6) Timocharis reported that he saw the northern cusp of the moon touch the star β Scorpii. The time was -294 December 21 at the beginning of the 10th hour of the night, which was 3 2/5 ordinary hours after midnight. The mean time was 3 1/6 hours at Alexandria. For this time, the geocentric position of the moon that Ptolemy calculates was 211° 15′ in longitude and +1° 20′ in latitude. The apparent position was 212° in longitude and +1° 05′ in latitude.

[†]The exclamation point is mine.

[‡]Ptolemy overestimates the time difference between Rome and Alexandria by slightly more than 10 minutes. In 10 minutes the moon travels about 5′.

Since the star was at the northern cusp of the moon, its
longitude was also 212° while its latitude was +1°05′ plus
the radius of the moon. That is, the latitude was +1° 20′,
very closely.

TABLE IX.3

STELLAR COORDINATES FROM CONJUNCTIONS
AND OCCULTATIONS

Star	Date	Longitude		Latitude	
		Calculated from star table	Deduced from the observations	From star table	Deduced from the observations
		° ′	° ′	° ′	° ′
Eastern Pleiades	137 Jul 20	33 40		+3 20[a]	
	-282 Jan 29	29 29	29 30	+3 20[a]	+3 40
	92 Nov 29	33 13	33 15	+3 20[a]	+3 40
α Virginis	137 Jul 20	176 40		-2 0	
	-293 Mar 9	172 22	172 20	-2 0	-2 0
	-282 Nov 9	172 29	172 30	-2 0	-2 0
	98 Jan 11	176 17	176 15	-2 0	-2 0
β Scorpii	137 Jul 20	216 20		+1 20	
	-294 Dec 21	212 01	212 00	+1 20	+1 20
	98 Jan 14	215 57	215 55	+1 20	+1 20

[a]Peters and Knobel [1915] say that the correct value from the
star table is +3° 40′. They are almost surely correct; the
value of 3° 20′ is probably a scribal error.

(7) Menelaos in Rome saw the southern cusp of the moon
lying in a straight line† with the stars δ Scorpii and π
Scorpii, with its center to the east of the line, and it was
as far from δ Scorpii as that star is from π Scorpii. The
moon had covered β Scorpii, for that star could not be seen.
This happened on 98 January 14, at the end of the 11th hour
of the night. This time, which is 5 hours of the night after
midnight, was 6 1/6 ordinary hours after midnight at Rome.
In Alexandria time, this was 7½ hours after midnight, in both
apparent and mean time. By Ptolemy's calculation, the center
of the moon was at 215° 20′ in longitude and at +2° 10′ in
latitude. The apparent position was 215° 55′ in longitude
and +1° 20′ in latitude. Hence (!)‡ this was also the position of the star.

---

†The configuration is shown in Figure IX.7, below.

‡Again the exclamation point is mine.

This observation is impossible. Instead of showing this now, I shall continue with Ptolemy's treatment, and I shall show the impossibility in a few moments.

In both observations, the latitude of β Scorpii is constant at +1° 20'. Between the two observations, which are about 391 years apart, the longitude increased by 3° 55' (= 3.917 degrees). Hence the rate is 1° per century.

Table IX.3 summarizes the coordinates involved in the preceding observations. Three stars are involved, namely the eastern Pleiades, α Virginis (Spica), and β Scorpii.† For each star, Table IX.3 compares the position calculated from the star table or catalogue, using a precession rate of 1° per century, with the position that Ptolemy infers from the observation. To illustrate the use of the table, consider the first star. The date 137 July 20 is the epoch of the star catalogue. For that date, the position in the catalogue is 33° 40' in longitude and +3° 20' in latitude. (However, we suspected above that Ptolemy meant to write +3° 40' for the latitude, as the footnote indicates.) The date -282 January 29 is about 419 years before the epoch of the catalogue. If the precession is 1° per year, the longitude on -282 January 29 is 4° 11' less than on 137 July 20, so that it is 29° 29'. The latitude remains unchanged at +3° 20'. Other entries in the table are obtained in the same way.

The table gives incontrovertible proof that the conjunctions and occultations have been fabricated. Except for the possible confusion about the latitude of the eastern Pleiades, the latitudes from the star catalogue agree exactly, to the minute of angle, with those deduced from the observations. Even for the eastern Pleiades, the two observations give exactly the same latitude. In the seven comparisons in longitude, there are three differences of 1' and four differences of 2', for an average difference of 1'.6.

There are many sources of error in the observations. To start with, there is the error in comparing the angular position of the moon and a star in some cases. When the star is hidden by the moon, it can usually be anywhere behind the disk of the moon. The times are rounded to the nearest half hour, so that the correct time may be as much as 15 minutes from the time given, even if there is no clock error. In 15 minutes, the moon moves about 7'; this contribution alone exceeds any difference in Table IX.3. In some cases, there are appreciable uncertainties in the time difference between Alexandria and some other place.

The most important source of error is probably that in Ptolemy's tables of the moon. At the times of the observations, the phase of the moon is well distributed over all phases. Specifically, there are phases near the quarters,

---

†β Scorpii actually consists of two stars so close together that they cannot be resolved with the naked eye. The one called β¹ Scorpii is the brighter and it is probably the one that the ancient observers saw.

the octants, and syzygy.† Taken over all phases, the standard deviation of the longitude calculated from Ptolemy's tables is 0°.581, about 35', as we saw in Section VII.6. Let us ignore all other sources of error. If the average difference in longitude is 1'.6, the values of longitude lie within a preassigned range of width 3'.2. For simplicity, let us change this to 3'.5, which is 0.1 standard deviations.

The probability that a single observation will lie within a preassigned range with a width of 0.1 standard deviations is less than 0.04, about 1 chance in 25. The probability that all seven observations meet this condition simultaneously is about 1.6×10^{-10}, or about 1 chance in 6 billion. We may safely say that the observations are fabricated.

Now the question is whether the fabrication is by miscalculation or whether the basic facts have been fabricated. I have not found a firm answer to this question, and I suspect that the observations contain a mixture of the two kinds of fabrication. I shall start the study of this question by showing that two of the observations are impossible.

Observation (2) says that the southern cusp of the moon was covering the southeastern part of the Pleiades. Figure IX.5 shows the configuration of the moon and the Pleiades at the time when the moon had about the same longitude as the eastern part of the Pleiades. The broken line shows the path of the moon as it went by the Pleiades. We see that the moon could not have occulted any of the stars in the eastern Pleiades.‡

Fotheringham [1915] suggests that the observer meant to write "northern and western" rather than "southern and eastern" part of the Pleiades. We see from Figure IX.5 that the moon could have occulted 19 Tauri and possibly 20 Tauri. These are the most northern, but not the most western, of the Pleiades, but Fotheringham's suggestion must nonetheless be considered as physically possible. Earlier, I suggested [Newton, 1970, p. 163] that the recorder wrote "over" the southern and eastern Pleiades, meaning that the moon was north of them, and that Ptolemy mistook this to mean that the moon was over them in the sense that it was hiding them from view.

---

†New moon does not occur, since the moon cannot be seen then. However, the moon was only about 5° from being full on -293 March 9.

‡Several people have studied the records of the occultations and conjunctions in attempts to solve the problems that they pose. Earlier studies include those by Fotheringham [1915], Britton [1967], and Newton [1970]. Britton cites a study by C.F.C. Schjellerup, Sur les conjunctions d'étoiles avec la lune rapportées par Ptolemy, Copernicus, 1, pp. 223-236, 1881, which I have not consulted. So far as I know, no one has approached the records from the viewpoint that they might be fabricated.

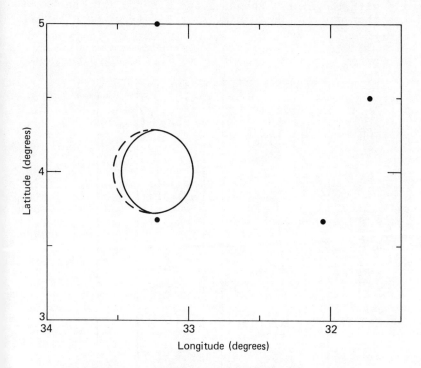

Figure IX.6   The configuration of the moon and of the Pleiades
on the night of 92 November 29 according to Ptolemy's data.
The four stars which he calls the Pleiades cannot be identified.
It is possible that the star to the southeast is the combination of
27 and 28 Tauri and that the star to the southwest is 17 Tauri.
The other stars do not seem to belong to the Pleiades at all, as
the term is used today. Ptolemy says that the southern edge
of the moon occulted the southeastern star.

If we use Ptolemy's data, however, we do not need to
assume any errors in writing or interpretation. The four
points in Figure IX.6 are the four stars that Ptolemy calls
the Pleiades in the star catalogue. Comparison of this
figure with Figure IX.5 shows why attempts to identify Ptolemy's Pleiades have been unsuccessful. If my speculation is
correct that the southeastern star is the combination of 27
and 28 Tauri, the star at the southwestern corner† may be
17 Tauri. The other stars do not seem to belong to the
Pleiades at all, as this term is used today. We see from
the figure that the southern edge of the moon is within a
minute or so of covering the southeastern star. Thus I
believe that Ptolemy fabricated the entire observation, with

―――――――――――
†It is easy to forget that east is to the left in these
pictures of the heavens.

the original intention of making the southern cusp coincide exactly with the star. Rounding and other slight numerical effects then caused him to miss by a minute or so of arc.

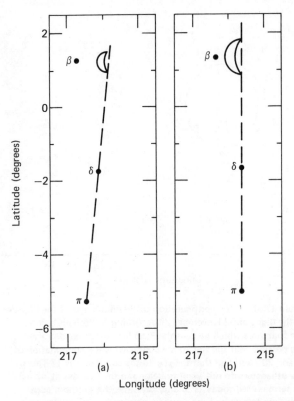

Figure IX.7   The configuration of the moon and of three stars in Scorpius on the night of 98 January 14. Part (a) is drawn using modern data, calculated back to the night in question, and part (b) is drawn using Ptolemy's data. Ptolemy says that the moon occulted β Scorpii at the same time that its southern cusp was in line with δ Scorpii and π Scorpii. This is impossible with either modern data or Ptolemy's data. The figures are drawn to scale.

Observation (7) is also impossible. It says that the southern cusp of the moon made a straight line with δ Scorpi and π Scorpii and that its center was east of this line. So far the observation is not only possible but accurate, as we can see from Figure IX.7. Part (a) of the figure shows the positions of β Scorpii, δ Scorpii, and π Scorpii as calcu-lated from modern data, but for the epoch of the observation When the southern cusp is on the line with δ and π Scorpii, its center is east of the line, as Ptolemy says. The rest

of the record, which says that the moon occulted β Scorpii
at this time, is impossible. The moon lacks a great deal
of being large enough to cover β Scorpii. In fact, it cannot occult β Scorpii if any part of it touches the line from
δ Scorpii and π Scorpii.

The observation is also impossible if we use Ptolemy's
data. Part (b) shows the configuration that results from
his numbers. In his catalogue, δ Scorpii and π Scorpii have
the same longitude. Thus, if the cusp of the moon lies on
the line, the center does also. However, if the cusp lies
on the line, the moon again cannot occult β Scorpii,† although it comes closer to the star than it does in part (a).

TABLE IX.4

PTOLEMY'S CALCULATIONS OF LUNAR COORDINATES

Date	Longitude				Latitude			
	Calculated by Ptolemy		Calculated by me[a]		Calculated by Ptolemy		Calculated by me[a]	
	°	′	°	′	°	′	°	′
-282 Jan 29	30	20	30	12	+3	45	+4	9
92 Nov 29	33	7	32	13	+4	50	+4	56
-293 Mar 9	171	21	171	15	-1	50	-2	1
-282 Nov 9	171	30	171	39	-2	10	-1	57
98 Jan 11	175	45	175	25	-1	20	-1	53
-294 Dec 21	211	15	211	24	+1	20	+1	17
98 Jan 14	215	20	215	19	+2	10	+1	33

[a]These values are calculated by me from Ptolemy's theory.

I suggest the following explanation of this observation:
The observer wrote that the southern cusp was on the line
from δ Scorpii and π Scorpii while its center was east of
the line; I do not see how Ptolemy could have fabricated
this statement, in view of part (b) of Figure IX.7. However,
Ptolemy wanted an observation in which the moon occulted β
Scorpii, so that he could combine this observation with observation (6) to calculate the precession. Therefore he
added the part about occulting β Scorpii and fabricated the
time to give him the position that he wanted. Through carelessness he forgot to reconcile the two parts of the observation. He could easily have done so by having the two
parts of the observation happen at different times; the occultation would have been a short time after the cusp was in

†I have drawn the moon larger in part (b) of the figure than
in part (a). The moon was slightly past the last quarter,
as the drawing shows, and its apparent size in Ptolemy's
theory is much larger than it should be.

line with δ Scorpii and π Scorpii.

Carelessness, perhaps we should even say slovenliness, is the dominant feature of these "observations". Consider observation (4). The record says that the time was 3 1/8 ordinary hours after midnight, in local apparent time. Ptolemy says, however, that the time when Spica rose was $2\frac{1}{2}$ ordinary hours after midnight. Actually, Spica rose at $2^h 47^m$ after midnight, very closely.† At the time that Ptolemy says, the moon and Spica would not be visible because they were below the horizon. However, he had to have $2\frac{1}{2}$ hours for the time in order to get the position of the moon that he needed. Hence he simply stated this time without checking to see whether Spica had yet risen.

We can also see Ptolemy's carelessness from Table IX.4. This table compares the positions of the moon that Ptolemy claims to have calculated with those that I calculate from Ptolemy's theory. The only observation for which Ptolemy's longitude has acceptable accuracy is 98 January 14. Otherwise, the discrepancies range from 6' to 54'. Matters are no better with the latitudes, which should be more accurate. The accuracy of Ptolemy's values is acceptable only for -294 December 21, where the discrepancy is only 3'. The greatest error is 37', not as large as the greatest error in longitude, but the average error is greater.

It is interesting to look at observation (4) again. The longitude that he needs for the moon is 171° 30'. At the time that he uses, the calculated longitude is 171° 39'. If had calculated the longitude carefully, he would have had to put the time even farther before the rise time of Spica than he did.

As to the basic facts of the records, we seem to have a mixture of genuine and fabricated statements. Observation (2) seems to be wholly fabricated. Observation (7) seems to have a genuine part and a fabricated part, with a fabricated time. In observation (4), we have two times. One is certainly fabricated, but the other may be genuine.‡

I do not urge my detailed interpretations of the record. It seems to me that only three firm statements can be made about the conjunctions and occultations: (1) There has been fabrication of the times given in the records. (2) Ptolemy' treatment of the records is careless, and stronger adjective are probably justified. (3) Since there has been fabricatio of the times assigned to the conjunctions and occultations, we cannot use any of the records in research about the motio

---

† I have calculated this time by using modern data and theory I believe that it is correct within a few minutes.

‡ If any part of Ptolemy's account can be trusted, Timocharis originally said that the time was $3\frac{1}{2}$ hours of the night after midnight. If Ptolemy's conversion to ordinary hours is correct (I have not checked it), this is about 20 minutes after the moon and Spica would have risen from a true horizon. Hence this time is plausible.

of the moon. Even though I have speculated that one of the
two times stated in observation (4) may be genuine, I would
not dare to use it because it occurs in such a suspicious
context.

We should notice that we have proved in two different
ways that Ptolemy fabricated the longitude of Spica (α
Virginis).

6. The Star Catalogue in the *Syntaxis*

The star catalogue in the Syntaxis was not the first
extensive star catalogue to be prepared, but it is the oldest that has survived. As such, it is a valuable document,
and it is unfortunate that it occurs in a fraudulent book.
Peters and Knobel [1915] count 1028 stars in it. Since
there are questions about the identifications of some of
the stars, and about whether some entries are duplicates,
other students of the matter have reached slightly different
counts.

Ptolemy says that the longitudes of the stars are for
the epoch 137 July 20. It has long been known that the
longitudes are all too small by about $1°.1$, on the average.
In just the same way, the longitudes of the sun and moon
calculated from Ptolemy's tables are also too small by about
$1°.1$. Now we believe that Hipparchus prepared a star catalogue whose date was about 2 2/3 centuries before Ptolemy's.
We know that Ptolemy, by means of the "observations" that
have been the subject of the three preceding sections, proved
that the precession between Hipparchus's time and his was
2 2/3 degrees. However, the correct precession rate is about
$1°.38$ per century, and the precession between Hipparchus and
Ptolemy is a little more than 3 2/3 degrees.

Thus there are two explanations of the errors in Ptolemy's
stellar longitudes. One explanation is that Ptolemy copied
Hipparchus's catalogue, keeping the latitudes unchanged but
adding 2 2/3 degrees to the longitudes. However, suppose that
Ptolemy did prepare the star catalogue by independent observations, as he claims that he did in Chapter VII.4. He claims
that he determined the positions of at least a basic set of
main reference stars by directly locating the star with respect to the moon, using an unspecified method of comparing
the coordinates of the two. In other words, he effectively
measured the difference in longitude between a star and the
moon. The longitude assigned to the star would then be the
measured difference plus the longitude of the moon as calculated from Ptolemy's lunar table. Since the calculated
longitude of the moon is too small by about $1°.1$, the measured longitudes of the stars would also be too small by
about $1°.1$.

Thus, by either hypothesis, the stellar longitudes would
be systematically too small by about $1°.1$. In order to decide how Ptolemy constructed his star table, we must use information other than the systematic error in longitude.

In Section IX.3 we studied the "observation" that Ptolemy used to establish the longitude of Regulus that is in the star catalogue. We found that it was fabricated. We also found that Ptolemy fabricated the longitude given for Spica (α Virginis). In Section IX.5 we studied the "observations" that were used for the coordinates of the eastern Pleiades, of α Virginis, and of β Scorpii, and we found that they were also fabricated. Thus we know that the coordinates of four stars found in the catalogue were fabricated, and that they were not found by independent observations, as Ptolemy claims. Since they were fabricated, there must have been pre-existing values that Ptolemy duplicated by means of his fraudulent observations. We cannot say rigorously what was the origin of the pre-existing values, † but the most likely source is Hipparchus's now lost catalogue.

We can say with complete confidence that Ptolemy did not find the stellar coordinates by the means that he claims. The proof is simple. The standard deviation of the random error in Ptolemy's theory of the moon is 35'. That is, this is the error, which can be of either sign, which we find in addition to the systematic error of about 1°.1 in longitude. Since the process of locating a star with respect to the moon also introduces random error, the random error in the stellar longitudes is necessarily greater than that in the longitude of the moon. However, by Table IX.1, the standard deviation of the random error in the stellar longitudes is only about 22', much less than the random error in the moon.

Hence Ptolemy did not observe the stars by the method that he claims to have used. Either Ptolemy used a different method, or he did not make the observations for the star catalogue at all.

The method by which Ptolemy "measured" the longitude of Regulus is in fact a different method (Section IX.3). There he aligned the astrolabe by means of the sun, just before sunset, and measured the longitude of the moon. Then, just after sunset, he measured the angle from the moon to Regulus By calculating the motion of the moon during the short wait, which would certainly be accurate enough, he thus found the longitude of Regulus. If he had used this method, however, he would not have a bias of 1°.1 in the stellar longitudes, as we saw in Section VII.2. The astrolabe in effect locates the sun in longitude by locating it in declination. This automatically refers the longitude of the sun, and hence of the stars, to the correct equinox. Since the stars in the catalogue are systematically referred to a wrong equinox, this is also not the method that Ptolemy could have used.

If Ptolemy measured the star coordinates, he did so by a method that is different from either of the two that he describes. There are methods, I am sure, that would yield a systematic error of about 1° while yielding a much smaller random error than the moon, but Ptolemy gives us no basis fo

---

†Except for Spica. Ptolemy tells us the longitude that Hipparchus gave for this star.

assuming that he used such a method. On the contrary, Ptolemy tells us in detail how he measured the coordinates of four stars in the table, and we know that these coordinates were fabricated. Since these are known to be fabricated, and since the method that he describes clearly did not lead to the claimed results, the most plausible conclusion is that Ptolemy fabricated the entire table.

We presumably know that Hipparchus prepared a star catalogue. If Ptolemy added 2° 40′, which he claims is the amount of precession between himself and Hipparchus, to the longitudes in Hipparchus's catalogue, we would have just the bias in longitude that we find. Before we accept the hypothesis that this is what Ptolemy did, we must ask if any facts conflict with this hypothesis or make it unlikely.

Many writers have said that all of the stars in Ptolemy's catalogue were visible at Rhodes, where Hipparchus observed. Since I have not seen a citation to quantitative results on this point, I have done an independent study. So far as I can find, the star that was farthest south in Hipparchus's time is α Crucis,† which had a south declination of 51°.36. This puts it 2°.2 above the horizon at Rhodes, or slightly more if we remember that refraction increases the apparent elevation. The next most southerly star was α Centauri, which was almost exactly 3° above the horizon.

Thus, so far as the stars themselves are concerned, all of the observations could have been made by Hipparchus at Rhodes. Since Alexandria is more than 5° farther south than Rhodes, many stars that are always below the horizon at Rhodes are visible in Alexandria. Since Ptolemy claims [Chapter VII.4 of the Syntaxis] that he tried to measure all of the stars down to the sixth magnitude, it would be surprising if he missed all such stars that were invisible at Rhodes. The fact that he did miss them increases the probability that the catalogue was in fact observed at Rhodes, and that Ptolemy did not make the claimed observations at all.

Most recent writers cite papers by Boll [1901] and Vogt [1925]. They say that Boll proved that Hipparchus's star catalogue contained only about 850 stars while Ptolemy's catalogue has 1028. They further say that Vogt further proved that the coordinates in Hipparchus's catalogue, in all the cases that can be checked, differ from those in Ptolemy's catalogue.‡ Thus Boll shows that about 170 stars in Ptolemy's catalogue necessarily represent observations that are independent of Hipparchus. Vogt's conclusion shows that Hipparchus's coordinates differ from those of Ptolemy

---

†Manitius does not identify this star. Peters and Knobel [1915, p. 118] identify it as α Crucis, as do all other sources whom they cite except Manitius. If we use the coordinates in Ptolemy's catalogue, we also conclude that the star was above the horizon in Rhodes.

‡Throughout the following discussion, I assume that 2° 40′ has been subtracted from Ptolemy's longitudes, so that Hipparchus's and Ptolemy's catalogues can be referred to an accurate equinox and have a common epoch.

for many of the 850 stars that the catalogues have in common.
Hence we have strong if not conclusive evidence that Ptolemy'
catalogue rests upon independent observations, just as Ptolemy claims.

Here, unfortunately, is a flourishing specimen of the
species error immortalis. The studies of Boll and Vogt do
not tell us anything at all about Hipparchus's star catalogue
and its relation to Ptolemy's catalogue. They cannot, becaus
Hipparchus's catalogue is no longer available for study. Instead, the studies of Boll and Vogt actually deal with Hipparchus's Commentary on Aratus and Eudoxus [Hipparchus, ca.
-135] and, to a much lesser extent, with some minor sources
possibly connected with Hipparchus. The conclusions of Boll
and Vogt rest upon the unlikely assumption that the Commentar
and the other sources are rigorously consistent with Hipparchus's lost catalogue.

Hopefully a scholar, as he advances in his career, ac-
quires more understanding of his subject and a greater maste
of the data. There is no reason I can see why different documents by Hipparchus, or by any other scholar, should be completely consistent. In particular, there is little reason
why the Commentary should be consistent with Hipparchus's
star catalogue. In the Commentary, Hipparchus seems to be
ignorant of the precession of the equinoxes. If Ptolemy is
right in putting Hipparchus's catalogue 265 years before his
own time, the catalogue was probably prepared after Hipparch
discovered the precession, as we see from two arguments.
Ptolemy's dating puts Hipparchus's catalogue in -128, within
a few years of his death. As a matter of probabilities, he
is more likely to have discovered precession in the long par
of his career before -128 than in the short part that remain
Further, still if Ptolemy is correct, Hipparchus found his
length of the year in -134. Neither argument is conclusive.
Hipparchus could have prepared the catalogue well before -12
and referred it to that year for some reason we do not know.
Likewise, he could have measured the length of the year in
-134 without realizing its significance for the precession
until some years later.†

I shall not rely upon the negative assertion that the
consistency of the catalogue and the Commentary (or other
sources) has not been proved. Instead, I first shall show,
using Boll's data and methods, that Hipparchus's catalogue
contained about 1000 stars, just about the number that Ptolemy has. Since we can show from the same data and reasoning
that the catalogue had both 850 and 1000 stars, we can obviously show nothing by this method. Then I shall show,

†In the Commentary, Hipparchus does not use latitude and
longitude as the coordinates. He uses declinations and
either right ascensions or the hypotenuse of the triangle
whose legs are the longitude and the distance from the
equator to the ecliptic; see page 185 of Neugebauer [1957].
These are poorly adapted to the phenomenon of precession.
If we could be sure that Hipparchus used latitude and longi
tude in his catalogue, this would be strong evidence that
he had discovered precession before he prepared the catalogue.

using Vogt's data and methods, that Ptolemy did not prepare the catalogue contained in the Syntaxis. Since we have shown that neither Hipparchus nor Ptolemy prepared the catalogue in the Syntaxis, where are we? Obviously, nowhere. We have shown nothing because all the arguments described rest upon an improbable and unverifiable hypothesis.† As we know from mathematical logic, a false proposition implies all propositions.

Boll uses several manuscripts that have been published within the past century which list the names of 42 constellations and the number of stars in each.‡ Some of the manuscripts are in Greek while the others are translations into what Boll calls barbarous Latin. From the fact that some of the manuscripts refer to Hipparchus in connection with the list, Boll concludes that the list is derived from Hipparchus's catalogue. If I have understood Boll correctly, Hipparchus's name appears only in the translations and in the later Greek manuscripts; it does not occur in the earlier Greek manuscripts. Hence the connection with Hipparchus seems tenuous. Instead of being authentic, it may have been the mistake of some medieval scribe.

In the 42 constellations for which the number of stars is usable, the total number of stars is 640. In the same constellations, there are 772 stars in the catalogue in the Syntaxis. Altogether there are 1028 stars in the Syntaxis. If the same ratio applies to the unlisted constellations as to the listed ones, there were (640/772) × 1028 = 852 stars in Hipparchus's catalogue.

Aside from the doubtful identification of the list with Hipparchus's catalogue, there are other difficulties. Perhaps the most serious one is that Hipparchus's Commentary [Hipparchus, ca. -135] gives the number of stars in six constellations that also appear in the list. For these there are 52 stars according to the list, but the Commentary gives 61. For each constellation, there are fewer stars according to the list than according to the Commentary. If we apply this ratio to all the constellations in the list, we conclude that Hipparchus assigned (61/52) × 640 = 751 stars to these constellations, almost the same as the number in the Syntaxis. Applying the ratio to the entire catalogue, Hipparchus would have had about 1000 stars.

These numbers apply to the Commentary rather than to the catalogue of Hipparchus. However, the Commentary is almost surely earlier than the catalogue and it is unlikely that Hipparchus would have stars in it that he did not have in his catalogue. Hence we conclude that the catalogue of Hippar-

---

†Fragments of lists of stellar coordinates have been found within the past century. Not enough has been preserved to let us identify the source or sources. However, since new discoveries continue to be made, it is possible that Hipparchus's catalogue may be found some day. See Neugebauer [1957], particularly page 68.

‡More accurately, 42 is the number of usable entries. There are fragmentary references to other constellations.

chus had as many stars as the catalogue in the Syntaxis, so far as we can tell. The list in the manuscripts used by Boll, then, must either be in error or it must refer to some catalogue other than Hipparchus's.

I do not particularly defend the reasoning by which I showed that Hipparchus's catalogue contained 1000 stars, more or less. However, I believe that it has as much justification as the reasoning used by Boll. The main conclusion is that the data are too fragmentary to let us reach any firm conclusion.

Vogt [1925] examined the data in the Commentary thorough In it he found 881 items of numerical data about 374 stars. these he adds 22 items of data attributed to Hipparchus that are not in the Commentary but that come to us through Ptolemy or Strabo.† Some of the items are direct coordinates, includ ing 64 declinations and 67 right ascensions. The only longi- tudes come through Ptolemy. Many of the items are the hypot- enuses in the triangle mentioned in a footnote above. The rest concern stars that rise just as another is setting, or similar matters.

For 122 of the 374 stars mentioned, the information is sufficient to allow calculation of the stellar position. For these, Vogt calculates the latitudes and longitudes and com- pares the results with the catalogue given in the Syntaxis. Let us look at the latitudes, for example. In 47 instances out of the 122, the discrepancy in latitude is greater than $0°.5$; in 11 instances, it is between $0°.4$ and $0°.5$, and so on In only 20 instances is the discrepancy less than $0°.1$. From this, Vogt concludes that Hipparchus did not make the measure ments used in the Syntaxis catalogue.

This conclusion obviously does not follow from the dis- crepancies. The discrepancies merely show that Hipparchus did not calculate the declinations, or other quantities in question, from the data in the catalogue. For the declina- tions, at least, there is no reason that he should have done so, even if the catalogue had existed at the time. The dec- lination of a star is simply measured by means of a meridian circle. If Hipparchus wanted a declination, it was probably as easy for him to measure it as to calculate it from a table of latitudes and longitudes. He may even have realized that measured declinations were more accurate than latitudes and longitudes.‡

Vogt's work suggests that we apply the same method to the stellar data in the Syntaxis. Ptolemy gives the declina- tions of 18 stars which he claims to have measured himself.

---

†We have seen most of the items that come through Ptolemy. They include the declinations of 18 stars that were dis- cussed in Section IX.4, and the longitude of Regulus that was discussed in Section IX.3. This leaves little informa- tion that comes through Strabo. For this, Vogt cites Chap- ter 5.41 of Strabo's Geography, which I have not consulted.

‡See Table IX.1.

TABLE IX.5

DECLINATIONS OF 18 STARS, AS STATED BY
PTOLEMY AND AS CALCULATED FROM HIS STAR CATALOGUE

Star	Longitude[a]	Latitude[a]	Declination Calculated[b]	Stated
η Tau[c]	33.67	3.67	16.39	16.25
α Aur	55.00	22.50	41.00	41.17
γ Ori	54.00	-17.50	2.12	2.50
α Vir	176.67	- 2.00	- 0.48	- 0.50
η UMa	149.83	54.00	59.24	59.67
α Boö	177.00	31.50	29.73	29.83
α Aql	273.83	29.17	5.36	5.83
α Tau	42.67	- 5.17	10.99	11.00
α Ori	62.00	-17.00	4.25	5.25
α CMa	77.67	-39.17	-15.74	-15.75
α Gem	83.33	9.50	33.17	33.40
β Gem	86.67	6.25	30.06	30.17
α Leo	122.50	0.17	20.11	19.83
ζ UMa	138.00	55.67	65.21	65.00
ε UMa	132.17	53.50	65.99	66.25
α Lib	198.00	0.67	- 6.56	- 7.17
β Lib	202.17	8.83	- 0.59	- 1.00
α Sco	222.67	- 4.00	-19.71	-20.25

[a]From the catalogue in the Syntaxis.
[b]Calculated from the longitude and latitude in the
Syntaxis.
[c]This identification is questionable.

For these 18 stars, I have calculated the declinations from
the latitudes and longitudes in the star catalogue, using
23° 51′ 20″ (the value that Ptolemy claims to have observed)
for the obliquity of the ecliptic. The calculated values
are compared with the values that Ptolemy claims to have
measured in Table IX.5.

The first six stars in Table IX.5 are the ones that
Ptolemy uses in finding the precession (Section IX.4),
listed in the order that Ptolemy uses them. The other
twelve are the ones that he does not use, listed in the
order that he lists them. The reader should remember that
Ptolemy mixes the ones he uses with the ones he does not use.
The identification of the first star is questionable. In

the description of the measurement, Ptolemy calls this "the middle of the Pleiades". Figure IX.5 shows that a star in the middle of the Pleiades is likely to be η Tauri, but Ptolemy does not list "the middle of the Pleiades" in his catalogue. In Table IX.5, I have used the latitude and longitude of the eastern Pleiades, even though this may not be correct.

The discrepancies between the calculated declinations and those which Ptolemy claims to have measured range from 0°.01 to 1°.00. If we accept Vogt's conclusion that Hipparchus did not measure the positions in the star catalogue, we must also conclude that Ptolemy did not measure them.

If Ptolemy actually measured the twelve declinations that he did not use, and if Hipparchus actually constructed the star catalogue, we do not expect the last twelve entries in Table IX.5 to agree. It is interesting that the stars which Ptolemy did use do not agree either. If they did, we would have to reject, as being fabricated, the declinations that Ptolemy attributes to Hipparchus. That is, in fabricating the results that give him the precession, Ptolemy could have started by calculating the declinations in his own time from the star catalogue. He could then have subtracted the necessary change in declination, thus fabricating the declinations that he attributes to Hipparchus.

The fact that the first six declinations in Table IX.5 do not agree indicates that Ptolemy did not do this. Apparently he started with genuine measurements by Hipparchus. To these, he added the changes in declination needed to give a precession of 1° per century, thus fabricating the measurements that he claims to have made.

Of course, the same remark applies to Table IX.5 as to the corresponding results of Vogt. Table IX.5 does not prove that Ptolemy did not measure the stellar coordinates. It only proves that he did not calculate the declinations from the latitudes and longitudes in the star catalogue.

It is now useful to summarize this section, which has established the following main points:

(1) The longitudes in the catalogue have a bias of about 1°.1, which could arise in either of two ways. It would happen if Ptolemy fabricated the table from Hipparchus' table, using a wrong rate of precession. It would also happen if Ptolemy measured the longitudes in the way that he claims to have done.

(2) Ptolemy describes in detail how he found the coordinates of four stars. All the observations involved, whether he claims to have made them himself or whether he attributes them to others, are fabricated. Thus four entries in the table are fabricated.

(3) Ptolemy did not measure the stellar coordinates in the way that he claims to have done, nor by any other way that he describes.

(4) The catalogue contains no stars that were visible
at Alexandria but invisible at Rhodes. It should contain
many such stars if Ptolemy based it upon independent observations as he claims to have done.

(5) There are no facts known which contradict the
hypothesis that Ptolemy fabricated his catalogue from Hipparchus's. Two published counterclaims prove not to deal
with Hipparchus's catalogue at all. They prove only that
Ptolemy did not fabricate his catalogue from certain sources
which are different from Hipparchus's catalogue.

7. Fractions of Degrees in Ptolemy's Star Catalogue

The earlier arguments about the stellar coordinates
have involved the coordinates themselves and their comparison with values that we calculate from either modern or ancient data. Another powerful class of argument involves
only the fractional parts of a degree and their statistical
distribution. Several recent writers say that this distribution shows that two different instruments were used in
measuring the coordinates that appear in Ptolemy's catalogue.
Two instruments then suggest two observers. This school of
writers concludes that Hipparchus measured some of the coordinates and that Ptolemy independently measured the others.

As we shall see, the distribution of fractions actually
proves a quite different conclusion. The distribution proves
clearly that we are dealing with a homogeneous body of coordinates. It also proves, with odds of 1 billion to 1, that
the entire table was fabricated, and it shows us how to reconstruct the fabrication.

TABLE IX.6

FRACTIONS OF A DEGREE THAT OCCUR
IN THE STAR CATALOGUE

Fraction, minutes	Number of cases with this fraction		
	In the longitudes	In the latitudes	Theoretical
0	226	236	171
10	182	106	128
15	4	88	86
20	179	112	128
30	88	198	171
40	246	129	128
45	0	50	86
50	102	107	128
Totals	1027	1026	1026

The fractions that appear in the star catalogue are multiples of either 1/6 of a degree or 1/4 of a degree. In the original Greek, no fraction at all appeared when the coordinate was an integer. This constrasts with modern practice; we specifically insert 0 in some form when the fraction is zero, thus showing that it has not been omitted by accident. In the original Greek, the multiples of 1/6 were written in the forms 1/6, 1/3, 1/2, 2/3, and 1/2 + 1/3.†️ The multiples of 1/4 were written in the forms 1/4, 1/2, and 1/2 + 1/4. We cannot discriminate the cases when 1/2 is a multiple of 1/6 and those when it is a multiple of 1/4. In discussing the fractions, it will be easier to use their equivalents in minutes of arc.

Table IX.6 shows the distribution of the fractions that occur in the star catalogue, both in the longitudes and the latitudes. I prepared the table by counting the cases in Peters and Knobel [1915]. In carrying out this rather tedious task, it is not possible to guarantee that there has been no error. However, the distribution in Table IX.6 agrees closely with the distribution given by Vogt [1925]. Hence there is no significant error in Table IX.6. I omitted one longitude and two latitudes where Peters and Knobel concluded that the original reading was questionable. I shall discuss the meaning of the column labelled "Theoretical" in a moment.

First let us look at the latitudes. The first property of their distribution to be observed is its symmetry. There are 106 10's,‡ 88 15's, and 112 20's, for a total of 306 cases. There are 286 values of 40's, 45's, and 50's. The difference between 286 and 306 is not statistically significant.‡

What most writers have concentrated on is the fact that multiples of 10′ (1/6 of a degree) and 15′ (1/4 of a degree) both occur in the latitudes. This proves, they say, the use of different graduated circles, one graduated by intervals of 10′ and one graduated by intervals of 15′. This does not follow at all. Suppose that a circle is graduated at intervals of 30′, but that the user does not intend to estimate any fraction less than 10′. Suppose further that a star appeared between some degree mark and the following 30′ mark. If the star appeared quite close to either graduation mark, the observer would write the fraction (0′ or 30′) appropriate to that mark. If a star was appreciably closer to one mark, but not quite close, he would write either 10′ or 20′, as

---

†️With few exceptions, only fractions with a unit numerator were used in Greek arithmetic; see Section II.1. The fraction 2/3 was one exception; it had a special symbol.

‡For simplicity in discourse, I shall omit the sign for minutes of arc when I am stating an element of this distribution. When I write 106 10's, for example, I mean that there are 106 cases in which the fraction is 10′.

‡If the total number of cases is N, the standard deviation of the difference is $\sqrt{N}$. Now $\sqrt{N} = \sqrt{286 + 306} = 24.4$, which is greater than the dissymmetry. Hence the lack of symmetry is not significant.

the case may be. Finally, if the star is close to the center, he would write 15'. Similar remarks apply if the star appeared between a 30' mark and the following degree mark.

To be quantitative, suppose that the observer's eye was capable of assessing the position accurately and that the observer rigorously followed this procedure. If the correct number of minutes was between 55 and 5, he would assign the value 0' (that is, he would not write any fraction). This would happen in 1/6 of the cases. If the correct number of minutes was between 5 and $12\frac{1}{2}$, he would assign 10'; this would happen in 1/8 of the cases. If the correct number of minutes was between $12\frac{1}{2}$ and $17\frac{1}{2}$, he would assign 15'; this would happen in 1/12 of the cases, and so on.

The column labelled "Theoretical" in Table IX.6 shows the number of times each fraction should occur on this basis. In calculating this column, I have multiplied the fractions just found by the number 1028 of stars in the catalogue, and I have then rounded to the nearest integer. We should first compare the "Theoretical" column with the distribution of the latitude fractions.

We note first that there are 38 more 0's than 30's, while the standard deviation of the difference for this number (236 + 198) of observations is 20.8. The probability that this would happen by chance is about 0.07, so that the difference is statistically significant but not overwhelmingly so. The difference indicates that the circle was probably graduated only in degrees, rather than in intervals of 30'. If it had been graduated in intervals of 30', there should be as many 30's as 0's. The theoretical division described above is the same whether the circle was graduated by degrees or by intervals of 30'. What seems to have happened, with division by degrees, is that the eye was attracted, so to speak, to the degree mark and assigned it more space than it assigned to the middle of the interval where there was no mark.

Similarly, there are more of both 0's and 30's than the strict theoretical division allows. This is not surprising. It means, as we just said, that the eye allotted more space to the region called 0', probably because of the presence of the graduation mark. It allotted the next largest amount of space to 30', probably because of the attraction of symmetry. The assignment of extra cases to 0 and 30 causes a depletion of the neighboring values, and it naturally depleted 10 and 50, which are next to 0, more than it depleted 20 and 40, which are next to 30.†

---

†The basis of this conclusion may be affected by copying errors. If a copier failed to copy a fraction, he would automatically alter it to a 0, since the 0 was not written in. This is one of the main reasons for the modern practice of explicitly writing 0 when it is meant. Error by omission tends to increase the number of 0's, so the true number of 0's may be less than 236 in the latitudes. However, the conclusion is supported by the fact that the 10's and 50's are depleted more than the 20's and 40's. If copying error were the explanation, all other values should be depleted almost equally.

After we allow for the extra attraction of the values 0 and 30, the distribution of the latitudes agrees well with the "Theoretical" column, with one exception. There are considerably fewer 45's than there should be, although the number of 15's is almost exactly right. This fact poses a problem whether we accept the contention that there were two graduated circles or not. Either way, there should be as many 45's as 15's.

The only explanation that I have thought of rests upon copying errors. The fraction that I have represented by 45' was actually written as $\frac{1}{2} + \frac{1}{4}$.[†] Accidental omission of either fraction would cause a loss of a 45 case, enriching either the 30's or the 15's. However, a copying error is unlikely to lead to a 45 case. This explanation is supported by the fact that the average error in the latitudes is decreased if we assume that some of the 15's actually represent 45's; I have tested this matter by calculation.[‡] However, it is contradicted by the 50's. The fraction that I am writing as 50' was written as 1/2 + 1/3 in the original, and 50 should be as badly depleted by copying error as 45. The table shows that this is not the case.

To summarize the latitude distribution, the distribution does not indicate the use of two or more instruments, or the participation of two or more observers. The distribution shows strongly that we are dealing with a homogeneous body of data, obtained with a circle graduated in degrees only, with fractions being estimated by eye. The number of 45's has been depleted by some process that we cannot explain satisfactorily. It cannot be explained by the assumption of multiple instruments or multiple observers.

The longitudes tell a quite different story from the latitudes. The distribution of the longitude fractions does not come from any possible body of observations, whether they were made by using one instrument or more, and whether they were made by one observer or more.

We note first that there are no 45's among the longitude and only 4 15's. This number is so small that we can dismiss it as the result of copying error, and I shall assume in the rest of the discussion that there are no 15's or 45's among the longitude fractions. This fact seems inexplicable if the longitudes came from measurement. Why would an observer or observers who recorded a nearly normal number of 15's and 45' in the latitudes not record any in the longitudes? The idea

---

[†]Except that it was written using the Greek numerals for $\frac{1}{2}$ and $\frac{1}{4}$, of course.

[‡]One cannot test this meaningfully by simply altering some 15's to 45's; it would always be possible to select some cases in which a 45 gave a smaller error than a 15. I have verified that the error is decreased if we replace 15 by any number from 16 to about 20. That is, the correct average size of the numbers where 15 appears is actually somewhat larger than 15. This is what we expect if some of the 15's were originally 45's.

that the latitude circles were graduated differently from
the longitude circles is possible a priori, but it is implausible. As we shall see, it also disagrees quantitatively
with the distribution of the fractions.

Now let us average the number of 10's, 30's, and 50's.
This average is 124. Then let us average the number of 0's,
20's, and 40's; this average is 217. The disparity of these
averages led Vogt [1925, column 41] to write: "I can explain
this unequal relation ... only by the assumption that Ptolemy's
astrolabe was graduated not by sixths but by thirds, ..."
Vogt failed to observe that the distribution does not support
this assumption. Instead, the distribution contradicts it
overwhelmingly.

If Vogt's assumption were correct, we would have equal
numbers of 0's, 20's, and 40's. Instead, we have many more
40's than either 0's or 20's. In fact, we have more 40's
than any other fraction. The difference between the number
of 40's and 20's is 67, but the standard deviation of the
difference is $\sqrt{179 + 246}$ = 20.6. If the longitudes come from
observation, with a circle graduated in any way whatsoever,
the probability that this difference could happen by chance
is about 1 in 1000. There are also many more 10's than 50's;
in fact, there are almost as many 10's as 30's and 50's put
together, whereas the number of 10's, 30's, and 50's should
be the same by Vogt's assumption. The difference between the
10's and the 50's is 80, while the standard deviation of the
difference is $\sqrt{182 + 102}$ = 16.8. The probability that this
could have happened by chance with any kind of graduations is
about 1 in 1 000 000. The probability that both maldistribu-
tions could happen in the same sample is about 1 in 1 billion.

If the reader will let me neglect this tiny probability,
we may say conclusively that the distribution of longitudes
cannot be the result of observation. Hence the longitudes
must be fabricated, and we can reconstruct most of the fabrication from the distribution. It must have been done by
adding an angle of the form $N° 40'$ to a set of observed
values; N is some integer. If this were done, values that
were originally 0's go into 40's, and this explains the abnormally large number of 40's. Values that were originally
50's go into 30's, and this explains the abnormally small
number of 30's. In fact, we can explain the entire distribution quantitatively, including the absence of 15's and 45's,
by the hypothesis that the longitudes were fabricated in the
manner described.

We start from the assumption that the distribution of
latitude fractions in Table IX.6 gives the way that the observer or observers naturally divided the interval between
degree marks. Thus the distribution of fractions in the
latitudes and longitudes should be nearly the same. This is
an assumption with a high degree of plausibility behind it.
Of course the original set of longitudes would not have
exactly the same distribution as the latitudes. Deviations
of around 10 or 20 for any specific fraction are to be
expected.

TABLE IX.7

THE ORIGIN OF THE DISTRIBUTION OF THE
FRACTIONS IN THE LONGITUDES

Fraction, minutes	Number of cases with this fraction			
	In original sample[a]	With 40 added, neglecting 15 and 45	With 40 added, after reassignment of 15 and 45	In the star catalogue
0	236	112	200	226
10	106	198	198	182
15	88	——	0	4
20	112	129	179	179
30	198	107	107	88
40	129	236	236	246
45	50	——	0	0
50	107	106	106	102

[a]This is actually the distribution of the fractions in the recorded latitude values. Since the observer would tend to interpolate in the same way in both the latitude and longitude circles, the original distribution of fractions should be the same in latitude and longitude, within ordinary statistical fluctuations.

Table IX.7 now shows what happened. In the table, the last column gives the distribution of longitudes taken from Table IX.6. The column labelled "In original sample" gives the distribution of latitudes from Table IX.6, and we assume that this is also the original distribution of the longitudes Now let us add 40 to each original fraction,[†] neglecting the 15's and 45's for the moment. This gives us the distribution in the column labelled "With 40 added, neglecting 15 and 45". This distribution already agrees satisfactorily with the last column except for the 0's and the 20's, when we remember that there is some statistical uncertainty in the column labelled "In original sample".

Now let us consider what would happen to the 15's and 45's in the original values. The fraction 15′ would become 55′ and the fraction 45′ would become 25′. However, the rules of the catalogue do not allow the appearance of 25's and 55's so Ptolemy had to eliminate them by changing them into neighboring values that are allowed. Thus an original 15 became either a 50 or a 0 while an original 45 became either a 20 or a 30.

[†]When we add 40 to 40, for example, we get 80. This would become 20, with 1 added to the whole number of degrees.

We could try to speculate about whether he would choose to change a 15 (which became 55 after addition of 40) into a 50 or a 0, and whether he would change a 45 into a 20 or a 30. Speculation of this sort would be inconclusive and, luckily, it is not necessary. We can see by direct inspection which choices he made. We see that the number of 0's is much too small when we neglect the 15's while the number

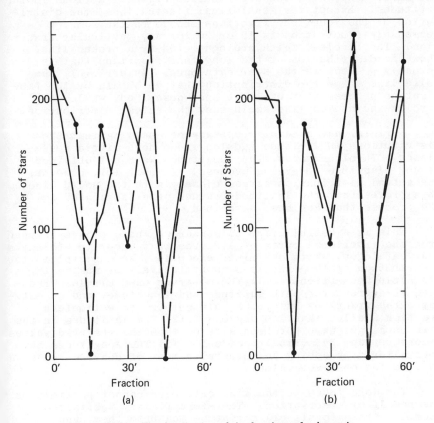

Figure IX.8   The distribution of the fractions of a degree in Ptolemy's star catalogue. In part (a), the solid line shows the distribution of the fractions in the latitude coordinates. This should also be the distribution of the fractions in the longitudes, except for normal statistical fluctuations. The circles, which are connected by a broken line, show the actual distribution in the longitudes. In part (b), the solid line shows the distribution of the longitude fractions if Ptolemy fabricated the stellar coordinates by the process discribed in the text. The circles and the broken line again show the actual distribution. There is no agreement of the two lines in part (a), but there is almost exact agreement in part (b). Hence Ptolemy fabricated the coordinates.

of 50's is right. Thus he added the original 15's to the 0's. Similarly, he added the original 45's to the 20's. When we make these additions, we obtain the distribution in the column labelled "With 40 added, after reassignment of 15 and 45". The agreement of this column with the distribution in the star catalogue is highly satisfying.

The reader can see this agreement more easily from Figure IX.8 than from Table IX.7. In part (a) of the figure, the solid line shows the distribution of the fractions in the latitudes. Except for statistical fluctuations, the distribution of the longitude fractions should be the same; a reasonable fluctuation is 15 or 20 for any particular fraction. The circles, which are connected by a broken line, show the distribution of the longitude fractions that is actually present in the star catalogue. In part (b), the solid line shows the distribution that we obtain by the fabrication process that has just been described, while the circles and broken line again show the actual distribution.

Part (a) has a chaotic appearance, showing that the distributions in latitude and longitude have little resemblance. In other words, the longitudes were not obtained by any plausible observing process. In part (b), however, the actual distribution agrees closely with the solid line. This proves that the longitude values were fabricated by the process that has been described.

This is a useful place to summarize what we have learned from the fractional parts of a degree that occur in Ptolemy's star catalogue. With one minor exception, we can explain the distribution of the fractions quantitatively by making three plausible assumptions: (1) The observer used circles graduated in degrees for both latitude and longitude, and he estimated the fractions by eye; he did not try to estimate a fraction smaller than 1/6 degrees. (2) The latitudes in the star catalogue have not been altered from the observed values except perhaps by scribal accident. (3) The longitudes have been altered by adding an integral number of degrees, plus 40', to the observed values.

The longitudes in the star catalogue cannot possibly be the result of observation. Therefore Ptolemy fabricated them, contrary to his claim that he measured them, and he must have fabricated them in the manner just described.

On the basis of the fractions alone, we cannot rigorously eliminate the possibility that the fabrication was done by the guilty associate rather than by Ptolemy. It is conceivable that Ptolemy told his hypothetical associate to make the measurements in the way that Ptolemy claims they were made. It is then conceivable that the associate deceived Ptolemy by fabricating the measurements instead of making them. However, the measurement of the stellar coordinates would be a long job and Ptolemy would have known this. During the long time when Ptolemy supposed his associate to be busy at the task, it is hardly imaginable that nothing aroused Ptolemy's suspicions. For example, we might expect him to show up at the observatory by accident some

night, perhaps to show it to a friend, and thus to find
that his associate was not there busily measuring stars
as he was supposed to be.†

The distribution of the fractions by itself cannot
tell us the integer part of the amount that Ptolemy added
to the original longitudes, but other circumstances can
tell us once we know that the fractional part was 40'. The
original catalogue had to be prepared about -130, so that
the amount added was 2° 40'. If its epoch differed appreciably from this, the bias in the longitudes would not be
what it is. Suppose, for example, that Ptolemy added 1°40'.
Since Ptolemy claims that the precession is 1° per century,
the epoch of the original catalogue would be -30 rather
than -130. Between -30 and +137, the epoch of Ptolemy's
alleged catalogue, the correct precession is close to 2° 20',
so that the bias would be only 40'. However, the bias is
slightly greater than 1°.

It is also possible that there was a catalogue other
than Hipparchus's with an epoch near -130, but there is no
suggestion that such a catalogue existed, so far as I know.‡
Further, the epoch would have to be less than 15 years from
the epoch of Hipparchus's catalogue. Suppose, for example,
that it was 15 years later, in -113. This is $2\frac{1}{2}$ centuries
before Ptolemy's epoch, so that Ptolemy would have added
2° 30', not 2° 40', to the original longitudes. However, we
know that the fractional part of his addition was 40'.

The reader may ask how firmly we know that the fractional part was 40' and not some other value, f say, where f is
a multiple of 10'.‡ The answer is that we can be quite sure.
The most conspicuous indicator of the added fraction comes
from the maximum in the final distribution. Since it is
virtually certain that 0 dominated in the original distribution, it is virtually certain that f must be the value that
dominates after the addition; hence f is 40. There are other
strong indicators as well, such as the fact that there are
fewer 30's than any other value. The possible fates of the
values that originally ended in 15 and 45 give interesting
indicators. For example, if Ptolemy had added an amount
with 30' as the fractional part, he would not have needed to
suppress the 15's and 45's as he did. Even if we assume that

---

†What we know of astronomy before modern times suggests, in
fact, that the place where an astronomer observed and the
place where he lived were frequently the same. If so, there
seems to be no way in which an associate could fabricate
data without the knowledge of his principal.

‡Neugebauer [1957, pp. 68-69] describes a fragment of a catalogue which gives only integral degrees for the coordinates
of 41 stars; the catalogue was discovered about 1936 by
W. Gundel, in a paper I have not consulted. From a study of
the longitudes, Neugebauer finds that the epoch of the
catalogue was between -130 and -60. Hence the fragment may
well have been prepared from Hipparchus's catalogue.

‡The added fraction f must be a multiple of 10', since only
multiples of 10' occur in the longitudes.

the original had no 15's and 45's for some reason, we still
know that 30 was not the value of f. If it had been, the
distribution in the catalogue would have a maximum at 30,
whereas it actually has a pronounced minimum there.

Ptolemy would surely be startled if he could know how
much we can learn about his fabrication simply from studying
the fractions in the star catalogue.

8. A Summary of Results About the Stars

It is useful to end this chapter with a summary of what
we have learned from a study of Books VII and VIII of the
Syntaxis, and to a lesser extent from Hipparchus [ca. -135].

(1) Ptolemy probably did not independently verify that
the stars are fixed with respect to each other, as he claims
that he did.

(2) Measured latitudes and longitudes in ancient Greek
sources have standard deviations of about 20'. Measured declinations have standard deviations of about 10'. There was
little if any improvement in accuracy after -300, at least
with respect to declinations. We cannot study latitudes and
longitudes before Hipparchus from lack of data; in fact, they
were probably not used in earlier times, before the discovery
of the precession. Among other things, this shows that declinations were measured with different instruments from those
used in measuring latitudes and longitudes. Tentatively,
declinations were measured with a meridian instrument while
latitudes and longitudes were measured with an astrolabe.

(3) The observation by which Ptolemy claims to have
measured the longitude of Regulus is fabricated. Some writers
have claimed that the error in the longitude can be explained
by experimental sources of error such as refraction. Actuall
the value that Ptolemy obtains cannot be explained by experimental sources, no matter what their size. The crucial point
is the exact agreement with preassigned values, and exact
agreement, occurring time after time, cannot be the consequence of errors in measurement.

,   (4) The observation by which Ptolemy claims to have
measured the longitude of Spica is also fabricated.

(5) Ptolemy uses declinations measured by Hipparchus
and allegedly by himself to prove that the precession is 1°
per century. The measurements that he claims to have made
and that he actually uses are fabricated. Luckily, the other
including the ones that he claims to have made but does not
use, prove to be genuine. These are the only observations of
any sort that he claims to have made which are probably genuine.† The genuine measurements of declination allow us to
calculate the rate of precession and the obliquity of the
ecliptic in ancient Greek times. The results agree accuratel
with modern theory.

---

†Some of the alleged observations of stellar configurations,
which he does not actually use, may also be genuine.

(6) Ptolemy uses seven conjunctions or occultations of
stars, all allegedly made before his own time, to prove that
the precession is 1° per century. These are fabricated, even
though they are observations which Ptolemy attributes to
other astronomers. Some of the fabrication may be by miscalculation only, but some of it is almost surely a falsification of the basic facts. Ptolemy's treatment of these conjunctions or occultations is extremely careless, even after
we allow for the possibility of fabrication by miscalculation.
I believe that we may say that there is in fact a lack of
competence.

(7) The coordinates in the star catalogue were not measured by the method which Ptolemy claims that he used. The
reason we know this is that the claimed method would have led
to much larger random errors than those which we actually
find.

(8) The coordinates in the star catalogue were not measured by the only other method which Ptolemy describes, namely
the method which he claims to have used with Regulus (item 3
above). We know this because the method would have led to
longitudes that are free of bias. Instead, the longitudes
have a bias of about 1°.1.

(9) The latitudes in the star catalogue were obtained
by measurement, almost surely made by a single observer using
a single instrument which was graduated in degrees but not in
fractions of a degree. The fact that the fractions contain
multiples of both 1/6 and 1/4 of a degree does not indicate
the use of two graduated circles, as some have concluded.
It results naturally from the use of a circle, graduated in
degrees only, by an observer who was accustomed to a non-
decimal number system.

(10) The distribution of the fractional degrees in the
longitudes, when combined with the bias in the longitudes,
proves conclusively that the longitudes were obtained by
adding 2° 40' to a set of longitudes that had been measured
2 2/3 centuries before the epoch that Ptolemy claims for his
catalogue. The original measurements were almost surely made
at a point several degrees north of Alexandria.†

(11) The epoch of the original catalogue which Ptolemy
used in fabricating his own catalogue was less than 15 years
from -128. This date is the one that Ptolemy assigns to the
catalogue which Hipparchus constructed, and it is consistent
with everything that we know about the work of Hipparchus.
Hence the original catalogue was almost surely that of Hipparchus. It is unlikely that some one else constructed an
independent catalogue so close in time to Hipparchus's.

(12) If we subtract 2° 40' from the longitudes given
by Ptolemy, leaving the other information unchanged, we

---

†I do not mean for this to imply that the point was necessarily close to the longitude of Alexandria, although it
probably was.

probably reconstruct a genuine catalogue prepared within a few years of -128, with two exceptions: (a) Ptolemy says that he has rearranged some of the assignments of stars to constellations from those used by earlier workers, and (b) we have irretrievably lost the longitudes that originally had fractions of 1/4 or 3/4 of a degree. Since these have been altered by only 5', however, this is not a serious matter.

The catalogue reconstructed in this way is the oldest extensive catalogue that is known, and it is thus a document of great interest. It tells us how the heavens appeared in ancient times. A careful study of the descriptions and magnitudes of the stars should be rewarding, although it would probably tell us more about people than about the stars themselves. Pannekoek [1961, p. 157] mentions an interesting point about Sirius (α Canis Majoris), for example. The Syntaxis describes it as reddish, whereas we think of it as bluish-white. Pannekoek suggests the following explanation: Ancient peoples in eastern Mediterranean regions were often interested in the first time Sirius becomes visible after the sun passes it. The first visibility occurs in the eastern sky as the star is rising. In the time of Hipparchus, this happened on a date close to July 18.† Interest in its first visibility meant that people tended to look at it when it was close to the horizon, when it would tend to be reddened by the long passage of its light through the atmosphere.

---

† Sirius is the brightest star in Canis Major (the Greater Dog). I have seen it stated that the "dog days" are called this because they begin when Sirius, the Dog star, first becomes visible. I do not know whether the statement is authentic.

CHAPTER X

THE MOTION OF MERCURY

1. Ptolemy's Model for the Orbit of Mercury

For Venus and the outer planets, Ptolemy uses the equant model that was described in Section IV.5. For the reason that was described there, this model cannot be expected to work well for Mercury. Figure X.1 shows the situation in more detail. This is the same as Figure XIII.9 in <u>APO</u>.

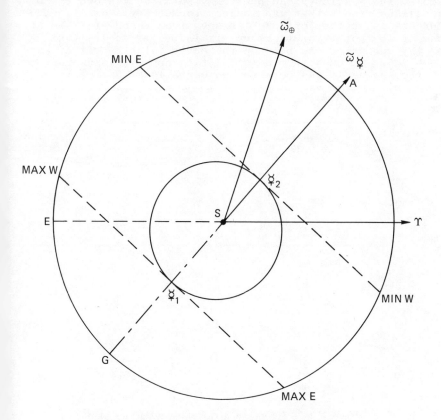

Figure X.1   The heliocentric orbits of Mercury and the earth. The orbits are drawn approximately to scale, using the orbital parameters that were correct in Ptolemy's time. The line from S to ♈ points to the vernal equinox, the line to $\tilde{\omega}_{\mathary{\yen}}$ points to the perihelion of Mercury, and the line to $\tilde{\omega}_{\oplus}$ points to the perihelion of earth. Point $\yen_1$ is the aphelion of Mercury and point $\yen_2$ is its perihelion. The broken lines are tangent to Mercury's orbit at its aphelion and perihelion. The reasons for the designations MAX E, MIN W, and so on, are explained in the text. This is the same as Figure XIII.9 of <u>APO</u>.

The figure shows the heliocentric orbits of Mercury and the earth drawn approximately to scale. The parameters used in drawing it are those that were correct in Ptolemy's time. The direction from the sun S to $\tilde{\omega}_\oplus$ points to the perihelion of the earth's orbit and the direction to $\tilde{\omega}_\mathrm{Ʉ}$ points to Mercury's perihelion. The line GSA is the major axis of Mercury's orbit. The point $Ʉ_1$ is its aphelion and the point $Ʉ_2$ is its perihelion. The line $S\Upsilon$ points to the vernal equinox. Angle $\tilde{\omega}_\oplus S\Upsilon$ is about $71°.04$ and angle $\tilde{\omega}_\mathrm{Ʉ}S\Upsilon$ is about $48°.39$.

When he developed his theory of Mercury, the main data that Ptolemy worked with were measurements of the maximum elongation. Suppose that the earth is at point E, for example, while Mercury can be at any point on its orbit. Draw the lines from E that are tangent to Mercury's orbit. If Mercury is to the left of S as seen from E and if it is at one of the points of tangency, it is as far to the east of the sun as it can be when the earth is at E. If it happens to be at the other point of tangency, it is as far west of the sun as it can be when the earth is at E.

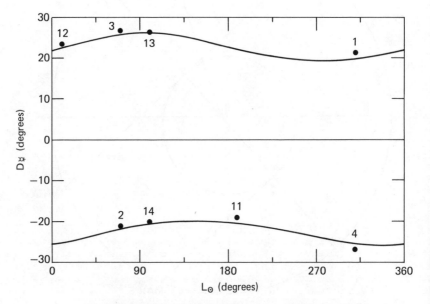

Figure X.2   The maximum elongation of Mercury, plotted as a function of the mean longitude $L_\odot$ of the sun; $L_\odot$ refers to the sun's geocentric position. The upper curve, where the maximum elongation $D_\mathrm{Ʉ}$ is positive, refers to east elongations, when Mercury is an evening star, and the lower curve refers to west elongations, when Mercury is a morning star. The parameters used in plotting the curves are the values that were correct in Ptolemy's time. The numbered points indicate observations that Ptolemy uses in deriving his theory of Mercury.

Now let $\lambda_{\mathrm{\mathchar"20A}}$ be the geocentric longitude of Mercury when it is at one of the tangent points, and let $L_\odot$ be the geocentric longitude of the mean sun at this time. I shall denote the maximum elongation by the symbol $D_{\mathrm{\mathchar"20A}}$ and define it by†

$$D_{\mathrm{\mathchar"20A}} = \lambda_{\mathrm{\mathchar"20A}} - L_\odot . \qquad (X.1)$$

When $D_{\mathrm{\mathchar"20A}}$ is positive, Mercury is east of the sun and it is an evening star. When $D_{\mathrm{\mathchar"20A}}$ is negative, it is west of the sun and is a morning star.

$D_{\mathrm{\mathchar"20A}}$ is a function of $L_\odot$, and this function is plotted in Figure X.2. The maximum east (+) elongation is shown in the upper part of the figure and the maximum west (−) elongation is shown in the lower part. The numbered points indicate observations that Ptolemy uses in finding the parameters of Mercury's orbit; they will be discussed later. Figure X.2 is calculated with the same parameters that were used in Figure X.1, so it is also correct for Ptolemy's time.

When Mercury was east of the sun, its greatest possible distance from the mean sun was about 26° 16′, and this came when $L_\odot$ was about 100°. The least positive value of $D_{\mathrm{\mathchar"20A}}$ was about 19° 23′, coming when $L_\odot$ was about 274°. On the other hand, the most negative value was about −25° 47′, and it came when $L_\odot$ was about 339°, while the least negative value was about −19° 50′, coming when $L_\odot$ was about 145°.‡

There are two interesting features about these numbers. First, the values of $L_\odot$ when $D_{\mathrm{\mathchar"20A}}$ has its greatest east and west values differ by about 120°, as do the values of $L_\odot$ when $D_{\mathrm{\mathchar"20A}}$ has its smallest east and west values. Second, the difference between the greatest east and west values is nearly 0°.5, as is the difference between the smallest values. We need to ask what would happen if the heliocentric orbit of Mercury happened to be a circle, as it nearly is for Venus. We shall see in Section XI.3 that the first feature is almost identical for Venus and Mercury, but that the second feature is different. The difference between corresponding east and west values is only about 0°.2 rather than 0°.5. That is, the lack of east-west symmetry is the most obvious problem presented by Mercury. However, most discussions that I have read seem to make out that the problem is the difference between the values of $L_\odot$, although this is actually the same problem for Mercury and Venus.‡

---

†This is the definition used by Ptolemy. In modern usage, the elongation is usually calculated with respect to the true position of the sun, as it tacitly is in Figure X.1, rather than to the mean position.

‡I have used the past tense here because I am describing what happened in Ptolemy's time. At the present time, the phenomena are qualitatively the same but the numbers are different.

‡The difference between the greatest and least values of $D_{\mathrm{\mathchar"20A}}$ is also considerably greater for Mercury than for Venus.

The problem of Mercury has a strong analogy to the problem of the lunar evection. $D_\mercury$ for Mercury plays a role similar to that of E, the maximum equation of the center, for the moon. In both cases, E or its analogue $D_\mercury$ varies strongly with position in the orbit. This means that either the epicycle radius or the deferent radius must vary with the mean longitude.

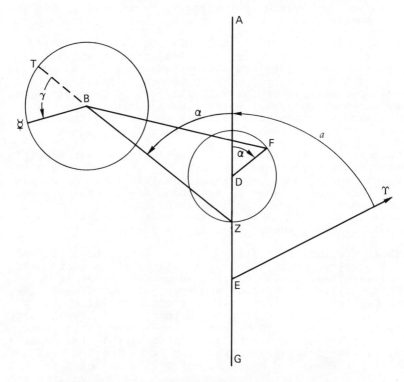

Figure X.3   Ptolemy's model for Mercury. B is the center of Mercury's epicycle and $\mercury$ is the position of Mercury. E is the earth and the line EΥ points toward the vernal equinox. A is the point that Ptolemy calls apogee, but it is the apogee of the deferent rather than of the planet itself. The longitude a of apogee is the angle from EΥ to EA, measured in the direction shown. Z is a point between E and A, and the epicycle center B rotates uniformly around Z rather than around E, in the direction shown. D is a point between Z and A, and point F rotates around D in the direction shown. F rotates at the same rate as B, so that the angle ADF is always equal to the angle α between ZA and ZB. During the motion, the distance BF, rather than the distance BZ, remains constant. The anomaly γ of the planet, which increases in the direction shown, is measured from point T, which lies on the extension of ZB.

Ptolemy claims to have solved the lunar problem by varying the radius of the deferent, using the "crank" model of Figure VII.2 to do so. Undeterred by the serious failure of this model, he proceeds to use a similar idea for Mercury. That is, knowing that he has to vary either r, the radius of the epicycle, or R, the radius of the deferent,† he again chooses to vary R.

The model which he uses to do this is shown in Figure X.3. As usual, E is the earth and ♈ is the vernal equinox. The line from E to A is in the direction that Ptolemy calls apogee. It is actually the apogee of the deferent rather than of Mercury itself, but Ptolemy generally uses apogee in this sense when he is speaking of a planet. The angle from E♈ to EA is called the longitude a of the apogee for Mercury.

A short distance from E toward A is a point Z. I shall use the term "first eccentricity" for the distance EZ, and I shall use the symbol e_1 for it. Point B is a point on the deferent, and the line BZ rotates uniformly around Z. There is no convenient name for the angle α between ZA and ZB; I shall call it the mean apogee distance of Mercury. The angle α increases in the direction shown. Point D is at a short distance e_2 from Z toward A; I shall call e_2 the second eccentricity. As B rotates around Z, point F rotates around D at the same rate but in the opposite direction, so that angle ADF also equals α. During the motion, the distance BF rather than the distance BZ remains constant. The radius DF will be called e_3, the third eccentricity. Finally, ☿ is the position of Mercury. The anomaly γ is measured from point T, which lies on the extension of line ZB.

The sum of the angles a and α is equal to the mean longitude $L_\odot$ of the sun:

$$L_\odot = a + \alpha. \qquad (X.2)$$

Ptolemy uses this relation explicitly in constructing his theory of Mercury.

When $\alpha = 0$, the point F is directly above D and B is directly above F. The distance from E to B is then the constant distance BF plus the sum of the three eccentricities. This is the greatest distance possible between B and E. When $\alpha = 90°$, the distance from B to D is a minimum, but the distance from B to E continues to decrease for awhile as α increases. B is actually closest to E when α is about 120°. Since the picture is symmetrical about the line EA, B is at the same distance when α is about 240°. Ptolemy describes the situation by saying (Chapter IX.8) that Mercury goes through perigee twice in each revolution. What he means is that the center of its epicycle has two perigee positions in each earth year. The planet itself goes through perigee once in each complete revolution of ☿ about B, which takes about 116 days.‡ That is, Mercury is at perigee when γ is

---

†See also Equation VII.1 in Section VII.3.

‡This is different from Mercury's period of revolution about the sun, which is about 88 days.

close to 180°, and it is at perigee again when γ is close to
180° + 360°. It takes about 116 days for γ to increase by 360

The anomaly γ increases linearly with time. If we let
$\gamma_{\Mercury 0}$ denote the value of γ at some epoch, such as Ptolemy's
standard epoch,† and if we let $\gamma_{\Mercury}'$ denote the amount by
which it changes in a day, $\gamma_{\Mercury 0}$ and $\gamma_{\Mercury}'$ are parameters that
must be found from observations. The position a of apogee,
the three eccentricities e_1, e_2, and e_3, and the radius r
of the epicycle must also be found from observations. In
addition, Ptolemy assumes that the value of a may change
with time, so that its rate of change a' is also a parameter
that must be found. Altogether, then, Ptolemy must find
eight parameters by the analysis of observations.

2. Five Planetary Conjunctions with the Moon

At a particular point in his study of Mercury, Ptolemy
uses an observation made when Mercury was nearly in conjunction with the moon. He does the same thing at a corresponding point in the study of every planet. It is convenient to
study the authenticity of all five conjunctions at once. We
shall see how Ptolemy uses the observations when we study
his analysis of the individual planets.

I shall now summarize the circumstances of the five conjunctions.

1. In Chapter IX.10 of the Syntaxis, Ptolemy says that
he used the astrolabe to observe Mercury and the moon at $4\frac{1}{2}$
ordinary hours before midnight on 139 May 17. Upon comparing
it with Regulus (α Leonis), he found Mercury to be at longitude $77\frac{1}{2}$, which was 1 1/6 degrees east of the apparent center
of the moon. Hence the moon was apparently at longitude
76;20 degrees. He found the time from the fact that the
point on the ecliptic with longitude 162° was at the meridian. From the time, he calculates that the true longitude
of the moon was 77;10 degrees, and the parallax in longitude
was 50', so that the apparent place of the moon was 76;20
degrees, in agreement with the observation. Ptolemy does no
state the time of the observation in mean time, but it is
clear that he did use the mean time in calculating the position of the moon. When I calculate the position of the moon
from his tables, I get 77;19 degrees if I do not correct for
the equation of time, but I get 77;09 degrees if I do correc
for it. The calculated position of the moon confirms that
the longitude of Mercury was $77\frac{1}{2}$ degrees. At the time of th
observation, the longitude of the sun was 53°, its mean long
itude was 52;34 degrees, the mean longitude of the moon was
72;14 degrees, and its anomaly was 281;20. These are all
calculated values. Mercury had not reached its maximum elon
gation.

2. In Chapter X.4 of the Syntaxis, Ptolemy says that h
observed Venus and the moon with the aid of the astrolabe.
The time was 4 3/4 ordinary hours after midnight on 138

---

†Noon, Alexandria time, on -746 February 26.

December 16.† By comparing Venus with α Virginis (Spica),
he found that the longitude of Venus was 216½ degrees. Venus
was also on the straight line between β Scorpii and the center of the moon, and its distance from the moon was 1½ times
its distance from β Scorpii. Ptolemy calculates that the
true longitude of the moon was 215;45 degrees and the parallax was 1 degree, so that the apparent longitude of the moon
was 216;45 degrees. Again he has used the equation of time
without saying so; I get 215;53 for the true longitude if I
do not use it and 215;43 if I do use it.

    In Ptolemy's star table, the longitude of β Scorpii is
216;20 degrees. Since the apparent longitude of the moon as
seen from Alexandria was 216;45 degrees, the longitude of
Venus was 216;30, in exact agreement with the measurement
made with the astrolabe. Ptolemy also calculates the latitudes, but we do not need them. At the time of the observation, the longitude of the sun was 263°, the longitude of
the mean sun was 262;09 degrees, the longitude of the mean
moon was 221;24, and its anomaly was 87;30 degrees. Venus
had passed the point of maximum elongation.

    3. In Chapter X.8 of the Syntaxis, Ptolemy says that
he observed Mars and the moon with the aid of the astrolabe
at 3 ordinary hours before midnight on 139 May 30; the point
of the ecliptic that was on the meridian had longitude 200°.
When he compared Mars with Spica (α Virginis), he found that
the longitude of Mars was 241 3/5 degrees. Mars was also
1 3/5 degrees east of the apparent center of the moon. Ptolemy
calculates that the mean longitude of the sun was 65;27 degrees, that the mean longitude of the moon was 244;20 (I get
244;25, so I probably used a different value of the equation
of time), and its true longitude was 239;00 degrees. The
parallax was 1°, so the apparent longitude was 240;00 degrees.‡ Hence Mars was at 241 3/5 degrees, in exact agreement with the measurement on the astrolabe.

    4. In Chapter XI.2 of the Syntaxis, Ptolemy says that
he observed Jupiter and the moon by means of the astrolabe
at 5 ordinary hours after midnight on 139 July 11. The mean
longitude of the sun was 106;11 degrees and the ecliptic
point with longitude 2° was on the meridian. When he compared Jupiter with α Tauri, he found that the longitude of
Jupiter was 75;45 degrees, and it was directly north of the
center of the moon. He calculates that the mean longitude
of the moon was 79 degrees, that its anomaly was 272;05 degrees, and that its true longitude was 74;50 degrees. I
believe that he made a sign error in calculating the equation
of time; when I make one, I agree closely with him in the
position of the moon. When I use what I believe to be the
correct equation of time, I get the lunar longitude to be

---

†The point on the ecliptic whose longitude was 152° was on
the meridian.

‡Since the moon was full, Ptolemy's calculation of the
parallax is fairly accurate. In the earlier observations,
the moon was not very close to a syzygy, and Ptolemy's
calculated parallax is too large.

74;56 degrees. He takes the parallax to be 55'. Since the moon is about midway between the last quarter and the new moon, his parallax is somewhat too large. With his parallax, he gets the apparent longitude of the moon to be 75;45, and the longitude of Jupiter is the same, in exact agreement with the value read on the astrolabe.

5. In Chapter XI.6 of the Syntaxis, Ptolemy says that he observed Saturn and the moon by means of the astrolabe. The time was 4 ordinary hours before midnight on 138 December 22. The ecliptic point whose longitude is 30° was in the meridian, and the mean longitude of the sun was 268;41 degrees. When he compared Saturn with α Tauri, he found that its longitude was 309;04 degrees.† Saturn was also $\frac{1}{2}$ degree east of the northern tip of the crescent moon. The moon was about midway between new moon and the first quarter. Ptolemy calculates that the mean longitude of the moon was 308;55 degrees, that its anomaly was 174;15 degrees, and that its true longitude was 309;40 degrees. I believe that he forgot to use the equation of time; I get 309;37 degrees when I do not use it and 309;28 degrees when I do. Ptolemy takes the parallax to be 1;06 degrees, and thus he calculates 308;34 degrees for the apparent longitude of the moon; this parallax is considerably too large because the moon is not close to syzygy. Since Saturn was $\frac{1}{2}$ degree farther east, it was at 309;04 degrees, in exact agreement with the astrolabe reading based upon the position of α Tauri.

I do not believe that it is necessary to tabulate these observations in order to show that they have been fabricated We note that Ptolemy claims to measure the longitude of each planet in two different ways. First he measures it by measuring the separation between the planet and a star. Although he does not say so explicitly, he takes the longitude of the star from his star table and adds the measured difference to get the longitude of the planet. Next, he measures the separation between the planet and the moon. He then calculates the longitude of the moon, including the parallax in longitude, from his lunar theory and adds the measured difference to get a second value for the longitude of the planet.

The two measurements of longitude agree to the minute of arc in every case. For the first four planets, we may grant him rounding to the nearest multiple of 5'. For Saturn however, the "measured" longitude is not a multiple of 5', there has been no rounding except to the nearest minute, and he still gets exact agreement. Thus we may say that four measurements agree within a zone that is 5' in width, while the last agrees within 1'.

The standard deviation of Ptolemy's lunar theory is about 35' (Section VII.6) and the standard deviation of the longitudes in his star table (Section IX.2) is about 22'. The standard deviation in a comparison of longitudes, in the

---

†This is a fraction of a degree that Ptolemy does not use in any other observation that I recall. It is impossibly precise.

way that Ptolemy does it, should thus be $\sqrt{(35)^2 + (22)^2}$, which is slightly more than 40', if we assume that there were no errors in the planetary observations themselves. If we make a reasonable allowance for errors in observation, the standard deviation of a comparison should be about 50'.

Thus four comparison values lie in a pre-assigned zone whose width is 0.1 standard deviations and the fifth lies in a zone whose width is 0.02 standard deviations. The probability that this could happen by chance is about 1 in 100 000 000. Hence the results did not happen by chance; they were fabricated.

Ptolemy seems to have become careless toward the end of the work. In the Saturn observation, he gives the longitude with an unlikely precision, and he seems to have forgotten about the equation of time. He also overlooked another circumstance which shows in an interesting and independent way that the Saturn observation was fabricated. Ptolemy says that Saturn was east of the northern tip of the crescent. Actually, at the time in question, the latitudes of Saturn and the moon were nearly the same [APO, pp. 488-489]. Hence, when the longitude of Saturn was 30' greater than the longitude of the moon, Saturn was within about half an hour of being occulted by the center of the moon's dark limb. The most accurate observation that Ptolemy could make would be to time the beginning of the occultation. Even if he did not want to wait the necessary half hour, I do not believe that he could even have looked at Saturn and the moon and written that Saturn was east of the northern tip. He should have written that it was east and slightly north of the center.

I have not thought of a reason why Ptolemy chose to find the planetary longitudes in two different ways for these observations. Perhaps he wanted to show that his planetary and lunar theories were consistent with each other, or that his lunar theory was accurate. If this was his intention, he accomplishes it by understatement; he does not point out this implication of his alleged measurements.

3.   Ptolemy's Parameters for Mercury

We saw in Section X.1 that Ptolemy's model for the motion of Mercury has 8 parameters that must be found from observations. He uses altogether 17 observations in order to find the parameters. One of these is the conjunction of Mercury with the moon that was studied in the preceding section. This is the 15th observation in the order that Ptolemy uses them, and it was made when Mercury was east of the sun but had not attained its maximum elongation. The last two of the 17 in order of use were made 4 days apart in the year -264, when Mercury was also east of the sun but not yet at maximum elongation. The first 14 observations were all made when Mercury was at maximum elongation, 7 when it was east of the sun and 7 when it was west.

I shall start by presenting the observations of maximum elongation. I do not believe that it is necessary to quote

-265-

TABLE X.1

OBSERVATIONS OF MERCURY AT MAXIMUM ELONGATION

Identifying number	Date			Observer	$L_\odot$, degrees[a]	$D_\yen$, degrees[a]	Error in measured longitude degrees[b]
1	132 Feb	2		Ptolemy	309;45	21;15E	+1.66
2	134 Jun	4		Ptolemy	70;00	21;15W	-1.21
3	138 Jun	4		Ptolemy	70;30	26;30E	+1.41
4	141 Feb	2		Ptolemy	310;00	26;30W	-1.05
5	-261 Feb	12		Dionysios	318;10	25;50W	+0.02
6	-261 Apr	25		Dionysios	29;30	24;10E	+0.80
7	-256 May	28		Dionysios	62;50	26;30E	+1.06
8	-261 Aug	23		Dionysios	147;50	21;40E	-1.80
9	-236 Oct	30		Anonymous	215;10	21;00W	+0.26
10	-244 Nov	19		Anonymous	234;50	22;30W	-0.20
11	134 Oct	3		Ptolemy	189;15	19;03W	+1.16
12	135 Apr	5		Ptolemy	11;05	23;15E	+1.16
13	130 Jul	4		Theon	100;05	26;15E	+0.45
14	139 Jul	5[c]		Ptolemy	100;20	20;15W	-0.23

[a]As stated by Ptolemy. E means that Mercury was east of the sun, W that Mercury was west of it.

[b]The measured value minus the value calculated from modern theory.

[c]The Egyptian date that Ptolemy states is equivalent to 139 July 8, but his calculations show that he meant to write the equivalent of 139 July 5.

the records of the observations from Ptolemy's text. The first ten observations are found in Chapter IX.7 of the Syntaxis, the next two are found in Chapter IX.8, and the remaining two are found in Chapter IX.9. Some records give directly the difference in longitude between Mercury and a particular star. Others describe a configuration that Mercury makes with two or more stars, with enough detail that we can infer the longitude of Mercury when we know the positions of the stars from a star catalogue. The records are discussed in detail in APO [pp. 182-190].

The data from the records are summarized in Table X.1. The first column gives an identifying number that will be used in discussing the observations. The next column gives the dates of the observations, and the third column gives the observer. The fourth column gives the mean longitude of the sun while the fifth column gives the elongation $D_{\u263F}$ of Mercury. $D_{\u263F}$ is defined in Equation X.1; it is based upon the mean longitude of the sun rather than upon its actual longitude. Both of these columns give the values that are stated by Ptolemy in the records. If a reader wants the longitude of Mercury that was measured, he can find it from the value of $L_{\odot}$ and the elongation.

We could compare the elongation that Ptolemy states with the value that we calculate from modern theory. Since this comparison would involve the error in Ptolemy's solar tables as well as the error in measuring the longitude of

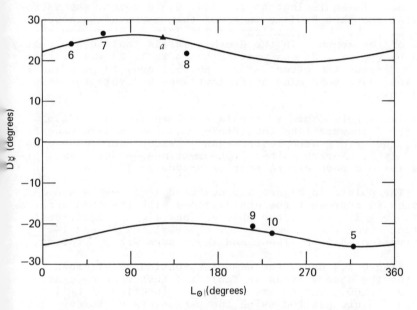

Figure X.4   The maximum elongation $D_{\u263F}$ of Mercury plotted as a function of the mean longitude $L_{\odot}$ of the sun, for the year −260. The points numbered 5 through 10 represent observations made near the year −260 that Ptolemy uses in developing his theory of Mercury. The point labelled a and marked with a triangle is involved in discussing Ptolemy's use of the observations.

Mercury, it is more instructive to study the error in the measured longitude rather than in the elongation of Mercury. This error is given in the last column of the table.

A few comments need to be made. 8 observations were made in Ptolemy's time, 7 allegedly by himself and 1 by his near contemporary Theon. 6 observations were made between -261 February 12 and -236 October 30. Of these, 4 were made by Dionysios. We know nothing about him except that he made these and a few other observations. The other two observations that were already ancient to Ptolemy are not attributed to any observer. It is chronologically possible that they were made by Dionysios, but the style of the records and the units used in the measurements make this unlikely.

In the record that is dated 139 July 5 in Table X.1, all texts and translations that I have seen say that the observation was made on the 24th of a certain Egyptian month. This date is equivalent to 139 July 8. However, the astronomical use that Ptolemy makes of this observation shows that he meant the equivalent of 139 July 5, which was the 21st of the month. In the Greek numerals that were probably used by Ptolemy, 21 was written as KA while 24 was written as $K\Delta$. Hence the error is one that was easy to make and it may well have been made by Ptolemy when he wrote the original text.

The errors shown in the last column are interesting. For the 7 observations that Ptolemy claims to have made himself, the errors are greater than 1° except for the one on 139 July 5. For the other 7 observations, 4 have small errors while 3 have errors near or exceeding 1°.

The points in Figure X.2 numbered 1 through 4 and 11 through 14 represent the observations with the same numbers in Table X.1. All of the observations lie a significant distance from the curves except observations 13 and 14. Observation 13 is by Theon and the others are by Ptolemy.

Figure X.4 shows the same information as Figure X.2, but for the observations in Table X.1 that were made near the year -260. The curves of $D_\yen$ as a function of $L_\odot$ in Figure X.4 are plotted using the parameters of Mercury that are correct for the year -260, and the points numbered 5 through 10 represent the observations with the same numbers in Table X.1. The observations lie close to the curves except observation 8. The triangular point marked a is not an observation but a point that will be used in discussion.

The parameters $\gamma_{\yen_0}$ and $\gamma_\yen{}'$ are not involved when Mercury is at maximum elongation, but all of the other six parameters are involved. Hence Ptolemy can find all of the parameters except $\gamma_{\yen_0}$ and $\gamma_\yen{}'$ from the observations in Table X.1, and we shall next see how he does so. Most of the following discussion is found in Section XIII.8 in APO.

Ptolemy starts from the fact that the model of Mercury is symmetrical about the line AE in Figure X.3.† In observation 1 in Table X.1 the maximum elongation is 21;15 degrees east while it is 21;15 degrees west in observation 2. Hence the symmetry line EA bisects the values of $L_\odot$. If we round the value of $L_\odot$ in observation 1 to 310°, we find that EA is the line that runs from longitude 10° to longitude 190°, approximately.

From these observations alone we cannot tell which value is the longitude a of apogee. In order to avoid awkward circumlocutions while we await this discovery, I shall anticipate a later result and say at once that a is 190° or close to it.

Ptolemy confirms this result by using observations 3 and 4, which he also claims to have made himself. With these, the elongations are 26;30 degrees east and west respectively, so he finds again that apogee is close to 190°. More precisely, the first pair of observations gives 189;52,30 degrees while the second pair gives 190;15 degrees; these values differ by less than half a degree. Ptolemy decides to use the value $a = 190°$ exactly.

Ptolemy finds these results in Chapter IX.7 of the Syntaxis. Still in this chapter, he says that he will show that apogee in ancient times‡ came at 186° rather than 190°. He first presents the record of observation 5, for which $L_\odot = 318;10$ degrees and $D_\yen = 25;50$ west. He then says that he cannot find, among the ancient observations, one for which $D_\yen = 25;50$ east, so he must use two observations that bracket this value. These are observations 6 and 7. For these, the values of $D_\yen$ are 24;10 degrees east and 26;30 degrees east while the values of $L_\odot$ are 29;30 and 62;50 degrees, respectively. Ptolemy interpolates between these observations to find the value of $L_\odot$ when $D_\yen = 25;50$ east, and he finds $L_\odot = 53\frac{1}{2}$ degrees.‡ Hence, at the time of the ancient observations, apogee was the average of 318;10 degrees and 53;30 degrees; this average is 185 5/6 degrees.

He then presents the record of observation 8, for which $L_\odot = 147;50$ degrees and $D_\yen = 21;40$ east. Since he cannot find an ancient observation with $D_\yen = 21;40$ west, he interpolates between observations 9 and 10. For these observations, $L_\odot = 215;10$ and 234;50 degrees while $D_\yen = 21;00$ degrees and 22;30 degrees west, respectively. By interpolation, he finds $L_\odot = 224;10$ degrees.* Hence apogee was at the average of 147;50 and 224;10, which is exactly 186 degrees.

---

†I do not mean that the model is symmetrical about AE at any particular instant; it is obviously not symmetrical at the instant for which Figure X.3 is drawn. I mean that the orbit generated by the model is symmetrical when α and γ go through their complete ranges of 360°.

‡Ancient to him, that is.

‡I find that the interpolation should give $L_\odot = 53;19$ degrees.

*The correct interpolation gives 223;54 degrees.

Hence, Ptolemy concludes, the position of apogee was almost exactly 186° in the time of Dionysios and it is almost exactly 190° in his own time, 4 centuries later. Therefore the apogee position precesses by 1° per century. This, as Ptolemy points out, is the same as the rate at which the equinoxes precess, so that the orbit of Mercury always retains the same position with respect to the star background. This contrasts with the behavior of the sun, whose orbit always maintains the same position with respect to the equinoxes in Ptolemy's theory.†

Ptolemy next wishes to find the radius r of Mercury's epicycle and, in the process, he not incidentally finds the sum $e_1 + e_2$ of two of the eccentricities in Figure X.3. This sum is the distance DE in the figure.

Let ρ denote the constant distance BF in Figure X.3. I noted in Section X.1, just below Equation X.2, that the distance from E to the center of the epicycle is $\rho + e_1 + e_2 + e_3$ when the mean apogee distance α is 0°, that is, when $L_\odot = 190°$ in Ptolemy's time. Let A be the point where the center of the epicycle is when $\alpha = 0°$; this is the condition that Ptolemy calls apogee. Let G be the point where the center of the epicycle is when $\alpha = 180°$. This is the condition when $L_\odot = 10°$, but it is not the condition that Ptolemy calls perigee. If the reader will sketch the configuration of the model when $\alpha = 180°$, he will see that the distance from E to G is $\rho - e_1 - e_2 + e_3$.

Ptolemy needs observations made when Mercury was at maximum elongation under two conditions, once when $L_\odot = 190°$ ($\alpha = 0°$) and once when $L_\odot = 10°$ ($\alpha = 180°$). He says at the beginning of Chapter IX.8 that there are no ancient observations made under these conditions. The reason he gives is that there are no stars bright enough to serve as references which are close to where Mercury would be on these occasions. Hence Mercury must be located by measuring a large angle between it and a bright star, and this can only be done with an astrolabe, which ancient observers‡ did not have. Hence he must use observations that he has made himself by means of the astrolabe.

As I have shown elsewhere,‡ Ptolemy's statement simply is not true. On 134 October 3, the star γ Virginis, which Babylonian observers frequently used as a reference star, was readily observable and it was within a few degrees of Mercury, and this configuration must have happened many times before. On 135 April 5, Mercury was within about 1° of η Tauri, which the Babylonians also used frequently, and η

---

†The precession of the equinoxes is about 1°.385 per century The sun's orbit precesses eastward by about 1°.719 per century with respect to the equinoxes, so that it moves eastward among the stars by about 0°.334 per century. The orbit of Mercury also moves eastward with respect to the stars, but at a rate of only about 0°.171 per century.

‡Ancient here again means ancient to Ptolemy.

‡APO, pages 465-466.

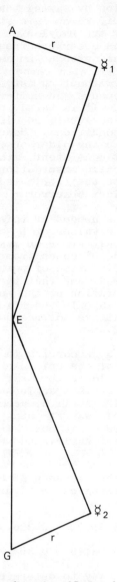

Figure X.5   The configuration of Ptolemy's model of Mercury
for the alleged observations of 134 October 3 and 135 April 5.
On 134 October 3, the center of Mercury's epicycle was at A and
Mercury was at the point ☿₁. On 135 April 5, the center of the
epicycle was at G and Mercury was at point ☿₂. Ptolemy claims
that he measured angle AE☿₁ to be 19;03 degrees and that he
measured angle GE☿₂ to be 23;15 degrees. The angles at ☿₁ and
☿₂ are right angles, since Mercury was at maximum elongation
on both occasions.

Tauri was readily visible if Mercury was. Further, Ptolemy states the observation by saying that Mercury was 8;20 degrees west of α Tauri, which is a star of the first magnitude. Inspection of the Babylonian observations in APO shows that Babylonian observers frequently measured separations larger than this without an astrolabe.

Thus observers without an astrolabe could readily make the required measurements, in spite of Ptolemy's statement that they could not. It is barely possible in spite of this that there were, by accident, no older observations made under the required conditions. However, I suspect that the real reason for not using older observations is that genuine observations are not consistent with Ptolemy's model. Hence he had to use fabricated observations and he chose on this occasion to attribute the fabricated observations to himself rather than to an older observer.

The observations needed at this point are numbers 11 and 12 in Table X.1. The values of $L_\odot$ are not exactly $190°$ and $10°$. However, the only error we make in assuming that they are exactly $190°$ and $10°$ comes from the trivial difference between $L_\odot$ and the direction of point B (Figure X.3) as seen from point E. If we assume that the values of $L_\odot$ are $190°$ and $10°$, the configuration of the model on the two occasions is shown in Figure X.5. In the figure, I have omitted all lines and points that are not needed in the analysis of observations 11 and 12.

On 134 October 3, according to Ptolemy's alleged observations, the center of the epicycle was at point A and Mercury was at maximum elongation west, at point $☿_1$. The angle at $☿_1$ was therefore a right angle and angle $AE☿_1$ equalled $19;03$ degrees. On 135 April 5, according to Ptolemy, the center of the epicycle was at point G and Mercury was at maximum elongation east, at point $☿_2$. The angle at $☿_2$ was therefore also a right angle, and angle $GE☿_2$ equalled $23;15$ degrees.

Let R_1 denote the distance AE and let R_2 denote the distance GE. The triangles in Figure X.5 then give the relations:

$$r = R_1 \sin 19;03 = 0.326\ 393\ R_1,$$

$$r = R_2 \sin 23;15 = 0.394\ 744\ R_2.$$

Since Ptolemy has no way to determine the actual distances involved in Figure X.5 (or in Figure X.3), he must assign some distance arbitrarily. He chooses to assign the value 120 to the distance AE, which we have called R_1. The first equation above immediately gives $r = 39.15$, which is $39;09$ in sexagesimal notation; the error in this calculation is insignificant. The second equation above then gives the value of R_2.

We saw a moment ago that $R_1 = \rho + e_1 + e_2 + e_3$ and that $R_2 = \rho - e_1 - e_2 + e_3$. Now that we have values for R_1 and R_2, we can find the combinations $\rho + e_3$ and $e_1 + e_2$:

-272-

$$\rho + e_3 = 109;35,$$

$$e_1 + e_2 = 10;25,$$

(X.3)

in sexagesimal notation. These are Ptolemy's values; the values that I find do not differ significantly. Repeating the value of r for convenience:

$$r = 39;09.$$

(X.4)

From the measurements of maximum elongation, Ptolemy has now found a, a', r, and the sum $e_1 + e_2$, and he has used all but two of the measurements. From the remaining two, he must find e_3 and either e_1 or e_2. When he has found either e_1 or e_2, he can then find all of the e's by using the second of Equations X.3. Until this point, he has used the symmetry of the situation, or at least the symmetry of

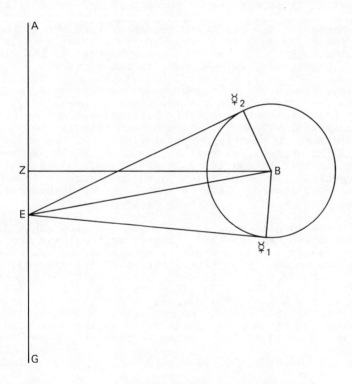

Figure X.6   The configuration of the epicycle in Ptolemy's model of Mercury for the alleged observations of 130 July 4 and 139 July 5. On both dates, we can take the line ZB to be perpendicular to AE, and we can assume that the center B of the epicycle was in the same place on both occasions. From this figure, Ptolemy finds the distance ZE.

-273-

the deferent curve. Now he specifically uses the circumstance in which symmetry is absent, by using observations
made when $L_\odot$ is midway between A and G in Figure X.5. That
is, he wants two measurements of maximum elongation, one
east and one west, for the same value of $L_\odot$, which should
be either 100° or 280°. The observations he can find that
come closest to these conditions are observations 13 and 14,
made by Theon and by himself, respectively. The fact that
$L_\odot$ is not exactly 100° for either observation causes a small
problem. Ptolemy solves this problem by using essentially
the principle given in the next paragraph, although he words
the matter in a considerably different fashion.

Suppose that ZB in Figure X.6 is perpendicular to line
AG,† and that Mercury is at maximum elongation east, so that
Mercury is at the point $\yen_2$. Since the distance ZE is small
compared with distance ZB, EB is nearly perpendicular to AG.
Because the elongation is a maximum, the radius $B\yen_2$ of the
epicycle can rotate through a moderate angle without changing
the elongation by an observable amount. Thus, as ZB rotates
through a small angle, the elongation angle $BE\yen_2$ does not
change. Hence, if the angle AZB decreases from 90° to, say,
89° 55′, the angle $AE\yen_2$ decreases also by 5′, while angle
$BE\yen_2$ remains unchanged. The same thing happens if the elongation is to the west, with Mercury at the point $\yen_1$.

In observation 13, $L_\odot$ = 100;05 degrees and $D_\yen$ = 26;15
degrees to the east. Hence the longitude of Mercury $\lambda_\yen$ =
126;20 degrees at the time of the observation. By the preceding paragraph, however, if $L_\odot$ had been exactly 100°,
which would make ZB exactly perpendicular to AG, $D_\yen$ would
still be 26;15 degrees, so that $\lambda_\yen$ would be 126;15 degrees.
Similarly, in observation 14, $L_\odot$ = 100;20 degrees, $D_\yen$ =
20;15 degrees to the west, and $\lambda_\yen$ = 80;05 degrees. Thus we
can pretend for present purposes that $L_\odot$ = 100;00 degrees,
$D_\yen$ = 20;15 degrees, and $\lambda_\yen$ = 79;45 degrees.

Since the longitude of A is 190 degrees, the angle
$AE\yen_2$ equals 63;45 degrees (190 - 126;15 degrees) and the
angle $AE\yen_1$ equals 110;15 degrees if we make ZB be exactly
perpendicular to AG. Since the center B of the epicycle
is in the same place for both observations under this condition, the angle $\yen_1 E\yen_2$ is 110;15 degrees minus 63;45 degrees, so that $\yen_1 E\yen_2$ = 46;30 degrees, and angles $BE\yen_1$ and
$BE\yen_2$ are both 23;15 degrees. We know that the radius of
the epicycle (Equation X.4) equals 39;09 in the convention
which makes AE equal to 120. Hence we find‡

$$EB = 39;09/\sin 23;15 = 99;9. \qquad (X.5)$$

Next we want to find the distance ZE, which is the
first eccentricity e_1. We remember that the two elongations

---

†The lettered points in Figure X.6 have the same interpretation as the corresponding points in Figure X.3. For
clarity, I have omitted the parts of the model that are not
needed in the immediate discussion.

‡This is Ptolemy's value. An accurate solution gives 99;11.

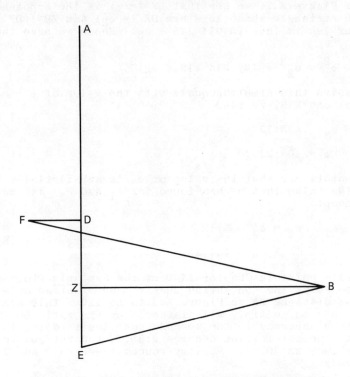

Figure X.7   The configuration of the crank radius DF and of
the connecting arm BF in Ptolemy's model of Mercury, on 130
July 4 and 139 July 5. From this figure, Ptolemy finds the lengths
of DF and BF, after he has found the distance ZE from Figure X.6.

in Figure X.6 are 26;15 and 20;15 degrees. Ptolemy states
the theorem that angle ZBE in the figure is half the difference between the two elongations, so that ZBE = 3 degrees
exactly. This makes e_1 = EB sin 3°. When we use the value
of EB from Equation X.5, we get e_1 = 5;12 in sexagesimal notation, according to Ptolemy.

Ptolemy now changes the value of $e_1 + e_2$ (Equations
X.3) from 10;25 to 10;24, and we can readily accept this
change as a reasonable one. He now has the result that
$e_1 = e_2 = 5;12$.

To get e_3 from the same observations, we use Figure X.7.
The letters that appear in this figure have the same meaning
that they have in Figure X.3, so that ZE equals e_1, DZ = e_2,
and DF = e_3, while BF is the constant distance that we have
called ρ. DF and ZB are perpendicular to AE. From the preceding results, we know that angle ZBE equals 3° and that
distance EB equals 99;9, which is 99.15 in decimal notation.
Hence distance ZB equals EB cos 3° = 99.014 118.

From Figure X.7, we see that BF (= ρ) is the hypotenuse of a right triangle whose legs are DZ (= e_2) and ZB + DF. The latter leg is thus 99.014 118 + e_3. Hence we have the relation

$$\rho^2 = e_2{}^2 + (99.014\ 118 + e_3)^2.$$

When we solve this simultaneously with the value of $\rho + e_3$ from Equations X.3, we find

$$e_3 = \quad 5;13,$$

$$\rho = 104;21,49.$$

Ptolemy points out that the value of e_3 is substantially the same as the value that he has found for e_1 and e_2, and he finally adopts

$$e_1 = e_2 = e_3 = 5;12,$$

$$\rho = 104;22. \tag{X.6}$$

At this point in Chapter IX.9 of the Syntaxis, Ptolemy decides to adopt the convention that $\rho = 60$, instead of taking the distance EA in Figure X.3 to be 120. This means that he wants to multiply all distances by the ratio 60/104; When we keep accuracy to the seconds position in doing this, each of the eccentricities becomes 2;59,22 and the epicycle radius becomes 22;30,26. Ptolemy rounds these to 3 and 22;3 respectively.

The distance from E to the apogee of the deferent curve is ρ plus the sum of the three eccentricities; this is 69 wi the new convention and the new values. Hence the greatest distance to Mercury is 69 plus the epicycle radius, and it comes out to be 91;30. A more complex calculation shows tha the distance to either of the two perigee points on the deferent curve is 55;34, which makes the least distance of Mercury equal to 33;4. These are the distances that I used in Section IV.8 in the discussion of Ptolemy's model of the universe.

Finally, in Chapter IX.10 of the Syntaxis, Ptolemy turn to finding the anomaly of Mercury at some epoch, and the rat at which the anomaly changes. To do this, he uses three observations. One of them is the conjunction of Mercury with the moon that was studied in Section X.2, and we have alread concluded that it was fabricated. The other two are observa tions that he attributes to Dionysios, and I shall summarize their circumstances briefly.

1. In the morning of -264 November 15, Dionysios observed that Mercury was 1 moon to the east of the line joinin β Scorpii and δ Scorpii, and that it was 2 moons north of β Scorpii in latitude. By a moon, he undoubtedly means an angle equal to the apparent diameter of the moon, which we can take as 30′ in interpreting the observation. I have ana lyzed the configuration on pages 190-191 of APO. Mercury wa

46' east and 1° north of β Scorpii. Dionysios does not state
the hour of the observation. From his calculations, I judge
that Ptolemy takes the time to be sunrise. However, β Scorpii and δ Scorpii are stars of the third magnitude according
to Ptolemy's star catalogue, and I think it is more likely
that the time was about 45 minutes or an hour before sunrise.

The longitude of β Scorpii in Ptolemy's star catalogue
is 216 1/3 degrees. Since the observation was almost exactly
4 centuries before the date of his catalogue, the longitude
of the star was 212 1/3 degrees at the time of the observation, according to Ptolemy. Hence the longitude of Mercury
was 212°20' + 46' = 213°06' according to the observation.
However, Ptolemy says that the longitude was 213°20'. It
is possible that there has been an error in transmitting the
text, but I think it is more likely that Ptolemy was careless
in inferring the longitude from the observation.

2. Four days later, in the morning of -264 November 19,
Dionysios reported that Mercury was $1\frac{1}{2}$ moons east of the line
joining β Scorpii and δ Scorpii, so that it had moved 15' to
the east during the four days. No statement is made about
the latitude.

From Figure X.3, it is clear that there are two values
of the anomaly which will give the same longitude of Mercury.
Ptolemy uses the second of Dionysios's observations only to
show that Mercury had not reached its maximum elongation at
the time of the first observation, and this tells him which
of the two possible values is the correct one. Calculating
the anomaly from the longitude is a tedious but obvious calculation, and I shall give no details.

Ptolemy finds that the anomaly was 212°34' on the morning of -264 November 15. From his own observation of the
conjunction of Mercury with the moon and the longitude of
Mercury that he deduces from it, he calculates that the anomaly was 99°27' at the time of his alleged observation on 139
May 17. The change in anomaly is 246°53' plus the number of
complete revolutions of anomaly that Mercury made between
the two times. Relatively simple observations show that this
number was 1268. The time interval was 402 Egyptian years
plus 283 days plus $13\frac{1}{2}$ hours. Hence, by division, the anomaly changes by 3.106 698 971 degrees per day, which is
3;6,24,6,58,39,51 degrees per day in sexagesimal notation.
However, this is not the value that Ptolemy uses in his
planetary tables. There he uses the value

$$\gamma_\u03a5' = 3;6,24,6,59,35,50 \text{ degrees per day.} \qquad (X.7)$$

An elementary calculation gives the result that

$$\gamma_{\u03a5 o} = 21;55 \text{ degrees} \qquad (X.8)$$

at the fundamental epoch, which is noon, Alexandria time,
on -746 February 26.

4.   The Accuracy of Ptolemy's Model for Mercury

In Section XIII.8 of APO, I prepared one computer program to calculate the geocentric longitude of Mercury from Ptolemy's theory, and another to calculate the geocentric longitude by combining Newcomb's theory of Mercury [Newcomb, 1895a] with his theory of the sun [Newcomb, 1895]. With these programs, I calculated the longitude at 51 times separated by 80 days. The center time in the calculations was taken to be noon on 137 July 20, which is an epoch that Ptolemy often uses and that is in the center of his astronomical activity. Since all of Ptolemy's longitudes are too small by about 1°.1, I added 1°.1 to his longitude of apogee. In the total time span of 4000 days, both Mercury and the earth make an integral number of revolutions about the sun, with reasonable accuracy.

TABLE X.2

A COMPARISON OF PTOLEMY'S PARAMETERS FOR MERCURY
WITH THOSE THAT GIVE THE BEST FIT

Parameter	Ptolemy's Value	Best Fit
Apogee, degrees	191.1[b]	219.013
Apogee rate, degrees per century	1.00	~ 1.6[c]
First eccentricity (e_1)	0.049 900	0.085 738
Second eccentricity (e_2)	0.049 900	-0.015 284
Crank radius (e_3)	0.049 900	0.012 025
Epicycle radius	0.375 090	0.376 704
Anomaly at the epoch,[a] degrees	189.430	187.195
Anomalistic rate, degrees per day	3.106 699 043	3.106 404 735

[a]Noon Alexandria time, 137 July 20.

[b]Ptolemy's value is 190° when it is referred to his erroneous equinox. I added 1°.1 in order to refer to a more accurate equinox.

[c]Not studied by the fitting process.

In the comparison of the values from Ptolemy's theory with those from Newcomb's theories, the greatest error in the sample of 51 values was 7°.84. It occurred on 137 July 20, which was the central date used in the comparison. The standard deviation of the error was 2°.99.

In Section VII.6 I quoted a statement by Dreyer [1905, p. 200] which says that Ptolemy's theories represented the motion of the solar system as closely as observers (in Greek times) could follow them. I find it hard to believe that

Dreyer ever studied the accuracy of Ptolemy's models. We
have seen that the statement is not correct for either the
sun or the moon. Now we see that the errors in Ptolemy's
model of Mercury can only be described as gross. The standard deviation in the stellar coordinates in the star catalogue, for example, was (Section IX.2) only about 20' in
each coordinate; we may take this as the accuracy with which
Greek observers could follow the planets. The standard deviation in Ptolemy's model of Mercury is almost exactly $3°$,
almost ten times the accuracy of observation.

I then varied the parameters in Ptolemy's model until
I found the values that make the standard deviation a minimum.
The results of this study are summarized in Table X.2.[†] The
first column in the table identifies the parameters and the
second column gives Ptolemy's values of them. Here I have
made two changes from the practice that Ptolemy follows in
most of his discussion of Mercury. The length of the connecting arm BF in Figure X.3 seems to me to be a more fundamental property of the model than is the distance AE. Thus,
instead of taking the distance AE as the reference to which
all other distances are referred, I take the length ρ of the
arm BF as the reference. Instead of calling it 120, I call
it 1. This is close to Ptolemy's second normalization.

The third column in Table X.2 gives the values of the
parameters that yield the greatest possible accuracy of the
model. When these values are used, the greatest error drops
from $7°.84$ to $3°.26$ and the standard deviation drops from
$2°.99$ to $1°.70$. However, even with the best values of the
parameters, the model cannot come close to representing the
motion of Mercury as closely as Greek observers could follow
it.

The new values of the parameters are also interesting.
The best value of apogee is almost $30°$ greater than the
value that Ptolemy finds. Now the large eccentricity of
Mercury's orbit (Figure X.1) makes it easier to find the
apogee of Mercury than it is to find the apogee of Venus's
orbit, which is nearly circular. Yet, as we shall see in
Section XI.4, Ptolemy locates the apogee of Venus with an
error of only about $4°$ while he misses the apogee of Mercury
by almost $30°$. We shall need to enquire into the source of
this peculiar and enormous error, but I shall defer this
enquiry to Section X.5.

However, we should notice here that Ptolemy's error in
the position of apogee is not his most serious error, so far
as the accuracy of his model is concerned. I tried the effect of letting apogee alone vary while all other parameters
keep the values that Ptolemy gives them. Under these conditions, the greatest accuracy is attained when apogee is at
$219°.6$, almost the value in Table X.2. The standard deviation under these conditions is still $2°.54$. Thus using the
best possible position of apogee does not help much unless
the other parameters are also changed.

[†]This is identical with Table XIII.4 in APO.

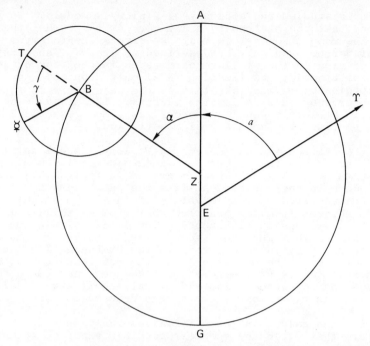

Figure X.8   The model of Mercury that Ptolemy should have used. If we consider all of the models that can be reached from the crank model of Figure X.3, this is almost exactly the one that gives the best accuracy. We note that the crank mechanism in Figure X.3 has shrunk to zero. That is, the crank does not improve the accuracy; it detracts from it.

The "best fitting" values of the epicycle radius, the anomaly at the epoch of 137 July 20, and the rate at which the anomaly changes are all close to the values that Ptolemy uses. Most of the error found when we use his parameters thus comes from his values of the three eccentricities. The three eccentricities should not be equal or even close to it. The best value of e_1 is considerably larger than Ptolemy's value while the best values of e_2 and e_3 are much smaller. In fact, the best value of e_2 is negative. This means that the point D in the model (Figure X.3) should be below point Z instead of being above it.

Since the best fitting values of e_2 and e_3 are so small, I tried the experiment of letting them be zero while keeping the other values in the third column of Table X.2. The standard deviation in consequence rose only from $1°.70$ to $1°.73$, a trivial change.

Thus Ptolemy should have used only the simple model of Figure X.8 for Mercury rather than the model he did use. This is the model that results from setting e_2 and e_3 equal to zero. The circle around which point F travels in Figure X.3 has collapsed to the point D and D has moved into coincidence with Z. Thus the model has become a simple eccentric on which the center B of the epicycle travels.

It is interesting to see that the simple model of Figure
X.8 still preserves one important feature that Ptolemy wanted
to describe with his model. This is the doubling of the values of $L_\odot$ where the greatest and least values of $D_\mercury$ occur;
this doubling was shown in Figure X.2. In that figure, we
saw that the greatest east value of $D_\mercury$ came when $L_\odot = 100°$
in Ptolemy's time and the greatest west value came when
$L_\odot = 339°$. Likewise, the least east value came when $L_\odot =
274°$ and the least west value came when $L_\odot = 145°$. These
are the values that apply in Ptolemy's time.

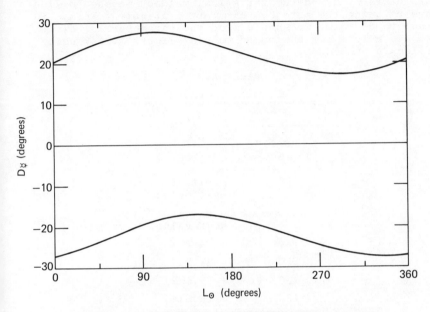

Figure X.9   The maximum elongation $D_\mercury$ of Mercury, considered as a function of the mean longitude $L_\odot$ of the sun, that results
from the model of Figure X.8. Figure X.9 agrees rather closely with
with Figure X.2, which is plotted using values calculated from
modern theory. The important point is that the crank mechanism of Ptolemy's model is not needed in order to give the most
striking feature of the situation. The occurrence of the double
maximum and the double minimum is a consequence of the
eccentric, not of the crank.

Figure X.9 shows the values of $D_\mercury$ given by the simple
model of Figure X.8, which uses the parameters shown in
Table X.2 except for setting $e_2 = e_3 = 0$. The greatest and
least values of $D_\mercury$ come at about the same values of $L_\odot$ in
both Figures X.2 and X.9, but the total swing in $D_\mercury$ is
greater in Figure X.9. This suggests that the best value
of e_1 is somewhat less than the value shown in Table X.2 if
we impose the restriction that e_2 and e_3 should both be
zero. It is not surprising that we should change the value
of e_1 somewhat if we impose restrictions that were not used
in deriving Table X.2.

Figure X.9 shows us that the crank mechanism, which is the most prominent feature of Ptolemy's model, is not the feature that produces the doubling of the greatest and least values of $D_{\mathbb{Y}}$, and that it is in fact irrelevant to the doubling phenomenon. The doubling comes from the eccentric, and it is present even if we omit the crank entirely.

The mathematical theory needed to derive Table X.2 did not exist in Ptolemy's time. Thus, if our object is to study what Ptolemy did, rather than how accurate his theory is, we should compare what Ptolemy did with what he should have done with the means that he had. That is, instead of finding the parameters by a modern statistical method, we should find them by using the methods that were available to him. In order to do this, we should find the parameters a, e_1, e_2, e_3, and r from the data used to plot Figure X.2 by using Ptolemy's procedure.

TABLE X.3

SOME MAXIMUM ELONGATIONS OF MERCURY

$L_{\odot}$, degrees	$D_{\mathbb{Y}}$, degrees
100	Greatest east value
340	Greatest west value
40	+ 24.277
220	- 21.327
130	+ 25.562
130	- 19.914

Table X.3 shows the sample of values that is exactly analogous to the sample that Ptolemy uses, but taken from the set of values used to plot Figure X.2. In the table, I have made the simplification that the greatest east value of $D_{\mathbb{Y}}$ comes when $L_{\odot} = 100°$ and that the greatest west value comes when $L_{\odot} = 340°$. These values are used only to locate apogee, and a moderate amount of approximation in them has a negligible effect upon the results. With these values, apogee comes at $220°$. After he locates apogee in his own time, Ptolemy uses the west (negative) value of $D_{\mathbb{Y}}$ at apogee ($220°$) and the east (positive) value of $D_{\mathbb{Y}}$ at the opposite point ($40°$) to find r and $e_1 + e_2$. His values of these quantities are given in Equations X.3 and X.4. He then uses both the east and west values at the point $90°$ before apogee ($130°$) to find e_1 and e_3; the values of e_1 and $e_1 + e_2$ then give him e_2.

TABLE X.4

PARAMETERS DERIVED FROM TABLE X.3

Parameter	Value
Apogee, degrees	220
First eccentricity (e_1)	0.04926
Second eccentricity (e_2)	0.01207
Crank radius (e_3)	0.00129
Epicycle radius	0.38646

Table X.4 shows the results of using Ptolemy's method with the values from Table X.3. These should be compared with the best fitting parameters from the last column of Table X.2. The values of r and of apogee agree closely. On the other hand, the value of e_1 agrees closely with the value that Ptolemy finds. The most interesting feature, however, is that both e_2 and e_3 are again small as they were in the last column of Table X.2. In other words, the use of accurate data in Ptolemy's method again leads us to the simple model of Figure X.8 rather than to the complicated crank model of Figure X.3.

Finally, we must note the direct reason that Ptolemy gives for needing the crank mechanism. In Chapter IX.8, he points out that the value of D_{V} is $23\frac{1}{4}$ degrees when $L_{\odot}$ is 10°. This is the position in which the center of the epicycle is at point G in Figure X.5. At this position, the value of D_{V} is the same whether the elongation is to the east or the west. Hence the sum of the two elongations at this position is $46\frac{1}{2}$ degrees. When $L_{\odot}$ is either 70° or 310°, according to the data from Table X.1, the sum of the two elongations is 47 3/4 degrees, greater than it is when $L_{\odot}$ is 10°.

From Figure X.6, we see that the sum of the two elongations is the angle subtended by the epicycle from E, the position of the earth. Since this angle is greater when $L_{\odot}$ is 70° or 310° than it is when $L_{\odot}$ is 10°, the center of the epicycle must be closer to the earth when $L_{\odot}$ is 70° or 310°, according to Ptolemy. This is the circumstance that Ptolemy brings about by means of his crank mechanism.

However, Ptolemy is mistaken. When $L_{\odot}$ is at 40°, which is the correct position of the point G in Figure X.5, the positive (east) elongation is 24°.277. Since the situation is not truly symmetrical, the west elongation is slightly smaller, and the sum of the two elongations is 47°.926. At the point 60° away, where $L_{\odot}$ is 100°, the sum of the two elongations is only 46°.788, and it is almost exactly the same where $L_{\odot}$ is 340°. Thus the deferent is closer to the earth at point G than it is at the points 60° away, just as it is in the simple model of Figure X.8.

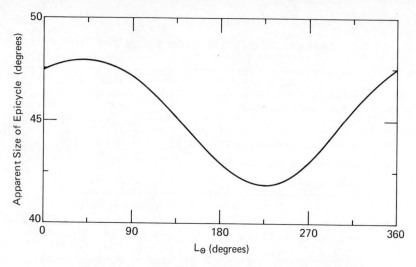

Figure X.10   The apparent size of the epicycle of Mercury, considered as a function of the mean longitude $L_\odot$ of the sun. The figure is based upon the parameters that were correct in Ptolemy's time. The apparent size is greatest when the deferent is closest to the earth (perigee) and it is least when the deferent is farthest from the earth (apogee). Contrary to Ptolemy's claim, there is only one perigee position.

In fact, the deferent is closer to the earth at the point G than it is at any other point whatsoever. We can see this with the aid of Figure X.10. In Figure X.10, I have calculated the sum of the maximum east elongation and the maximum west elongation, and I have plotted the sum as a function of $L_\odot$. The sum of the two elongations is the angle subtended by the epicycle as seen from E, which I have called the apparent size of the epicycle in the figure. We see that the curve has a single minimum, near the point where $L_\odot = 220°$. This must be apogee in Ptolemy's sense, that is, it is the point where the deferent is farthest from the earth. The curve also has a single maximum, near the point where $L_\odot = 40°$. This must be perigee in Ptolemy's sense, that is, it is the point where the deferent is closest to the earth.

Contrary to Ptolemy's claim that there are two perigee positions, then, we find that there is a single perigee position and a single apogee position. Once again we can find no justification for Ptolemy's crank mechanism.

From whatever point we start, then, we are unable to find a justification for the crank model. It does not give accurate results, it is not needed to give the doubling of the maximum and minimum values of $D_\text{ĝ}$, it does not result from analyzing data of reasonable accuracy, and it requires a doubling of the perigee position, a doubling that is contrary to fact. I can find only one explanation for it. This is that Ptolemy does not understand the situation, as I shall show in the next section.

5. The Fabrication of the Data

We can start our enquiry into the authenticity of the Mercury data with the simplest point, which concerns the anomaly and its rate of change. In his discussion of the anomaly, Ptolemy uses three observations, which were discussed in Section X.3. One of these was an observation that he claims to have made himself on 139 May 17. The others were made by Dionysios on -264 November 15 and -264 November 19. Ptolemy uses the last observation only to show that Mercury had not reached maximum elongation on -264 November 15, so we can ignore the last observation for the moment.

In Chapter IX.3 of the Syntaxis, Ptolemy says that Hipparchus has determined the rates at which the anomaly of each planet changes, and that he has corrected Hipparchus's rates by studies that he has made himself. He then says that Mercury makes 145 complete revolutions of its anomaly in 46 solar years plus 1 1/30 days, but he does not say how this rate was determined. He immediately makes the incompatible statement that this means a change of 52 200° in 16802; 24 = 16802.4 days. The number of degrees is correct, but the number of days should be 16802.38. However, if we use a change of 52 200° in 16 802.4 days, we get the value of $\gamma_{\mathbb{Q}}'$ in Equation X.7.

Thus the way in which Ptolemy finds the rate of change of the anomaly for Mercury is a mystery, and we find a similar mystery in his treatment of every other planet. With the moon, we saw that Ptolemy gives a value[†] which, he says, comes from certain Greek observations that he quotes. However, this is not so. The observations do not lead to the value that he states, which turns out instead to have a Babylonian origin. It may be that the planetary rates also have a Babylonian origin.

In Chapter IX.10 of the Syntaxis, Ptolemy says that the change in anomaly deduced from Dionysios's observation of -264 November 15 and his own observation of 139 May 17 agrees well with the value that he has found earlier, in Chapter IX.3. Thus he does not use his own observation. He finds the anomaly at the fundamental epoch (Equation X.8) by using the rate from Chapter IX.3 and the anomaly found from Dionysios's observation of -264 November 15.

By analyzing the lunar part of the observation in Section X.2 above, I showed that the alleged observation of 139 May 17 was fabricated. We can prove independently that it was fabricated by analyzing its planetary part. I shall defer this proof until Section XI.8, where I shall study the planetary part of the observation along with analogous observations of the other planets.

The two observations attributed to Dionysios on -264 November 15 and 19 may be genuine. The errors in the longitude of Mercury are about 0°.36 and 0°.57 [APO, Table XII.3]. The second error seems somewhat large, but not suspiciously so.

[†]See Section VI.4 above.

Now we can turn our attention to the 14 observations of $D_{\varnothing}$ that were summarized in Table X.1. These are the observations that Ptolemy uses to find apogee, its rate of precession, the three eccentricities, and the radius of the epicycle in his model of Mercury. We remember that the observations numbered 5 through 10 were made about 4 centuries before Ptolemy's time, that Ptolemy's contemporary Theon made another one, and that Ptolemy claims to have made the rest himself. We should also remember that he gives a false explanation of the necessity for using his own observations for numbers 11 and 12 in the table.

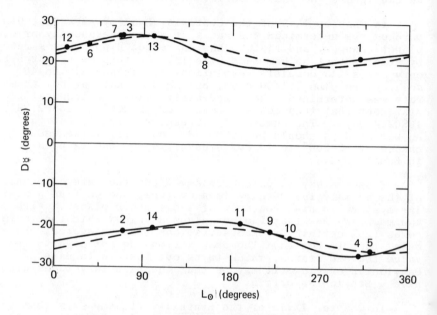

Figure X.11   The maximum elongation of Mercury in Ptolemy's time. The broken lines show the correct dependence of the maximum elongation $D_{\varnothing}$ on the mean longitude $L_\odot$ of the sun; these curves are the same as those in Figure X.2. The solid curves show the dependence that is obtained from Ptolemy's model. The points represent the 14 observations of $D_{\varnothing}$ that Ptolemy uses in finding the parameters of Mercury's orbit. In accordance with his deductions about the precession, I have added 4^0 to to the values of $L_\odot$ for the observations that were made 4 centuries before Ptolemy's time.

The broken lines in Figure X.11 are the curves of $D_{\varnothing}$ as a function of $L_\odot$ that were originally plotted in Figure X.2. That is, they give the correct variation of $D_{\varnothing}$ with $L_\odot$ for Ptolemy's time. The solid curves in Figure X.11 show the variation of $D_{\varnothing}$ with $L_\odot$ that is required by Ptolemy's model, and the numbered points represent the observations that Ptolemy uses in finding the six parameters listed in the preceding paragraph. That is, they represent the observations listed in Table X.1.†

We note that all 14 observations agree closely with the behavior of D_x that is required by Ptolemy's model. In most cases, the agreement is as close as the accuracy with which we can plot the points and curves. It happens that points 13 and 14 lie at places where the sets of curves touch or cross,‡ and these points agree with both sets of curves about equally well. Point 5 lies near a crossing point, but it agrees better with the correct curve than with the curve derived from Ptolemy's model. Point 6 is about midway between the two curves.

The behavior of the points in Figure X.11 is so striking that I think we can draw a conclusion immediately: The observations in Figure X.11 have been fabricated, with the possible exception of four points. The possible exceptions are points 5, 6, 13, and 14.

The evidence in favor of this conclusion is quite powerful. It comes from the facts that Ptolemy made four determinations of apogee which agree within a small fraction of a degree, and that he found all three eccentricities to be equal. Both findings conflict violently with reality.

First let us look at the apogee, which Ptolemy found to lie at 190° in his time and at 186° at a time four centuries earlier. As we saw in both Tables X.2 and X.4, the correct value is about 220°, and even a small body of data shows this. Specifically, Ptolemy's four determinations of apogee yield, in decimal notation, 189°.875, 190°.25, 189°.833, and 190°.000.‡ The standard deviation of these values from 190° is 0°.163. If we really believe that Ptolemy could find apogee with this precision, we must believe that he made four determinations that agree extremely closely but that are all in error by about 180 standard deviations in the same direction. The odds against this happening by chance are more than astronomical; they are so large that I am not sure they could be written in a reasonable space even with the aid of the so-called scientific notation that is designed to cope with large numbers. I have not even tried to calculate the odds.

The three eccentricities also agree with suspicious accuracy. Yet, as we saw in both Tables X.2 and X.4, they should not be the same. e_2 and e_3 should both be so small

---

†For the observations that were made 4 centuries before Ptolemy's time, I have added 4° to values of $L_\odot$ in Table X.1. This is the total amount of precession required by Ptolemy's theory.

‡The upper curves cross twice, once near $L_\odot = 100°$ and once near $L_\odot = 260°$. The lower curves also cross twice, once near $L_\odot = 330°$ and once near $L_\odot = 225°$. They nearly touch where $L_\odot = 95°$, but they do not cross there.

‡I have added 4°, the amount of precession according to Ptolemy, to the last two values, which were determined for an epoch 4 centuries before the first two values.

that we can replace them by 0 with only a trivial effect upon the accuracy of the resulting model.

However, the data used to find the three eccentricities depend upon the value used for apogee. In order to be as fair to Ptolemy as possible, let us grant his value of apogee and ask what we should then find for the eccentricities if we use correct values of $D_\math{y}$. The values that we should use on this assumption are the values found when $L_\odot$ has the values 10^u, 100°, and 190°, rather than the values used in Table X.3.

I did this analysis in Section XIII.9 of **APO**, and I shall only summarize the results here. If we use the convention that $\rho = 1$, the values of the epicycle radius r and of the three eccentricities are

$$r = 0.372\ 615, \qquad e_1 = 0.054\ 308,$$

$$e_2 = 0.006\ 154, \qquad e_3 = 0.047\ 231.$$

Thus e_3 is fairly close to e_1, although not nearly as close as Ptolemy makes it, but e_2 is still quite small. I then estimated the effect of changing the observations by a standard deviation. The result was that the value of e_2 found by Ptolemy is about 10 standard deviations away from what he should have found. The odds against this result happening by chance are approximately 10^{23} to 1.

Both formal analyses confirm powerfully what our instinct tells us when we look at Figure X.11: Most of the observations have been fabricated. This applies to the observations that Ptolemy attributes to other astronomers as well as to the observations that he claims to have made himself.

Before he fabricated the observations, Ptolemy had to choose the parameters that he used in the fabrication. There are six such parameters.† We may be sure that he chose to make apogee precess at the rate of 1° per century because this is his rate of precession of the equinoxes. We may also be sure that he chose to make all three eccentricities equal to each other, perhaps for esthetic or numerological reasons and perhaps for reasons that we have not thought of. In considering what he did, then, we may set $e_1 = e_2 = e_3 = e$. Thus there are really only three parameters involved in fabricating the data plotted in Figure X.11. These are a, the longitude of apogee, e, the value of each eccentricity in Figure X.3, and r, the radius of the epicycle.

In addition, we have the question of why Ptolemy chose to introduce the crank model at all. Put another way, why did he have three eccentricities, in view of the results that we have found? Why did he not use the simple model of Figure X.8?

---

†The parameters $\gamma_{\mathrm{y}Q}$ and $\gamma_\mathrm{y}{}'$, which relate to the anomaly, are not involved in the values of the maximum elongation.

His error with regard to the position of apogee may be explained by three of the observations of Dionysios, the observations that are numbered 5, 6, and 7 in Figures X.4 and X.11 and in Table X.1. We have already concluded that observations 5 and 6 may be authentic, although observation 7 probably is not. Let us assume that observations 5 and 6 are indeed authentic, and let us further assume that the value of $L_\odot$ in observation 7 is also authentic but that Ptolemy has altered the value of $D_\text{☿}$ by about 1°. This assumption explains Ptolemy's enormous error in the position of apogee.

It is easier to see the explanation by referring back to Figure X.4 in which there are not so many points and curves as in Figure X.11. We start with observation 5, for which $D_\text{☿}$ = 25;50 degrees west and $L_\odot$ = 318;10 degrees.[†] Ptolemy cannot find an observation made near the same time for which $D_\text{☿}$ = 25;50 degrees east, but he does find observations 6 and 7 that bracket the desired value of $D_\text{☿}$. Thus he interpolates to find a point between 6 and 7 that is not plotted; let us call it point $6\frac{1}{2}$ for brevity. For point $6\frac{1}{2}$, $D_\text{☿}$ = 25;50 degrees east and $L_\odot$ = 53;30 degrees.[‡] Hence, by symmetry, Ptolemy concludes, apogee is at the midpoint between the values of $L_\odot$ for observations 5 and $6\frac{1}{2}$. This is the point at longitude 185;50 degrees.

Ptolemy does not seem to understand the fundamental fact that a particular value of $D_\text{☿}$ occurs for two different values of $L_\odot$, except for the places where one of the curves has an extreme value. For the value $D_\text{☿}$ = 25;50 degrees east, one of the values of $L_\odot$ is symmetrical to observation 5 and one of the values is not. The value of $L_\odot$ for observation 5 is smaller than the value for which $D_\text{☿}$ has its largest west value. Hence the symmetrical point on the upper curve must be larger than the value for which $D_\text{☿}$ has its largest east value. That is, the symmetrical point is not the point $6\frac{1}{2}$, between points 6 and 7. It is the point a, on the other side of the extremum. For this point, $L_\odot$ = 124 degrees.

Thus the position of apogee is the midpoint between points 5 and a in Figure X.4. The midpoint is at 221°, quite closely.[‡] This is close to the value that we have found in Tables X.2 and X.4.

In using observations 5 and $6\frac{1}{2}$, Ptolemy uses the values of $L_\odot$, and he uses the fact that the values of $D_\text{☿}$ are equal and opposite, but he does not explicitly use the values of $D_\text{☿}$. After he misuses the observations in the fashion we have

---

[†] In this paragraph, the values of $L_\odot$ are those appropriate to the year -260. These are 4° smaller than the values used in plotting Figure X.11.

[‡] This is the value of $L_\odot$ that Ptolemy finds by interpolation. As I remarked in Section X.3, accurate interpolation gives $L_\odot$ = 53;19 degrees.

[‡] This does not agree exactly with the value that we have found by more elaborate methods. The discrepancy comes primarily from the fact that the geocentric orbit of Mercury does not have a line of symmetry, as Ptolemy assumes.

just described,† it is possible that he altered the values of $D_{\Ysmall}$ to make them accord with the parameters that he finally adopted. In Figure X.11, it looks as if he left observation 5 unaltered, and it may be that he also left observation 6 alone. It is likely that he did change the value of $D_{\Ysmall}$ for observation 7.

I said that he changed the value of $D_{\Ysmall}$ for a reason connected with the problem of fabricating observations of maximum elongation. It is difficult to calculate the value of $L_{\odot}$ for which a maximum elongation occurs. Even with modern mathematical methods, the calculations can be made only by trial and error methods, and each trial would have been laborious with the methods available to Ptolemy. Hence it is likely that he took the values of $L_{\odot}$, or at least the times of the observations, from genuine observations. However, once the value of $L_{\odot}$ is known, it is easy to fabricate the value of the elongation itself.

Anyone, even a scholar of high competence, can make an occasional error by inadvertence, and he can make such an error even on the simplest matters. Hence, if his misuse of points 5 and $6\frac{1}{2}$ were his only error with regard to the apogee of Mercury, we could readily understand and forgive. Unfortunately, this is not his only error, because he chooses to confirm the erroneous apogee that he has just found.

In a manner parallel to his use of points 5, 6, and 7, Ptolemy first quotes observation 8. Since he cannot find an old observation of a west elongation with this value, he uses observations 9 and 10 which bracket the desired value. Let me use point $9\frac{1}{2}$ for the point between points 9 and 10 that has the same elongation‡ as point 8.

Unlike points 5 and $6\frac{1}{2}$, the points 8 and $9\frac{1}{2}$ do have the correct symmetry relation. Thus, if they were valid observations, they would give a value of apogee near 220°, in violent disagreement with the value that Ptolemy finds from points 5 and $6\frac{1}{2}$ and from his lack of understanding of the basic symmetry. I think that there is little question about what happened next.

When he found a serious disagreement, whether by using valid forms of observations 8, 9, and 10 or by using other valid observations, a good scholar would have studied his procedures carefully until he found the reason for the discrepancy. Ptolemy did not do this. Instead, when he found that additional observations did not confirm his first value, he simply fabricated observations that did. Only thus, so far as I can see, can we explain the enormous error in observation 8, which agrees exactly with Ptolemy's model in

---

†Speaking rigorously, we can only conclude that Ptolemy misused some set of observations with the same basic symmetry properties as observations 5, 6, and 7. Since he does not mention any other possible set of observations, we may as well speak as if this were the set that he misused.

‡Except for the direction of the elongation.

spite of its error.

From this point on, it is easy to understand Ptolemy's actions with regard to apogee. From "observations" 5 through 10 he finds that apogee is at 186° at a time 4 centuries earlier. He adds 4°, obtaining 190° for apogee in his own time, and he proceeds to fabricate observations 1 through 4 to yield this value of apogee.

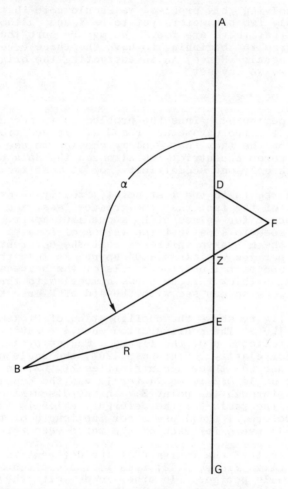

Figure X.12   The configuration of Ptolemy's model of Mercury when α = 120º. The epicycle has been omitted for clarity. The distances ZE, DZ, DF, and FZ are equal. Thus the connecting arm BF passes through the point Z. The distance BE is close to its minimum when α = 120º.

It is a little harder to see how Ptolemy handles observations 11 through 14, and the parameters r, e_1, e_2, and e_3 that he derives from them. We saw above that e_2 certainly was derived from fabricated data, and e_2 depends upon all of the observations 11 through 14. From Figure X.11, we see that observations 11 and 12 are surely fabricated, but we cannot tell about observations 13 and 14. However, Ptolemy has to have values of at least some of the parameters in order to fabricate the data. Our problem then is to find the origin of the parameters that Ptolemy used in the fabrication.

In solving this problem, we should note that there are really only two parameters yet to be found, although Ptolemy pretends that there are four. We may be sure that Ptolemy intends from the beginning to have the three eccentricities come out equal. Hence, in investigating the origin of the parameters, we may set

$$e_1 = e_2 = e_3 = e$$

from the beginning. Thus the problem is to account for the values of the two parameters r and e. In addition, we must account for the fact that Ptolemy chooses to use a model that requires three eccentricities although the data demand a model with only one eccentricity, or at most two.†

With regard to the last point, Ptolemy seems to have some idea of the fact that the greatest east and west elongations occur for values of $L_\odot$ about 120° apart, although he does not seem to know what the values of $L_\odot$ are. From this, he makes the mistaken inference that the deferent has to have two perigee positions 120° apart, so that the two perigees and apogee trisect the circle of the heavens. This leads him to think that he needs a model with three eccentricities, as we can see with the aid of Figure X.12.

The figure shows the configuration of Ptolemy's model when $\alpha = 120°$. The configuration when $\alpha = 240°$ is the left-to-right reflection of the figure. The epicycle has been omitted for clarity. Since $\alpha = 120°$, the angle ADF is also 120°. Since the three eccentricities ZE, DZ, and DF are equal, triangle DFZ is equilateral, and the connecting arm BF passes through the point Z. In the discussion, I shall use R for the radius of the deferent, which is the distance BE. As before, I shall use ρ for the length of the arm BF, and I shall use e for each of the three eccentricities.

We see that the radius R of the deferent is close to its minimum in Figure X.12; this is the configuration that Ptolemy calls perigee. In other words, with three equal eccentricities, the perigee positions occur about 120° apart Suppose, however, that points D and Z coincided, so that

---

†In the tests made with accurate data that have been describ both e_2 and e_3 come out essentially zero when we put apogee in the right place, and e_2 still comes out essentially zero when we put apogee in the wrong place at 190°.

there are only two eccentricities.† Then the perigee positions would occur close to the points where α equals 90° and 270°, which would make the perigee positions be 180° apart.

As we have seen, the separation of the positions of greatest east and west elongations depends upon the distance ZE, and we do not need the other eccentricities DZ and DF at all. However, Ptolemy does not seem to understand this point any more than he understands the implications of symmetry when he is trying to locate apogee. He seems to think that a separation of 120° between the points of greatest east and west elongation implies a model with three eccentricities.

There is an objection to concluding that this is the basis for Ptolemy's decision to use three eccentricities. As we shall see in the next chapter, the maximum elongation of Venus behaves in a similar manner to that of Mercury, but Ptolemy does not decide to use three eccentricities for Venus. Hence the origin of his model for Mercury may not be the separation of the greatest east and west elongation, and another possibility must be considered.

The following may be the origin of the triple eccentricity: After he found the wrong position for apogee, at 190°, Ptolemy looked at the opposite position, at 10°, and found the apparent size of the epicycle there. Then, perhaps by accident, he saw some observations which gave him the size of the epicycle at 70°. We see from Figure X.10 that the size is the same at 10° and 70°, to the accuracy that we can read the figure. However, it is possible that the data available to Ptolemy showed that the size was greater at 70° than at 10°, because of measurement error. This caused him to think that perigee was about 120° from apogee instead of being opposite to it. If he had checked this by looking at the symmetrical position, namely 310°, he would soon have found that this was not so. The apparent size of the epicycle at 310° is much smaller than at 10°, and it is not credible that measurement error would have made it seem larger.

Thus if Ptolemy studied the size of the epicycle at 70°, and if the errors in the data happened to make it look larger there than at 10°, and if Ptolemy did not check this finding at 310°, he might have been led to the model with three eccentricities. This hypothesis explains why he did not use the same model for Venus, but it has corresponding disadvantages. It requires an improbable set of events. Further, we notice from Table X.1 that Ptolemy claims to have made the measurements at 10° and 70° himself. If he had been led to his model by authentic measurements, it is hard to see why he would suppress them and substitute his fabricated data.

The hypothesis requires Ptolemy to act in a rather unsystematic manner, but this is not a disadvantage. He had to act in such a manner already before concluding that apogee was at 190°.

---

† A short time ago, we saw that D and Z become coincident if we use accurate observations for $L_{\odot}$ equal to 10°, 100°, and 190°.

Both explanations of the origin of the three eccentricities have advantages and disadvantages, and we cannot settle between them at present. Let us therefore leave this problem and look for the origin of the numerical values that Ptolemy chose for the two parameters r and e.

In order to find these parameters, Ptolemy needs only two independent observations. For example, let us consider observations 1 and 4 in Table X.1, for which the mean apogee distance α equals 120°. This is the situation for which Figure X.12 is drawn. Let δ denote the angle ZBE in the figure. Since BZ = ρ - e, it is simple to show the relation

$$e/(\rho - e) = \sin \delta/\sin (\alpha - \delta). \qquad (X.9)$$

We also know that the distance AE is the sum of ρ and the three eccentricities. Since each eccentricity equals e, and since Ptolemy uses the convention that AE = 120, we have

$$\rho + 3e = 120. \qquad (X.10)$$

It is trivial to solve Equations X.9 and X.10 simultaneously. Elementary trigonometry then gives us the value of R in the figure. Now the sum of the two elongations for observations 1 and 4 is 47;45 degrees, and half of this is 23;52,30 degrees. Since 47;45 is the total angle that the epicycle subtends at the distance R, we have r/R = sin 23;52,30, and this gives us r. The results are

$$e = 5;7,50, \qquad \rho = 104;36,30,$$
$$r = 39;15,57. \qquad (X.11)$$

These are not quite the same as the values in Equations X.4 and X.6. Even so, this may still have been Ptolemy's procedure. We are often unable to duplicate Ptolemy's calculations even when we know that we start from the same point.

The difficulty in assuming that Ptolemy found his parameters from observations 1 and 4, or from the pair 2 and 3 which he thought were symmetrical, lies in accounting for the values of elongation that he used. The two values of elongation were $26\frac{1}{2}$ and $21\frac{1}{4}$ degrees. At $L_\odot$ = 70°, the correct elongations are 25°.745 and 21°.869, while at $L_\odot$ = 310° the correct values are 25°.291 and 19°.897. It is possible to imagine that Ptolemy might have found a value of $21\frac{1}{4}$ degrees from some genuine observation with $L_\odot$ near 70°, but it is hard to imagine that he found a value of $26\frac{1}{2}$ degrees from a genuine observation with $L_\odot$ near either 70° or 310°.

Observations 13 and 14 are well suited for finding r and e. Since observation 13 is by Theon, and since it has normal accuracy, it may well be genuine, but observation 14 is by Ptolemy. By now, we are entitled to assume that any observation that Ptolemy claims to have made is fabricated. Still, we must admit the possibility that he departed from his usual habits and made a genuine observation on this

occasion, unlikely though it seems.† It is also possible
that Ptolemy used observations 5 and 6, which may well be
genuine, to find r and e. Of the explanations that the
evidence makes available to us, this is perhaps the most
likely.

We can now summarize the situation. Most of the observations that Ptolemy uses in finding the parameters of Mercury's orbit are fabricated. This applies to the observations that he attributes to other astronomers as well as to
the observations that he claims to have made himself. However, there is not enough information to let us reconstruct
Ptolemy's procedures with certainty, and we cannot be certain which observations, if any, are genuine. The observations with the best chances of validity are probably those
numbered 5 and 6 in Table X.1, which are attributed to Dionysios, and the observation of -264 November 19, which is
also attributed to Dionysios.

It is clear that Ptolemy has little understanding of
the problems posed by the motion of Mercury. He is aware
that the value of $L_\odot$ when Mercury has its greatest east elongation is about 120° from the value when Mercury has its
greatest west elongation, and this seems to be about all
that he knows. He does not know within 30° what the values
of $L_\odot$ are. He thinks that the separation of 120° requires
the complex crank mechanism of Figure X.3 when all it requires is the simple eccentric-epicycle mechanism of Figure
X.8. He also thinks that the deferent curve of Mercury
must have two perigee positions, whereas Figure X.10 shows
that it has only one perigee position. Finally, he does not
understand the symmetry properties of $D_\ogamma$, and in consequence
he misplaces the symmetry axis by about 30°.

6.   Two Other Possible Models for Mercury

One basic problem posed by the motion of Mercury is the
wide variation of the maximum elongation $D_\ogamma$ with the position
$L_\odot$ of the mean sun. As Ptolemy points out (see Section X.1
above), this means that either r, the radius of the epicycle,
or R, the radius of the deferent, must depend upon $L_\odot$. Ptolemy tries a model in which R is made to vary and, as we have
seen, his model does not work very well. Instead of recog-
nizing this fact and trying the effect of letting r vary,
Ptolemy pretends that his model does work and he uses fabricated data to maintain this pretense. It is natural to see
what happens when we do let the radius r vary.

One way to do this is to use the double epicycle model
of Figure X.13. We specifically take the distances DZ and
ZE to be equal, although we allowed the possibility that
they might be different when we originally discussed the
model in Section IV.6. The line SD rotates uniformly about
the point D although the center of its circle is at Z. The
earth is at E. The symbol ♈ denotes the vernal equinox.

────────────

We must remember that Ptolemy finds two of the e's by using
observation 14 among others, and that some of the observations used to find the e's are certainly fabricated.

The line SC rotates uniformly. If we measure its rotation from the line SD, its period of rotation is the synodic period of Mercury. That is, it rotates at the same rate as the anomaly γ in Figure X.3. Mercury is at the point M. The line from C to M rotates very slowly if at all.

In finding the parameters of the model, Greek astronomers would have had to resort to the analysis of observation. However, we can find most of them readily from modern theory. The point S is the sun, so that the direction from E to A is that of the sun's apogee. If the distance AZ is taken as 1, the distances DZ and ZE both equal the eccentricity of the sun's orbit, and the angle SDγ is the mean longitude of the sun. The radius SC of the larger epicycle is the semi-major axis of Mercury's heliocentric orbit, and the angle between SC and Dγ is the mean longitude of Mercury in its orbit about the sun. The direction from C to M is the direction of the aphelion of Mercury. The distance CM is harder to estimate.

Still in modern terms, the purpose of the two epicycles is to represent the heliocentric motion of Mercury. We saw in Section IV.2 that an epicycle cannot represent the motion accurately in both longitude and distance. If we choose the epicycle radius† to represent the motion accurately in longitude, the variation in distance is twice the correct amount. If we choose the radius to represent the variation in distance, the variation in longitude is half the correct amount.

When we look at the heliocentric orbit of Mercury from the earth, both the distance of Mercury from the sun and its longitude about the sun affect what we see. Therefore the accuracy of the double epicycle is limited. In studying the double epicycle model for Mercury, we should leave the radii of both epicycles as parameters to be adjusted until we have the best accuracy.

If we let r_1 and r_2 denote the two radii, the values that give the best fit to the motion of Mercury are

$$r_1 = 0.375\ 106, \qquad r_2 = 0.116\ 156. \qquad \text{(X.12)}$$

I found these in the same way that I found the best parameters for Ptolemy's model. Now the semi-major axis of Mercury is 0.387 098 6 and its eccentricity (in the modern definition) is 0.205 614 2. If we chose r_2 to give the correct variation of distance, r_2 would be the product of these values, namely 0.079 593 0. If we chose r_2 to give the correct variation of longitude, it would be twice as large.

The fitting program gives a value of r_1 that is close to the semi-major axis, as we would expect. The value of r_2 is about midway between the value that gives the longitude accurately and the value that gives the sun-Mercury distance correctly. This is also about what we expect.

The standard deviation of the error in the double epicycle model is 1°.69. This is only slightly better than the

---

†This means the radius of the smaller epicycle in the present context.

performance of the simple eccentric-epicycle model of Figure X.8, and the extra accuracy is probably not worth the extra complication.

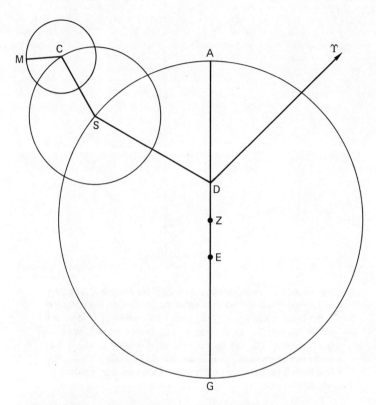

Figure X.13   A double epicycle model for Mercury. This is the model of Figure IV.5 specifically adapted to the motion of Mercury. The distances DZ and ZE are equal. In modern terms, A is the direction of the apogee of the sun. The distances DZ and ZE are equal to each other and to the eccentricity of the sun's orbit. Line SD rotates uniformly at the rate of 1 revolution per year. Line SC rotates uniformly in such a way that its direction is that of the heliocentric mean longitude of Mercury. Line CM is always parallel to the direction of the aphelion of Mercury.

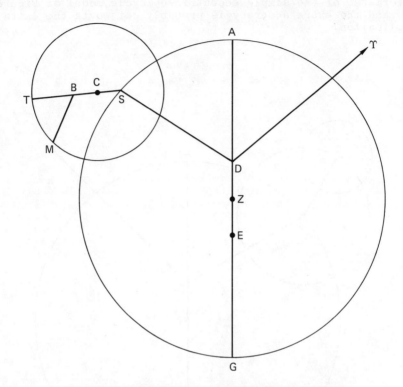

Figure X.14   A double equant model for Mercury. The part
of this that involves the motion of the point S on the deferent
is the same as in Figure X.13. Here, the motion of M around S
is also represented by an equant. The circle on which M moves
has its center at C, but the rotation of M is uniform about the
point B. Distances BC and CS are equal. In modern terms, the
line from S to T always points toward the aphelion position
of Mercury.

        Figure X.14 shows still another possible model.   It is
suggested by strong but simple theoretical reasons that were
available to Ptolemy, although I shall defer the explanation
of the reasons to Section XII.5.   The way in which the motio
of S is generated is the same in Figure X.14 as in Figure
X.13.   For the two epicycles of Figure X.13, I have substi-
tuted another figure that is obviously just like the mechanism that controls the motion of S.   For this reason, we may
call Figure X.14 the double equant model.   In modern terms,
the direction from S to T is the direction of Mercury's aphelion.   M is the position of Mercury, and the line BM rotates uniformly about B; the angle between BM and DƳ is the
heliocentric mean longitude of Mercury.   The distances BC
and CS both equal 0.079 583 0, and C is the center of the
circle on which M moves.

The standard deviation of the double equant model is
0°.32, or 19'. This is a dramatic improvement over other
models; it is better than Ptolemy's model by almost a factor
of 10. This is the accuracy that results from using the
parameters suggested by modern theory. Since the double
equant does not give the same detailed motion as that which
we get from gravitational theory, the accuracy could probably
be improved somewhat by letting the parameters vary. However,
it does not seem worthwhile to undertake the variation of the
parameters.

It is often said that the motion of Mercury is too complicated to be represented adequately by the methods available to Greek astronomers. However, the double equant, which
was available to Greek astronomers, gives an accuracy of 19',
and this is probably better than the accuracy of observation.
If this is the criterion of adequacy, the double equant must
be considered adequate. Further, as I have said and as I
shall demonstrate in Section XII.5, the reasons that lead to
the double epicycle were well within the horizons of the
Greek astronomers. Why they did not develop it, and why
medieval Islamic and European astronomers did not develop
it, are questions for the psychologists, philosophers, and
historians of science.

CHAPTER XI

VENUS AND THE OUTER PLANETS

1. The Basic Property of the Equant

In the equant model of celestial motion, which was described in Section IV.5 and pictured in Figure IV.4, the distinguishing feature is the use of the three separate points E, Z, and D in controlling the motion of the point B. The fact that B is taken as the center of an epicycle is not itself the important point. If we omit the epicycle, the equant model still gives an interesting model for the motion of B.

The angle BDϒ is the mean longitude of the point B and the angle BEϒ is the true longitude of B. The difference between them is the angle that has been called the equation of the center. The equation of the center is denoted by the symbol e_c in this work. Let δ denote the angle EBD in Figure IV.4. Elementary geometry gives the result that angle AEB equals angle ADB minus δ, so that δ is the negative of e_c.

The angle δ is the sum of two angles that are not drawn in, namely angles ZBD and EBZ. Let these angles be denoted by δ_2 and δ_1, respectively. As before, let e_1 be the first eccentricity ZE, let e_2 be the second eccentricity DZ, and let α be the angle ADB. So far as the motion of B is concerned, α is its mean anomaly measured from the apogee position A, and we have called this the mean apogee distance. We can calculate e_c by using the following equations in turn:

$$\sin \delta_2 = e_2 \sin \alpha,$$
$$\theta_2 = \alpha - \delta_2,$$
$$\tan \delta_1 = e_1 \sin \theta_2 / (1 + e_1 \cos \theta_2),$$
$$e_c = - \delta_1 - \delta_2 .$$

(XI.1)

In these expressions, and in those that immediately follow, the distance ZB is 1 and all angles are in radians.

Let us assume that e_1 and e_2 are so small that we can ignore their cubes or higher powers. By using Equations XI.1, we can write e_c in the simpler form:

$$e_c = -(e_1 + e_2) \sin \alpha + \tfrac{1}{2} e_1 (e_1 + e_2) \sin 2\alpha .$$

(XI.2)

Equation IV.8 gives the equation of the center for an elliptical orbit whose eccentricity is e. If we change the symbol for the anomaly from γ to α,

$$e_c = -2e \sin \alpha + (5/4)e^2 \sin 2\alpha .$$

(XI.3)

Equation XI.3, like Equation XI.2, is correct only if we can ignore the cubes or higher powers of e.

There are three interesting cases.

1. Let e_1 = 2e and let e_2 = 0. Equation XI.2 becomes

$$e_c = - 2e \sin \alpha + 2 e^2 \sin 2\alpha. \qquad (XI.4)$$

With these values, the model is the same as the simple eccentric model described in Section IV.3. When we compare Equations XI.3 and XI.4, we see that the error is $(3/4)e^2$ sin 2α. This is a result that we have obtained before, and I give it here only as a check on Equation XI.2.

2. Let e_1 = (5/4)e and let e_2 = (3/4)e. Then

$$e_c = - 2e \sin \alpha + (5/4) e^2 \sin 2\alpha. \qquad (XI.5)$$

If it is legitimate to ignore e^3 and higher powers, this variant of the equant model gives the same longitude of point B as the elliptical model.

3. Let e_1 = e_2 = e. Then

$$e_c = - 2e \sin \alpha + e^2 \sin 2\alpha. \qquad (XI.6)$$

The error in longitude is $\frac{1}{4}e^2$ sin 2α. This is not as good as Case 2, but it is only a third of the error present in the simple eccentric model.[†]

Now let us consider the variation of the distance of B from the earth E that the various models give. For the correct elliptic orbit, the distance varies from 1 + e at apogee to 1 - e at perigee.[‡] In Case 1, the simple eccentric, the distance varies from 1 + 2e at apogee to 1 - 2e at perigee; as we have seen, this is twice the correct variation. In Case 2, which gives the greatest accuracy in calculating the longitude, the distance varies from 1 + (5/4)e at apogee to 1 - (5/4)e at perigee, so that the greatest error in distance is $\frac{1}{4}e$ if there is no error in longitude. In Case 3, the distance varies from 1 + e at apogee to 1 - e at perigee. For this case, the distance is correct and the greatest error in longitude is only $\frac{1}{4}e^2$. Hence this case agrees closely with an elliptic orbit in both longitude and distance.

In fitting a model to observations, the nature of the observations determines whether Case 2 or Case 3 gives the best results. If only the longitude is important, Case 2 will be preferred. If distance is highly important, Case 3 will be prefered. In some actual cases, both longitude and distance are important. When this happens, we expect

---

[†]The epicycle model gives the same result as the simple eccentric.

[‡]With the convention that the average distance is 1.

e_1 to be between e and $(5/4)e$, if we take $e_1 + e_2$ to be 2e.

When we are using the equant model to describe the motion of a planet, we are interested in the longitude of the point B, but we are also interested in its distance. In the theory of Mercury, the angle that the epicycle subtends at the earth E plays an important role in determining the parameters of the motion. It will play an equally important role in the theories of Venus and Mars, our closest neighbors. Its role in the theories of Jupiter and Saturn is also important, but not as important as in the close planets. Since the angle in question is directly related to the distance from E to B, and since the longitude of the planet is directly related to the longitude of B, the theories of the planets require attention to both longitude and distance.

2. The Balance of Two Eccentricities

On the basis of the preceding section, we can look at Ptolemy's study of Mercury from a new viewpoint that will also be useful in looking at Venus and the outer planets. We remember that Ptolemy used the first ten observations of Mercury simply to locate the position of apogee at different times. In the ensuing discussion, let us ignore the fact that he did a poor job of locating it. Let us also ignore the fact the geocentric orbit of Mercury does not have a symmetry axis, as Ptolemy assumed. Instead, let us accept his idea of a symmetry axis and his location of apogee, and concentrate on what followed.

First he used an observation made when B was at the point which Ptolemy claims was the position of apogee. Then he used an observation made when B was at the opposite point. The configuration is shown in Figure X.5. In order to avoid the continuing use of awkward circumlocutions, I shall say that the observations were made at apogee and perigee, even though this is far from the truth. I shall also speak as if there were a symmetry axis. I hope that the reader will not be misled by this usage.

In Figure X.5, Mercury is at maximum elongation in both observations. For the observation at apogee A, for example, Mercury was at maximum elongation to the west of the sun. Since A is on the symmetry axis, the maximum east elongation would be the same size. Hence the total angle subtended by the epicycle is twice the maximum elongation in the observation. Thus the subtended angle is 38;06 degrees at apogee. By a similar argument, the subtended angle is 46;30 degrees at perigee.

The subtended angle is twice the angle whose sine is r/R. R is the distance from E to either A or G, as the case may be, and r is the radius of the epicycle. Since r is the same in both observations, the observations give the difference between the distances EA and EG. Half of this difference is the interval from E to the midpoint between A and G. I shall call this interval the "distance eccentricity"; it is determined from considerations of distance and not from the oscillations of the true longitude about the mean longitude.

-303-

After he used the observations illustrated in Figure X.5, Ptolemy used two observations made when the center of the epicycle was at the same point. These observations are illustrated in Figure X.6. They yield the difference between the mean longitude of the point B and its true longitude at the point where the angle α is $90°$.[†] At this point, the angle 2α is $180°$, so that $\sin 2\alpha = 0$. This point is midway between apogee and perigee.

Let us write the equation of the center in the form

$$e_c = \Lambda \sin \alpha + \kappa \sin 2\alpha.$$

The second term on the right is zero since $\sin 2\alpha = 0$ at the point where the observations were made. The observations then give us the coefficient Λ of $\sin \alpha$. I shall call Λ the "longitude eccentricity"; it is determined from the oscillation of the true longitude about the mean longitude.

Since we know that the deferent curve should really be an ellipse, we know that the distance eccentricity should be half of the longitude eccentricity. Near the beginning of Chapter X.6 of the Syntaxis, Ptolemy writes: "For the other three planets Mars, Jupiter, and Saturn, we have found that the same theory of motion fits all three of them, and this is the same theory that we have found for Venus. That is, the eccentric circle which carries the center of the epicycle is described about a center which is the midpoint between the center of the zodiac and the point which makes uniform the revolution of the epicycle. Because, for each of these stars the inequality that one finds from the deviation relative to the zodiac is double the inequality that one finds from the retrograde motions at the greatest and least distances of the epicycle."

For Mercury, we found a distance eccentricity by studying the maximum elongation at apogee and perigee, and we shall do the same for Venus. The outer planets Mars, Jupiter and Saturn do not have maximum elongations. However, they do exhibit the interesting phenomenon of retrograde motion that was described in Section I.5,[‡] and the arc over which the motion is retrograde, like the maximum elongation, depends upon how far away is the center of the epicycle. Thus, when Ptolemy refers to the "inequality that one finds from the retrograde motions", he is apparently referring to what I have called the distance eccentricity. Similarly, the "inequality that one finds from the deviation relative to the zodiac" is apparently what I have called the longitude eccentricity. If these interpretations are correct, Ptolemy is saying that the longitude eccentricity is twice the distance eccentricity for Venus and the outer planets. This is closer than taking them to be equal, but it is not highly

---

[†]More accurately, α is $270°$, which is the same as $-90°$. Here, the size of the angle and not its sign is the major consideration.

[‡]The inner planets also exhibit retrograde motion, but it is not as important a phenomenon for them as maximum elongation is.

accurate, as we shall see.

Equation XI.2 shows us that the longitude eccentricity, in the equant model, equals $e_1 + e_2$; this is the total distance from E to D in Figure IV.4. The figure also shows us that the distance from E to A is $1 + e_1$ while the distance from E to G is $1 - e_1$. Hence e_1 equals the distance eccentricity. The requirement that this be half of the longitude eccentricity obviously means that we must have $e_1 = e_2 = e$.

We saw in Section IV.5 that the equant model has eight important parameters. They are

a, the longitude of apogee,

e_1, the first eccentricity,

e_2, the second eccentricity,

r, the radius of the epicycle,

L_0, the mean longitude at some epoch,

n, the rate at which the mean longitude L_p changes,

γ_0, the anomaly at some epoch,

γ', the rate at which the anomaly changes.

When we know these parameters, we calculate L_p and γ by the relations

$$L_p = L_0 + n(t - t_0),$$
$$\gamma = \gamma_0 + \gamma'(t - t_0).$$

(XI.7)

These are the same as Equations IV.9, and they are repeated here for the convenience of the reader. t_0 denotes the epoch for which we have found the constants L_0 and γ_0.

We also saw that L_p, $L_\odot$, and γ are related. For Venus,

$$L_p = L_\odot.$$

(XI.8)

For the outer planets,

$$L_p = L_\odot + \gamma.$$

(XI.9)

Equations XI.8 and XI.9 reduce the number of parameters by two, because $L_\odot$ is known from the solar theory and is not to be evaluated again for the planetary theory. This means that we take γ_0 and γ' as parameters. Once they are known, we can find L_0 and n from Equation XI.9 for an outer planet. Of course, we can find L_0 and n for Venus from Equation XI.8 whether we already know γ_0 and γ' or not.

Thus the equant model has only six parameters that must be found by the analysis of planetary observations. These are the six that remain after we drop L_0 and n from the list above. If we imposed the condition that $e_1 = e_2 = e$, we would have only five parameters. However, Ptolemy does not impose this condition at the beginning of his study of Venus and the outer planets. He leaves it as a result that is to be established by the analysis of observations.

3. Ptolemy's Parameters for Venus

Ptolemy's derivation of the parameters in his model of Venus parallels closely his treatment of Mercury. He first uses observations of maximum elongation in order to find all the parameters that do not involve the anomaly; there are four such parameters for Venus. He then uses observations made when Venus was not at maximum elongation in order to study the anomaly.

We remember that Ptolemy allows the longitude of apogee to increase with time. For Mercury, he uses six observations of maximum elongation in order to show that the longitude of apogee increases by 1° per century. He then assumes without proof that the apogees of the other planets move at the same rate. For Venus, then, he does not use old observations in order to find how its apogee changes. He uses eight observations in order to find a, the eccentricities, and r, and he uses them for Venus in just the same way that he used eight observations for Mercury.

He defines the maximum elongation of Venus in the same way that he defined the maximum elongation of Mercury. That is,

$$D_{\venus} = \lambda_{\venus} - L_{\odot},$$ (XI.10)

in which $\lambda_{\venus}$ is the longitude of Venus when it is at maximum elongation and $D_{\venus}$ is the value of the maximum elongation. The observations of maximum elongation are found in Chapters X.1, X.2, and X.3 of the Syntaxis. I shall summarize two typical records.

In Chapter X.1, Ptolemy says that Theon observed Venus in the evening of the day that we call 132 March 8. It was at maximum elongation, and it was west of the center of the Pleiades by a distance equal to their width. It was also slightly south of them. From this, Ptolemy concludes that $\lambda_{\venus}$ = 31;30 degrees. From his theory of the sun, he calculates $L_{\odot}$ = 344;15 degrees. Hence $D_{\venus}$ = + 47;15 degrees;[†] the plus sign means that Venus was east of the sun.

In Chapter X.3, Ptolemy says that he observed Venus at maximum elongation in the morning of the day that we call 134 February 18. When he compared Venus with the star α Scorpii, he found $\lambda_{\venus}$ = 281;55 degrees, and he calculates $L_{\odot}$ = 325;30 degrees. Hence $D_{\venus}$ = -43;35 degrees.

[†]I attach a sign to $D_{\venus}$ for convenience but Ptolemy does not use one. He says that $D_{\venus}$ was so many degrees east or west, as the case may be.

TABLE XI.1

OBSERVATIONS OF VENUS AT MAXIMUM ELONGATION

Identifying number	Date	Observer	$L_\odot$, degrees[a]	Elongation, degrees[a]	Error in measured longitude, degrees[b]
1	132 Mar 8	Theon	344;15	47;15 E	+0.31
2	140 Jul 30	Ptolemy	125;45	47;15 W	-1.08
3	127 Oct 12	Theon	197;52	47;32 W	+0.47
4	136 Dec 25	Ptolemy	272;04	47;32 E	+0.91
5	129 May 20	Theon	55;24	44;48 W	-0.05
6	136 Nov 18	Ptolemy	235;30	47;20 E	+1.49
7	134 Feb 18	Ptolemy	325;30	43;35 W	+1.12
8	140 Feb 18	Ptolemy	325;30	48;20 E	+0.51

[a]As stated by Ptolemy. E means that Venus was east of the sun, W that Venus was west of it.

[b]The measured value minus the value calculated from modern theory.

The eight observations of $D_\male$ that Ptolemy uses are summarized in Table XI.1, which has the same arrangement as Table X.1 for Mercury. They are listed in the order that he uses them.

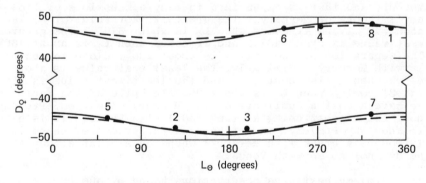

Figure XI.1   The maximum elongation of Venus. The broken curves show how the maximum elongation $D_\male$ depends upon $L_\oplus$, the mean longitude of the sun, according to modern theory. These curves are calculated using the orbital parameters that are appropriate to Ptolemy's time. The solid curves show how $D_\male$ depends upon $L_\oplus$ according to Ptolemy's theory of Venus. The numbered points are observations of $D_\male$ that Ptolemy uses in finding the orbital parameters of Venus.

Figure XI.1 shows how $D\female$ depends upon $L_\odot$, the mean longitude of the sun; it is like Figure X.11 for Mercury. The broken curves show the dependence that is calculated from modern theory using the orbital parameters that were correct in Ptolemy's time. The solid curves show the dependence that we calculate from Ptolemy's theory of Venus, using the parameters that he derives. The numbered points represent the eight observations of $D\female$ listed in Table XI.1.

Ptolemy starts with observations 1 and 2. Observation 1 was made by Theon and observation 2 was made by himself. Since the elongation was 47;15 degrees east for the first one and 47;15 degrees west for the second, the line connecting perigee and apogee bisects the values of $L_\odot$. Hence perigee and apogee are at $55°$ and $235°$, but he cannot tell yet which is which.

Ptolemy confirms this result with observations 3 and 4, which also form a matched pair. Observation 3 is by Theon and observation 4 is one that Ptolemy claims to have made himself. The longitudes indicated by this pair of observations are 54;58 and 234;58 degrees. I think we can agree that Ptolemy is justified in rounding these to $55°$ and $235°$, in confirmation of the first result.

Before we go further, we should notice two important points. First, there is no breakdown of symmetry here as there was in Ptolemy's treatment of Mercury. Observation 1 is east of the point where Venus has its greatest positive elongation while observation 2 is west of the point where Venus has its greatest negative elongation. Observations 3 and 4 likewise are symmetrically located. As a result, the position of apogee that Ptolemy finds is rather accurate.

Second, Venus shows the same phenomenon that we saw with Mercury and that is emphasized in many discussions of Ptolemy's theory of Mercury. The greatest east value of $D\female$ is about $48°.1$ and it comes when $L_\odot$ is about $310°$. The greatest west value is about $48°.3$ and it comes when $L_\odot$ is about $180°$. The separation of the values is about $130°$, close to what it is with Mercury. Similarly, the least east value is about $44°.5$ when $L_\odot$ is about $130°$ and the least west value is about $44°.3$, coming when $L_\odot$ is about $0°$. In spite of the fact that we have almost a duplicate of the Mercury situation, Ptolemy does not suggest using a crank model with Venus. It is hard to know why Ptolemy, in defiance of the evidence, chose to introduce a crank with Mercury† when, on similar evidence, he did not do so with Venus.

Ptolemy next uses observations 5 and 6, one by Theon and one by himself, which were made at perigee and apogee. Since the elongation is smaller in size when $L_\odot$ = 55;24 degrees than it is when $L_\odot$ = 235;30 degrees, we see that $55°$ is apogee and that $235°$ is perigee. We analyze this pair of observations just as we analyzed the pair of Mercury

---

†However, as I said in the preceding chapter, I suspect it is connected somehow with his error about symmetry and his consequent error in locating apogee.

observations in Figure X.5, except for what we call the distance from E to the center. With Mercury, that distance was $e_1 + e_2$; here it is simply e_1. The observations give[†]

$$e_1 = 1\tfrac{1}{4}/60 = 0.020\ 833,$$

$$r = 43\tfrac{1}{6}/60 = 0.719\ 444.$$

(XI.11)

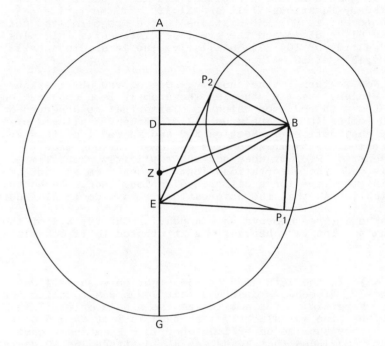

Figure XI.2   The configuration of Ptolemy's model of Venus for the alleged observations of 134 February 18 and 140 February 18. On both occasions, according to Ptolemy, point B was in the same place, and line DB is nearly perpendicular to line AG. Venus was at P_1 on 134 February 18 and at P_2 on 140 February 18. This figure is the same as Figure XIII.6 of <u>APO</u>.

Ptolemy completes the use of the maximum elongations with observations 7 and 8, both made by himself and both made when $L_\odot$ is midway between perigee and apogee, to high accuracy. The situation is illustrated in Figure XI.2, which is the same as Figure XIII.6 in <u>APO</u>. The total angle P_2EP_1 that is subtended by the epicycle equals 91;55 degrees, the sum of the two elongations. Since we now know the radius of the epicycle, we find the distance EB is 1.000 8333.[‡]

[†]The values that Ptolemy finds are close to the ones that come from solving the situation accurately, and I shall give only Ptolemy's values in the rest of the discussion.

[‡]I am using the convention that the radius of the deferent circle is 1. Ptolemy uses the convention that it equals 60, and I am thus dividing his values by 60.

The angle DBE is half of the difference between the elongations, or 2;22,30 degrees, so that the distance DE is $2\frac{1}{2}/60$
= 0.041 667, exactly twice the value of e_1 in Equation XI.11,
as Ptolemy solves the triangle. Since DE is the sum $e_1 + e_2$,
we have immediately that

$$e_1 = e_2 = 0.020\ 833. \hspace{3cm} (XI.12)$$

We should note the greatest and least distances that are
implied by Equations XI.11 and XI.12. If the radius of
the deferent is 60, the distance to the apogee of the deferent is $61\frac{1}{4}$ = 61;15 and the greatest distance is this plus
$43\frac{1}{8}$ = 43;10, or 104;25. Similarly, the least distance is
60 - 1;15 - 43;10 = 15;35.

Now we turn to the anomaly. One of the observations
that Ptolemy uses in studying the anomaly is a conjunction
of Venus with the moon which he claims that he observed on
138 December 16. I analyzed all of Ptolemy's alleged planetary conjunctions in Section X.2 and showed that they are
fabricated. Ptolemy also uses two observations made by
Timocharis. Ptolemy does not say explicitly where Timocharis made the observations, but he uses them without ad-
justing the time for a difference in longitude. Hence the
implication is that the observations were made in Alexandria.

These three observations appear in Chapter X.4 of the
Syntaxis. The two observations attributed to Timocharis are
as follows:

(1) In the 12th hour of the night between -271 October
11 and 12, Timocharis observed that Venus was touching the
star η Virginis. If the time was the middle of the 12th
hour, the time was slightly less than half of an ordinary
hour before sunrise on -271 October 12. From the record,
Ptolemy concludes that Venus was at longitude 154;10 degrees,
that the longitude of the mean sun was 197;03 degrees, and
that the anomaly of Venus was 252;07 degrees. The elongation
was 42;53 degrees to the west.

(2) Four days later, Timocharis found that Venus was
4;40 degrees east of η Virginis, so that it had moved 4;40
degrees in 4 days. Its longitude was now 158;50 degrees,
the mean sun was now at 200;59 degrees, and the elongation
was now 42;09 degrees. Since the elongation decreased between the two observations, Venus was already past maximum
elongation on the morning of -271 October 12. This tells
Ptolemy that the anomaly was 252;07 degrees and not the
other value that would give the same longitude. Ptolemy
makes no other use of this observation.

Ptolemy calculates that the anomaly of Venus was 230;32
degrees at the time of the conjunction of 138 December 16.
He then says that the change in anomaly since -271 October
12 agrees closely with the change calculated from the tables

that he has already constructed.[†] The value of γ' that we
find from these observations is in fact close to the value
that Ptolemy uses in his tables, but it is clearly not the
same. Thus Ptolemy does not tell us how he found γ' for
Venus. His practice here is the same as it was for Mercury.

4. The Accuracy of Ptolemy's Model for Venus

In order to study the accuracy of Ptolemy's model for
Venus, I calculated the position of Venus from modern theory
at 51 times separated by intervals of 44 days. I took the
center time to be 137 July 20. The interval was chosen so
that both Venus and the earth would make nearly an integral
number of revolutions around the sun in the total span of
2200 days. I also calculated the position of Venus from
Ptolemy's theory for the same times. In order to compensate
for the error in Ptolemy's position of the equinox, I added
$1°.1$ to his value of apogee. In order to avoid the systematic error in his expression for $L_\odot$, I used Newcomb's theory
for this quantity in Ptolemy's model of Venus.

TABLE XI.2

A COMPARISON OF PTOLEMY'S PARAMETERS FOR VENUS
WITH THOSE THAT GIVE THE BEST FIT

Parameter	Ptolemy's Value	Best Fit
Apogee, degrees	56.1^b	60.202
First eccentricity (e_1)	0.020 833	0.012 883
Second eccentricity (e_2)	0.020 833	0.014 711
Epicycle radius	0.719 444	0.722 804
Anomaly at the epoch,[a] degrees	273.825	274.675
Anomalistic rate, degrees per day	0.616 508 734	0.616 595 195

[a]Noon, Alexandria time, 137 July 20.

[b]Ptolemy's value is $55°$, but this is referred to an
erroneous equinox. I added $1°.1$ in order to refer
to an accurate position of the equinox.

The maximum error in the 51 positions of Venus calculated from Ptolemy's theory was $4°.29$, which came on 137
March 10, and the standard deviation of the error in his
theory is $1°.01$. These are considerably better than the errors that we found with Mercury, but they are far from just-
ifying Dreyer's statement[‡] that Ptolemy's theories are as
accurate as Greek observations. We have yet to find a body
for which the theory is as accurate as observation, even
approximately.

[†]In Chapter IX.4 of the Syntaxis, Ptolemy gives tables for
calculating the anomaly, the mean longitude, and the position of apogee for each planet.

[‡]See Sections VII.6 and X.4.

I then varied the parameters in the model, except for the rate of motion of apogee, until the standard deviation was as small as possible. In doing this, I took the motion of apogee from modern theory. With the best fitting values of the parameters, the largest error fell from $4°.29$ to $0°.32$ and the standard deviation fell from $1°.01$ to $0°.14$. Both improvements are by about an order of magnitude.

Ptolemy's parameters and those which give the best fit are compared in Table XI.2, which is the same as Table XIII.1 in APO. We see that the two values of apogee differ by only about $4°$, so that Ptolemy did a rather good job of finding it. I expected that the most important difference between Ptolemy's parameters and the best fitting ones would be in the eccentricities. However, when I used the best fitting eccentricities along with Ptolemy's values for the other parameters, the standard deviation fell only from $1°.01$ to $0°.70$. Changing apogee, the radius of the epicycle, and the rate of change of the anomaly had only a trivial effect. Almost all of the remaining change in accuracy from $0°.70$ to $0°.14$ came from the change in the anomaly at the epoch.

We should notice an important contrast between Ptolemy's models of Mercury and Venus. As we saw in Section X.4, it is not possible to achieve reasonable accuracy with Ptolemy's model of Mercury. Even when we use the best-fitting parameters, the standard deviation of the Mercury model falls only to $1°.70$. Ptolemy's model of Venus, on the other hand, gives a standard deviation of $0°.14$ when we use the best-fitting parameters, and this is better than the accuracy of observation. The basic reason is that the epicycle of Mercury that is required by the physical situation (its heliocentric orbit) has a high eccentricity while the required epicycle of Venus is nearly circular. Ptolemy's models, however, use a circular epicycle for both.

While Ptolemy's model of Venus is capable of high accuracy, Ptolemy did not come close to realizing its potential. Thus we may doubt that he really understood the problems involved, even if we word them in ancient rather than modern terms.

5. The Fabrication of the Venus Data

Figure XI.1 shows how the observations of maximum elongation compare with accurate values found from modern theory and with the values required by Ptolemy's theory. It is analogous to Figure X.11 for Mercury. The situations shown by the two figures are quite different. In Figure X.11, most of the observations agree strikingly well with the values from Ptolemy's theory and few of them agree well with the accurate theory. This allows us to say immediately that most of the observations of Mercury in Figure X.11 are fabricated, whether they are observations that Ptolemy claims to have made himself or whether he attributes them to other observers.

In Figure XI.1, in contrast, only a few of the observations obviously agree better with Ptolemy's theory than with the accurate theory. Observation 4, which is one that Ptolemy claims to have made himself, and observation 3, which is attributed to Theon, agree better with the accurate theory than with Ptolemy's theory, while observation 1, due to Theon, does not agree well with either. This does not mean that the observations are authentic, but it does mean that Ptolemy's procedures are different for the two planets.

Let us first look at the matter of finding apogee. It is doubtful that the time when maximum elongation occurs can be found with an accuracy as good as a day from naked eye observations. Hence the expected error in a value of $L_\odot$ in Table XI.1 is more than $1°$. When two values of $L_\odot$ are averaged in order to find a position of apogee, the error in the average should be somewhat less than the error in a single observation. We should be fairly close to the truth if we say that the expected error in a position of apogee, found in the way that Ptolemy finds it, is of the order of a degree or so.

Ptolemy finds two values of apogee, which disagree by only $2'$, although the expected disagreement is a degree or so, and although the actual error in both values is close to $4°$, as we infer from Table XI.2. The probability that Ptolemy's agreement could have happened by chance is obviously so tiny that we do not need to estimate it.

This means that the agreement in the two values of apogee was the product of fabrication, not of observation. Since the agreement of the two values can be produced by the fabrication of a single value, all we can say at the moment is that at least one of observations 1 through 4 in Table XI.1 is fabricated, but we cannot say which one. The errors exhibited in Figure XI.1 do not give us a basis of choice.[†]

Now let us turn our attention to the inference of the other parameters that can be found from the maximum elongations. These are the two eccentricities e_1 and e_2, and the radius r of the epicycle. We should notice a remarkable fact. Ptolemy finds the sum $e_1 + e_2$ from the analysis of observations 7 and 8 in Table XI.1. The value that he finds is exactly $2\frac{1}{2}$, on a scale in which the deferent has a radius of 60. This is exactly the eccentricity of the sun in Hipparchus's theory of the sun. I do not believe that Ptolemy points out this exact agreement, and I do not believe that it is a coincidence.[‡]

[†]The reader may have noticed that the errors shown in the last column of Table XI.1 do not necessarily agree with the errors shown in the figure. The reason is that the errors in Table XI.1 are the errors in the measured longitudes of Venus. If there is an error in judging when maximum elongation occurs, the error in the measured longitude is not the same as the error in the elongation.

[‡]According to Equation XI.2, the "longitude eccentricity" is the sum $e_1 + e_2$. This is the eccentricity that we find by studying the longitude of the point B in the equant model. That is, the sun and point B both exhibit the same eccentricity in their longitudes.

Pannekoek [1961, p. 142] points out this agreement and
explains it as follows: " ... one has to remember that
Ptolemy's observations were elongations, i.e., distances
from the real sun, from which the longitudes of Venus were
computed by means of reductions taken from solar tables based
on the erroneous eccentricity of 1/24†; so that he got as
'result' what he had put in under another name." That is, in
Ptolemy's results, the equation of the center for point B is
automatically equal to the equation of the center for the sun,
as a consequence of Ptolemy's methods of analyzing the data.

It seems to me that there are two errors in this explanation: (1) The equation of the center for B is not the
same as that for the sun. The longitude eccentricity (which
is the same as the maximum value of the equation) is the same
for both, but the apogees differ by more than 10°. That is,
Ptolemy does not get back as result what he puts in under another name. (2) In no case is the observed quantity an elongation from the real sun. In every case, at least as Ptolemy gives the records, the observed quantity is the longitude
of Venus, measured against the star background without reference to the position of the sun. Ptolemy then finds the
elongation $D_{\mathcal{Q}}$ by subtracting the calculated longitude of the
mean sun from the observed longitude of Venus. The position
of the real sun is not involved in this process, and hence
the equation of its center is not involved.

This means that an error in measuring the longitude of
Venus directly affects the values found for the eccentricities
If Ptolemy finds $e_1 + e_2 = 2\frac{1}{2}$ by using genuine observations,
the coincidence must come from accidental errors in observing
the longitude of Venus against the star background. Let us
next look at the process by which Ptolemy finds the value of
$e_1 + e_2$.

The most important number in the process is half of the
difference between the elongations in observations 7 and 8 in
Table XI.1. This is the angle DBE in Figure XI.2, which
Ptolemy finds to be 2;22,30 degrees. The value of $e_1 + e_2$ is
the product of the sine of this angle by a length that is
very close to 60, and Ptolemy finds that $e_1 + e_2$ is exactly
2;30. This means that he finds the value to lie within the
range from 2;29,30 to 2;30,30, a range of length 0;01.

Now let us look at the effect of measurement error. If
the standard deviation in a single measurement of longitude
is 0°.25, the standard deviation of the difference between
the elongations is 0°.25 × √2, the standard deviation of the
angle DBE is half of this, and the standard deviation of the
sine of the angle is about 0.00308. Thus the standard deviation of $e_1 + e_2$ is about 60 × 0.00308 = 0;11.

Ptolemy finds that $e_1 + e_2$ lies in a preassigned range
whose width is 1/11 standard deviations. If the value that
he finds were approximately correct, the probability that
such a close agreement could happen by chance is about 0.036,

---

†This is an eccentricity of $2\frac{1}{2}$ if the deferent has a radius
of 60.

-314-

about 1 chance in 28. However, we see from Table XI.2 that
the correct value of $e_1 + e_2$ is 0.027 594 for a deferent
radius of 1, or about 1;39 if the deferent radius is 60.
Thus the error in Ptolemy's value is 0;51, about 4.5 standard deviations. The probability that $e_1 + e_2$ would lie in
a preassigned range whose width is 1/11 standard deviations
and whose center lies at an error of 4.5 standard deviations
is about 7 out of 1 000 000. Thus there is little doubt
that Ptolemy fabricated either observation 7 or observation
8 or both. Personally, I have no doubt that he fabricated
both.

Ptolemy uses observations 5 and 6 to find e_1 and r. He
finds that e_1 is 1;15, exactly half of $e_1 + e_2$. That is, e_1
lies in a preassigned range whose length is again 0;01. When
we analyze the effect of an error of $0°.25$ in a single measurement of longitude, the probability that e_1 should agree
exactly (within rounding error) with half of $e_1 + e_2$ is about
0.0009, less than 1 chance in 1000. There is little doubt
that one of these observations is also fabricated. Since one
of the observations is by Theon and the other is by Ptolemy,
the one by Ptolemy must be fabricated. It is possible that
the observation by Theon is genuine, but it is also possible
that Ptolemy fabricated it.

As we have said, Ptolemy does not tell us how he finds
the value of $\gamma_\female{}'$; he does not find it from the observations
that he quotes. However, we know that the observation that
he claims to have made on 138 December 16, which he uses to
confirm the value of $\gamma_\female{}'$, is fabricated. The observations
made by Timocharis may be genuine. The errors in them [APO,
Table XII.5] are only about $10'$ and $5'$, and there is no apparent reason why Ptolemy should fabricate them.

I shall show in Section XI.8 how Ptolemy fabricated the
observation dated 138 December 16. We must ask here how he
fabricated the other false observations, which is the same
as asking how he chose the parameters that he uses in his
model. I think there is little doubt that he chose a priori
to make e_1 and e_2 equal to each other and to half of the
solar eccentricity. This does not mean that he subscribed
to a heliocentric theory of Venus. In fact, we see that he
did not subscribe to this theory, because the apogee that
he uses in the theory of Venus is not the same as the solar
apogee. Rather, it is likely that he started with the solar
eccentricity because it was a convenient value that works
fairly well.

If he had decided upon the values of the e's before he
started to work, he had to find only apogee a and the radius
r of the epicycle from the observations of maximum elongation
in Table XI.1. He could find these from any two genuine observations. Since there are three observations by Theon in
Table XI.1, I tentatively conclude that he found a and r by
using two of these. When I try using observations 1 and 5,
for example, I find that apogee is $55°$, when rounded to the
nearest degree, and I find r = 42;45. Apogee agrees exactly
with Ptolemy's value, and r is not far from Ptolemy's value
of 43;10. I think it is possible that the difference comes
simply from differences in computational technique.

Thus it is possible that Ptolemy did find his parameters by using two of the three observations attributed to Theon, and there is no fundamental difficulty in concluding that his own observations are fabricated. However, he has not left enough information to let us reconstruct his procedures uniquely.

6.   Ptolemy's Model for the Outer Planets

Although Ptolemy uses the equant model of Figure IV.4 for Mars, Jupiter, and Saturn as well as for Venus, there are three differences between his treatment of Venus and his treatment of the outer planets.

First, Ptolemy "proves" that the apogee of Mercury moves at the rate of 1° per century, and he uses this value for all of the other planets without further ado. Similarly, he "proves" that the two eccentricities in the model are equal for Venus, and he applies this equality to the outer planets without discussion.

Second, the mean longitude of an outer planet is related to the mean longitude of the sun by Equation XI.9, whereas Equation XI.8 applies to Venus.

Third, maximum elongation for an outer planet and maximum elongation for Venus mean different things. The maximum for Venus is about 48°, but the maximum for an outer planet is 180°. The condition when the elongation is 180° is usually called opposition; this means that the planet and the mean sun are on opposite sides of the earth.

Let λ_P denote the geocentric longitude of an outer planet and let D_P denote its elongation. Ptolemy's definition of D_P is analogous to his definition of $D_\smallsun$ and $D_\female$. That is,

$$D_P = \lambda_P - L_\odot. \tag{XI.13}$$

When $D_P = 180°$, I shall say that the planet is at mean opposition.

There are some immediate consequences of these differences. Ptolemy started his treatment of Venus by assuming that the eccentricities e_1 and e_2 might be different, and thus he had to find six parameters from the analysis of observations. For the outer planets, he assumes $e_1 = e_2 = e$ from the outset. This means that only five parameters must be found for each outer planet.

Another consequence is that an outer planet is truly at perigee when it is at mean opposition. In Figure IV.4, which shows the equant model, this means that the planet P is at the point on the epicycle that is closest to the earth E at the same time that it is in mean opposition. In other words, when P, E, and the mean sun all lie on a straight line, the point B also lies on the same line at the same instant. This statement is not obvious, I think. Ptolemy proves it as a theorem in Chapter X.6 of the Syntaxis, and I have outlined the proof elsewhere [APO, p. 445].†

Since the points B, P, and E lie on a straight line at
mean opposition, the radius of the epicycle points directly
at the earth when the planet is in mean opposition. This
means that we cannot find the radius of the epicycle from
mean oppositions, although we could find it from maximum
elongations of Venus. On the other hand, the condition of
mean opposition is sensitive to the anomaly and, conversely,
we can find the anomaly at the time of a mean opposition.

Hence the parameters that can be found from mean oppositions are not the same as those that we find from maximum
elongations of Venus. From maximum elongations of Venus,
the parameters found are apogee, the eccentricity, and the
radius of the epicycle. From mean oppositions of an outer
planet, we find apogee, the eccentricity, and the anomaly.

Ptolemy handles each of the outer planets in exactly
the same way. For this reason, I shall describe only his
treatment of Mars.

TABLE XI.3

MEAN OPPOSITIONS OF THE OUTER PLANETS
ALLEGEDLY OBSERVED BY PTOLEMY

Planet	Date	Hour	Longitude, degrees
Mars	130 Dec 15	01	81;00
	135 Feb 21	21	148;50
	139 May 27	22	242;34
Jupiter	133 May 17	23	233;11
	136 Aug 31	22	337;54
	137 Oct 8	05	14;23
Saturn	127 Mar 26	a	181;13
	133 Jun 3	16	249;40
	136 Jul 8	12	284;14

[a]Ptolemy does not give the hour in the
record, but his calculations show that
the time was 18 hours.

---

†In a heliocentric theory, P is really the earth and E is
the outer planet. The statement that must be proved as a
theorem in a geocentric theory becomes a tautology in the
heliocentric theory. See APO, Section XIII.6.

He starts with three observations of mean opposition, all allegedly made by himself. He gives the record of the observations in Chapter X.7 of the Syntaxis, apparently in a single sentence: † "We observed the first in the 15th year of Hadrian, at 1 equinoctial hour after midnight of the 26th and 27th of the Egyptian month Tubi, at the 21st degree of Gemini; the second in the 19th year of Hadrian, at 3 hours before midnight of the 6th and 7th of the Egyptian month Pharmuthi, at 28;50 degrees of Leo; and the third in the 2nd year of Antoninus, at 2 hours before midnight of the 12th and 13th of the Egyptian month Epiphi, at 2;34 degrees of Sagittarius."

Stated in modern terms, the mean oppositions occurred as follows: (1) at 01 hours on 130 December 15, at longitude 81 degrees, (2) at 21 hours on 135 February 21, at longitude 148;50 degrees, and (3) at 22 hours on 139 May 27, at longitude 242;34 degrees.

The data from these observations, and from the corresponding observations for Jupiter and Saturn, are summarized in Table XI.3. Only a few comments need to be made. Ptolemy says that the observations of Jupiter and Saturn were made by means of the astrolabe, but he does not say this for Mars; there is probably no significance to this omission. Ptolemy omits the hour of the Saturn observation on 127 March 26, but his calculations show that he meant it to be 18 hours. For most of the observations, Ptolemy does not say whether the hour is an hour of the night or an hour as we use the term (an ordinary or equinoctial hour), but his calculations show that ordinary hours are meant.

We should consider how such observations could be made. The observer needs to have an idea about the time when an opposition should occur. At a safe interval before the expected time, he measures the longitude of the planet, calculates the mean longitude of the sun at the time of the observation, and finds the elongation D_p. If he has been careful, the planet is not yet at opposition. If it is not, he makes additional observations until he finds that the planet is past opposition. From the times of observations before and after opposition, he interpolates to find the time and the circumstances of the opposition. Ptolemy in fact implies that he found the time and longitude of the Saturn opposition of 133 June 3 by using observations made before and after the stated time.

Suppose that the observations of longitude were rounded to the nearest multiple of 5'. The longitude found by interpolation would usually not be a multiple of 5'. Thus, although most of the longitudes in Table XI.3 are given to a precision that is impossible for observations, this does not of itself imply fabrication. However, it does not imply that the observations are genuine, either. If Ptolemy fabricated the observations, he would still have to use the same basic

---

†I do not know how accurately the division of modern editions into sentences, paragraphs, and chapters reflects the original text.

technique. The time of an opposition cannot be calculated by formula. In fabricating, Ptolemy would have to guess a time, calculate λ_p and $L_\odot$, and thence find D_p. He would then have to repeat the process until he found times lying on either side of opposition, and finally estimate the fabricated time, λ_p, and D_p by interpolation.

Ptolemy uses the three oppositions of each planet to find apogee a, the eccentricity e, and the anomaly γ_3 at the time of the third opposition. Worded this way, the matter is impossible. If we can find γ_3, we must also be able to find γ_1, the anomaly at the time of the first opposition. This is so because it does not matter, so far as astronomical theory is concerned, which way time flows, and there can be no basic difference between the first and last observations.[†] Thus, if we can find a, e, and γ_3, we can find γ_1 also. Since it is mathematically impossible to find four parameters from three observations, it is not possible to find a, e, and γ_3 unless we have information beyond the three observations.

What Ptolemy actually does is to use a value of γ' without calling attention to the fact. As I discussed in Section X.5, Ptolemy describes how the angular velocity γ' can be found by observing the lengths of time required for the longitude and the anomaly to repeat approximately. He does this in Chapter IX.3 of the Syntaxis, and the values of γ' that he gives there are rather accurate. Actually it would be satisfactory, in analyzing the oppositions, to use a crude value of γ', because the time span of the observations for a given planet is less than 10 years.

Now let t_1, t_2, and t_3 denote the times of the oppositions in Table XI.3 for a particular planet, and let γ_1, γ_2, and γ_3 be the values of anomaly at these times. Then

$$\gamma_1 = \gamma_3 - \gamma'(t_3 - t_1),$$

$$\gamma_2 = \gamma_3 - \gamma'(t_3 - t_2).$$

In analyzing the three oppositions, there are originally five unknowns, namely a, e, γ_1, γ_2, and γ_3. By using the equations just given, we can eliminate γ_1 and γ_2, leaving only three unknowns to be found from the three observations.

It is a difficult calculation to find a, e, and γ_3 from the circumstances of the observations, and I shall not take the space to describe the method here.[‡] I carried out the calculations for Mars in order to be sure that I understood the method, with the results

---

[†] Rigorously, there must be a small amount of friction in the solar system, and so there is a determined direction of time flow. However, the accumulated effects of friction on the planets since the time of Greek astronomy is about equal to the standard deviation of a Greek observation. Thus we can certainly neglect friction in this discussion.

[‡] It is given in APO, Section XIII.6.

$$\alpha = 115;29,37 \quad \text{degrees},$$

$$e = 0.100\ 003,$$

$$\gamma_3 = 171;24,57 \quad \text{degrees}.$$

In giving the value of e, I use the convention that the deferent has a radius of unity. Ptolemy does a more accurate job of solving the observations than he does in most circumstances. His values are

$$\alpha = 115;30 \quad \text{degrees},$$

$$e = 0.1, \qquad\qquad\qquad\qquad\qquad (XI.14)$$

$$\gamma_3 = 171;25 \quad \text{degrees}.$$

From the values of γ_3 and γ', Ptolemy then calculates the anomaly of Mars at his fundamental epoch.

Only the radius r of the epicycle remains to be found. Because of the difference between the inner and the outer planets that has been discussed, finding the radius of an outer planet is analogous to finding the anomaly of Venus, and Ptolemy does this by using the alleged planetary conjunctions that were discussed in Section X.2. Ptolemy can calculate the anomaly at the time of the observation from γ_3 in Equation XI.14, along with the value of γ' from his planetary tables. It is then a trivial calculation to find r from the longitude of the planet and the mean longitude of the sun. Ptolemy finds that r = 39;30 if the deferent radius is 60, or

$$r = 0.658\ 333$$

if the deferent radius is 1.

Ptolemy does one more thing for each planet. He cites an ancient observation and uses it in a further study of the anomaly. The observations are as follows:

(1)   In Chapter X.9 of the <u>Syntaxis</u>, Ptolemy says that Mars was observed to be touching the star β Scorpii in the morning of -271 January 18. He does not say where the observation was made nor by whom, and he does not give the hour. In his calculations, he clearly assumes that the observation was made at sunrise in Alexandria. From the observation, he deduces that the longitude of Mars was 212;15 degrees, that the mean longitude of the sun was 293;54 degrees, and that the anomaly of Mars was 109;42 degrees.

(2)   In Chapter XI.3, Ptolemy says that Jupiter occulted δ Cancri on the morning of -240 September 4. Again he does not give the observer, the place, or the hour. He assumes that the observation was made at sunrise in Alexandria. Ptolemy concludes that the longitude of Jupiter was 97;33 degrees that $L_\odot$ was 159;56 degrees, and that the anomaly of Jupiter was 77;02 degrees.

(3)   In Chapter XI.7, Ptolemy says that Saturn was 2
digits† south of γ Virginis in the evening of -228 March 1.
He takes this to be at sunset in Alexandria.   He deduces
that the longitude of Saturn was 163;10 degrees, that the
mean sun was at 336;10 degrees, and that the anomaly of
Saturn was 183;17 degrees.‡

I shall use Jupiter to illustrate Ptolemy's use of
these observations. At the beginning of Chapter XI.3, Ptolemy says that he has chosen (2) from among the ancient observations of Jupiter as one of the most accurate, and that
he will use it to find the parameter that we are calling γ'.
After he finds the anomaly at the time of the observation,
he compares it with the anomaly at the time of the third
opposition (137 October 8) of Jupiter in Table XI.3.   He
finds that there were 377 Egyptian years‡ plus 127 days plus
23 hours between the two observations, and that the anomaly
of Jupiter changed by 345 complete revolutions plus 105;45
degrees in this time. He then says that the change in anomaly agrees closely with the value found from the tables that
he has already given.

By his wording, both here and in Chapter IX.3, Ptolemy
intends to give the impression that he finds γ' from the
observations that he quotes, I believe. We have already
seen that the observations do not lead to his values of γ'
for Mercury and Venus, and the same thing is true for the
outer planets. For Jupiter, for example, the change in anomaly is 124 305 3/4 degrees, the time interval is 137 732 23/24
days, and the rate obtained from these values is 0.902 512 7430
degrees per day. In sexagesimal notation, this is

$$\gamma' = 0;54,9,2,45,8,57 \text{ degrees per day.}$$

However, the value that Ptolemy uses in constructing his
tables of Jupiter is

$$\gamma' = 0;54,9,2,46,26,0 \text{ degrees per day.}$$

It is possible that Ptolemy made an error in division in each
of the five cases, but this does not seem likely.

---

†In records like this one, a digit means 1/12 of a degree,
or 5'.

‡The observations of Jupiter and Saturn agree with the results of modern calculations about as well as we can expect.
At the time stated for the Mars observation, however, Mars
was about 50' from β Scorpii [APO, Section XII.4], and it
does not seem possible that this condition was described as
touching. Since Mars and the star were nearly touching in
the morning of -271 January 16, it is possible that the recorded date was wrong, or that Ptolemy made a mistake in
reading it. Ptolemy definitely uses -271 January 18 as the
date in his calculations.

‡The Egyptian year contains exactly 365 days.

7. The Accuracy of Ptolemy's Model for the Outer Planets

As I did with Mercury and Venus, I prepared a computer program to calculate the geocentric longitude of an outer planet from Ptolemy's model for the outer planets, and I

TABLE XI.4

A COMPARISON OF PTOLEMY'S PARAMETERS FOR MARS
WITH THOSE THAT GIVE THE BEST FIT

Parameter	Ptolemy's Value	Best Fit
Apogee, degrees	116.6[b]	117.977
First eccentricity (e_1)	0.1	0.097 536
Second eccentricity (e_2)	0.1	0.099 953
Epicycle radius	0.658 333	0.657 767
Anomaly at the epoch,[a] degrees	219.189	219.140
Anomalistic rate, degrees per day	0.461 575 567	0.461 504 019

[a]Noon, Alexandria time, 137 July 20.

[b]Ptolemy's value is 115°.5, but this is referred to an erroneous equinox. I added 1°.1 in order to refer to an accurate position of the equinox.

TABLE XI.5

A COMPARISON OF PTOLEMY'S PARAMETERS FOR JUPITER
WITH THOSE THAT GIVE THE BEST FIT

Parameter	Ptolemy's Value	Best Fit
Apogee, degrees	162.1[b]	162.197
First eccentricity (e_1)	0.045	0.047 089
Second eccentricity (e_2)	0.045	0.043 153
Epicycle radius	0.191 667	0.192 401
Anomaly at the epoch,[a] degrees	110.936	111.031
Anomalistic rate, degrees per day	0.902 512 842	0.902 523 801

[a]Noon, Alexandria time, 137 July 20.

[b]Ptolemy's value is 161°. I have added 1°.1 in order to refer apogee to the correct equinox.

TABLE XI.6

A COMPARISON OF PTOLEMY'S PARAMETERS FOR SATURN
WITH THOSE THAT GIVE THE BEST FIT

Parameter	Ptolemy's Value	Best Fit
Apogee, degrees	234.1[b]	235.963
First eccentricity (e_1)	0.056 944	0.064 977
Second eccentricity (e_2)	0.056 944	0.052 820
Epicycle radius	0.108 333	0.104 871
Anomaly at the epoch,[a] degrees	173.779	173.552
Anomalistic rate, degrees per day	0.952 146 738	0.952 137 765

[a]Noon, Alexandria time, 137 July 20.

[b]Ptolemy's value is 233°. I have added 1°.1 in order to refer apogee to the correct equinox.

TABLE XI.7

ERRORS IN PTOLEMY'S MODEL OF VENUS AND
THE OUTER PLANETS

Planet	Greatest error, degrees		Standard deviation, degrees	
	With Ptolemy's parameters	With best fitting parameters	With Ptolemy's parameters	With best fitting parameters
Venus	4.29	0.32	1.01	0.14
Mars	0.89	0.91	0.44	0.36
Jupiter	0.47	0.21	0.17	0.09
Saturn	0.79	0.19	0.38	0.10

compared the longitudes calculated in this way with those
calculated from modern theory. After doing this, I varied
the parameters in Ptolemy's model until the standard deviation of his model was a minimum. The results of the comparisons are summarized in Tables XI.4, XI.5, and XI.6.
These tables show Ptolemy's values of each parameter and the
values that give the minimum error (best fit), using the convention that the radius of the deferent is unity.

The greatest errors and the standard deviations are compared in Table XI.7. I have included Venus in the table so
that we may contrast Ptolemy's treatment of Venus with that
of the outer planets. We saw in Section XI.4 that Ptolemy

did a poor job with Venus; his errors are almost an order
of magnitude worse than those that we get with the best
fitting parameters. For Mars, in contrast, he does almost
as well as it is possible to do. The greatest error that I
found with the best fitting parameters is actually slightly
greater than with Ptolemy's parameters, but this is probably
a sampling accident. Ptolemy's performance with Jupiter and
Saturn is of intermediate quality.

In the discussion of the equant model in Section XI.2,
we introduced the "distance eccentricity" and the "longitude
eccentricity". The former is determined by the variation of
distance between the earth E and the point B on the deferent
in Figure IV.4. The latter is determined by the difference
between the actual longitude and the mean longitude of point
B. The distance eccentricity is the distance ZE in the figure, which I have denoted by e_1. The longitude eccentricity
is the distance DE, which is the sum $e_1 + e_2$. If the distance variation affects the observations much more than the
longitude variation, we expect to have $e_1 = e_2$. If the longitude variation is much more important, we expect to have
$e_1 = (5/3)e_2$. In the first case, $e_1/(e_1 + e_2) = 0.5$. In
the second case, $e_1/(e_1 + e_2) = 0.625$. We expect an actual
case to lie between these extremes.

TABLE XI.8

BEST FITTING VALUES OF THE ECCENTRICITIES IN
PTOLEMY'S MODEL OF VENUS AND THE OUTER PLANETS

Planet	e_1	$e_1 + e_2$	$e_1/(e_1 + e_2)$
Venus	0.012 883	0.027 594	0.466 88
Mars	0.097 536	0.197 489	0.493 88
Jupiter	0.047 089	0.090 242	0.521 81
Saturn	0.064 977	0.117 797	0.551 60

Table XI.8 summarizes the information about the best
fitting eccentricities for Venus and the outer planets.
Interpreting the e's is complicated somewhat by the fact
that the equant model has a circular epicycle, while the
orbit that forms the natural epicycle always has a measurable eccentricity. This fact means that we cannot interpret the eccentricities of Venus in a ready fashion, since
Venus is so close to the earth. For the outer planets, however, as we go to more distant planets, the apparent size
of the epicycle decreases and the importance of representing
its size correctly also decreases. Since the variation in
its apparent size depends upon the variation of distance,
the importance of the distance eccentricity should decrease
steadily as we go to the more distant planets.

This is just what happens in Table XI.8 for the outer
planets. The ratio in the last column is almost exactly 0.5

for Mars, meaning that the variation of distance is quite
important in determining its geocentric longitude. The ratio
moves steadily toward 0.625 as we go to Jupiter and Saturn,
which shows that the variation of distance has less effect in
determining their geocentric longitudes. Ptolemy should not
have found $e_1 = e_2$ for Jupiter and Saturn.

8.   The Fabrication of the Data for the Outer Planets

For each outer planet, Ptolemy quotes only five observations. Four of these are observations that he claims to
have made himself, and the other is one of the ancient observations (one for each planet) described in Section XI.6,
with dates between -271 and -228. Thus, if Ptolemy has fabricated his observations of the outer planets, we have few
ways to test the matter.

Since Ptolemy takes the two eccentricities to be equal,
there are only five parameters to be found from observation.
They are a, e, r, γ_0, and γ'. Further, as we saw in Section
XI.6, Ptolemy does not find γ' from the five observations
that he quotes, although his language is apparently designed
to give the impression that he does. This means that he
finds only four parameters from five observations. The re-
dundant observation gives us an opportunity to test for fabrication.

Ptolemy never says explicitly that dividing the change
in anomaly by the time interval between two observations
gives γ'. Instead, he first says that he will use the observations to find γ'. Then, when he has found the change
in anomaly between two observations, he says that the change
agrees well with the change calculated from the tables that
he has already given. Here we have a situation exactly analogous to Ptolemy's equinox and solstice observations in Section V.4.

We remember from Section V.4 that Ptolemy quotes some
earlier observations of the solstices and equinoxes. He
then says that he will verify the length of the year that
Hipparchus had found by using some observations that he has
made himself with great care. Instead, what he does is to
take Hipparchus's length of the year, multiply it by the
appropriate number of years, and add the result to the time of
an earlier observation. Finally he rounds this fabricated time
to the nearest hour and claims that this is the time he has
observed.

We should see if Ptolemy has done the same thing with
regard to the anomalies of the planets. For each planet, we
start with the time and the anomaly at the old observation,
and the time that Ptolemy chooses for the observation that
he claims to have made himself. We multiply the time interval between the observations by the rate at which the anomaly
changes according to Ptolemy's tables,[†] and we add the change

[†]Ptolemy says [Syntaxis, Chapter IX.3] that he starts with
the values determined by Hipparchus, and that he will correct them slightly by using his own observations. As we
have seen, he does not actually use his own observations,

in anomaly calculated this way to the original value of the anomaly. Using this value of the anomaly, and the other parameters a, e, and r of the planetary model, we also calculate the longitude of the planet.

TABLE XI.9

OBSERVATIONS THAT PTOLEMY USES TO CONFIRM THE
ANOMALISTIC RATES OF THE PLANETS

Planet	Date	Longitude, degrees		Anomaly, degrees	
		Stated by Ptolemy	Calculated[a]	Stated by Ptolemy	Calculated[a]
Mercury	139 May 17	77;30	77;29,12	99;27	99;27,36
Venus	138 Dec 16	216;30	216;29,12	230;32	230;34,48
Mars	139 May 27	242;34	242;33,36	171;25	171;24,54
Jupiter	137 Oct 8	14;23	14;22,54	182;47	182;47,48
Saturn	136 Jul 8	284;14	282;14,00	174;44	174;44,00

[a]Calculated from Ptolemy's theories, not from modern theori

The results of this test are given in Table XI.9. Sinc Ptolemy does the same thing for Mercury and Venus that he does for the outer planets, I have included the relevant ca culations for Mercury and Venus in the table. The first column gives the name of the planet and the second column gives the date of the observation which Ptolemy claims that he made of that planet. The third column gives the longitu that Ptolemy claims to have observed, while the fourth column gives the longitude calculated in the manner just described. The fifth column gives the anomaly that Ptolemy claims to derive from the observation, while the last colum gives the anomaly calculated in the manner just described.

The "observed" longitude agrees with the calculated longitude within less than 1' in every instance, although the standard deviation of an observation is 15' or more. The "observed" anomaly agrees with the calculated anomaly within less than 1' in every instance except Venus. We may safely say that Ptolemy's observations have been fabricated I shall not even bother to calculate the odds that this is so.

The discrepancy in the anomaly of Venus is 2;48 minut of angle. This amount of disagreement is not surprising a it does not contradict the conclusion that the observation fabricated. When Ptolemy fabricated the longitude, he had

---

in spite of his claim to do so. This suggests that the values which he uses may actually be those of Hipparchus, or they may be of Babylonian origin.

[†]Logically, we can say only that either Ptolemy's observat or the ancient observation has been fabricated, unless bo have been.

round it to the nearest minute in order to preserve verisimilitude. The rounding means that the calculations made in
his pretended analysis of the observation are not identical
with those made during the fabrication. Because the eccentricity of Venus is quite small, the anomaly must change
by a relatively large amount after the rounding of the longitude.

We saw in Section X.2 that Ptolemy claims to have observed a conjunction of the moon with each planet, and I
proved there that all of these conjunctions are fabricated.
Ptolemy uses the conjunctions of Mercury and Venus in his
pretended determinations of γ' for those planets; that is,
the observations of Mercury and Venus in Table XI.9 are the
same as the corresponding conjunctions in Section X.2. Thus
we have proved that these observations are fabricated in two
independent ways.

On the other hand, Ptolemy uses the conjunctions of the
moon with the outer planets to find the radius of the epicycle. That is, for Mars, Jupiter, and Saturn, the alleged
observations in Section X.2 are not the same observations as
those in Table XI.9. Thus, for each outer planet, we have
been able to prove fabrication for two of the four observations which Ptolemy claims to have made for that planet. I
have no personal doubt that he fabricated all of them, but
I see no way to demonstrate this directly for the outer
planets.

CHAPTER XII

SOME MINOR TOPICS

1. Turning Points of the Planets

When he has finished developing his models that give
the longitude of the planets, Ptolemy has finished the
major work of the Syntaxis. In this chapter, I shall describe briefly the minor topics which Ptolemy treats in the
last two books of the Syntaxis, which are books XII and XIII.

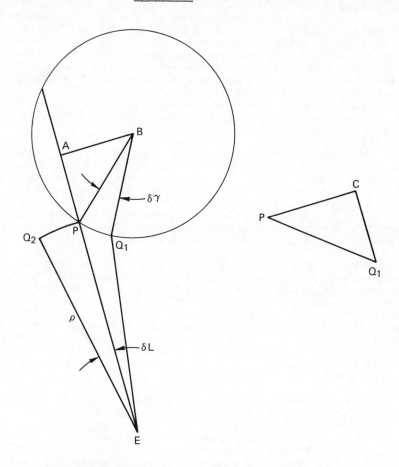

Figure XII.1   The theorem of Apollonius on turning points.
The circle whose center is B is the epicycle of the planet P. In
the simple epicycle model, B moves uniformly on a circle whose
center is the earth E. If the anomaly γ changed by $\delta\gamma$ with no
change in the position of B, P would move to Q_1. If the mean
longitude changed by δL with no change in γ, P would move to Q_2.
In the actual case, both γ and L change, and P is at a turning point
if the angles PEQ_1 and PEQ_2 are equal. The triangle at the right
shows a construction used in proving the theorem.

With one exception, these topics were important in Babylonian astronomy but received relatively little attention in Greek astronomy.

In this chapter, I shall also take up an item of unfinished business. When I presented the "double equant" model for Mercury in Section X.6, I promised to give the reasoning that leads to the model in a later section, and to explain why this reasoning was within the intellectual horizons of the Greek astronomers. The last section of this chapter provides a convenient place to do this.

Ptolemy ends Book XI with the study of the equant model as applied to Saturn. He devotes the first eight chapters of Book XII to the turning points of the planets, which were discussed briefly in Section I.5 of this work. During most of the time, a planet moves eastward through the heavens. However, there is a time when the eastward motion stops and the planet starts to move westward. After a relatively short time, however, the westward motion stops and the planet resumes its usual eastward motion. The reversal of direction is possible because the velocity of motion on the epicycle is greater than the velocity of the epicycle center in its motion on the deferent.

Ptolemy starts by giving, in Chapter XII.1, an important theorem on turning points that he attributes to Apollonius of Perga.[†] Perga is a town on the central part of the southern coast of Asia Minor. Apollonius is a major figure in the study of conic sections, he apparently introduced the eccentric model into astronomy, and he may have been the first to demonstrate the equivalence of the epicycle and eccentric models (see Section IV.3).

I shall demonstrate Apollonius's theorem using modern terminology, with no attempt to reproduce Ptolemy's demonstration, although the main outlines of both demonstrations are the same. We consider first the simple epicycle model of the planets. In Figure XII.1, B is the center of the epicycle. It moves uniformly on a circle whose center is at the earth E, and the longitude of B is the mean longitude L of the planet. The anomaly γ is the angle between the radius BP of the epicycle and the line from E to B. Both L and γ increase in the counterclockwise direction.

P is the position of the planet at some instant of time Suppose it were possible for the anomaly to increase by the amount $\delta\gamma$ while the longitude did not change. P would then move to Q_1, and the longitude of P would decrease by the angle PEQ_1. On the other hand, suppose that it were possibl for the mean longitude to increase by δL while the anomaly did not change. The entire epicycle would then move by the angle δL, and P would move to Q_2. The longitude of P would increase by the angle PEQ_2, which is the same as δL.

[†]It is interesting that he specifically attributes this theorem to someone else, although he fails to attribute the theorem of Menelaos (Section II.4 above).

In the actual case, the changes δL and $\delta\gamma$ occur simultaneously, and the longitude of the actual planet P does not change if the angles PEQ_1 and PEQ_2 are equal.

Now look at the small figure to the right of the main figure, which is an enlargement of the part of the main figure around the displacement PQ_1. PQ_1 is actually an arc of a circle, but it is so short that we can consider it as a straight line. When P moves to Q_1, part of the displacement changes the distance from E, but this part is not perceived by the eye. If CQ_1 is parallel to the line of sight EP, P appears to move to C rather than to Q_1. The angles PEQ_1 and PEQ_2 are equal if PC, rather than PQ_1, is equal to PQ_2.

Triangles PCQ_1 and PAB are similar, so that PC/PQ_1 = AP/BP; BP is the radius r of the epicycle. Thus PC = $(AP/r)PQ_1$. If the angles $\delta\gamma$ and δL are measured in the appropriate units, PQ_1 = $r\delta\gamma$ and PQ_2 = $\rho\delta L$, in which ρ is the distance from E to P. Thus PC = $AP\delta\gamma$. The ratio of PQ_2 to PC is then

$$PQ_2/PC = \rho\delta L/AP\delta\gamma = (\rho/AP)(\delta L/\delta\gamma).$$

However, the ratio of δL to $\delta\gamma$ is the same as the ratio of their rates. L increases at the rate of n degrees per day and γ increases at the rate of γ' degrees per day, in the notation that we have been using. Hence

$$PQ_2/PC = (\rho/AP)(n/\gamma').$$

P is at a turning point, that is, the longitude of P does not change, if the ratio PQ_2/PC equals 1. In other words, P is at a turning point if

$$AP/\rho = n/\gamma'. \tag{XII.1}$$

This is the theorem of Apollonius.

In the equant model, the angular velocity γ' is constant, but the angular velocity n and the distance ρ are not constant. Hence the position of P on the epicycle (that is, the anomaly) at a turning point depends upon the value of L, the mean longitude. After he proves Apollonius's theorem in Chapter XII.1, Ptolemy devotes the next five chapters to working out how γ at a turning point depends upon L for each of the five planets in turn, basing the treatment upon his models of planetary motion. Chapters XII.7 and XII.8 are devoted to tables of the turning points.

2. Maximum Elongations

We discussed the maximum elongations of Mercury and Venus at considerable length in earlier chapters, because measurements of maximum elongation play a leading part in finding the parameters of those planets in Ptolemy's theory. In particular, we saw that the maximum elongation is not uniform around the orbit; that is, the maximum elongation depends upon the value of $L_\odot$ at the time that the maximum occurs.

-331-

Ptolemy turns to this topic in Chapters XII.9 and XII.10. Here, however, he deals with the elongation from the actual sun rather than from the mean sun as in his earlier work. The reason is that the elongation from the mean sun is used in his theoretical work, but it is the elongation from the actual sun that is interesting as an observed phenomenon. He closes this part of his work with a table which gives, for both Mercury and Venus, the maximum elongation east and the maximum elongation west as functions of $L_\odot$.

I did not attempt to follow this part of Ptolemy's work, but his table shows an interesting feature. The sum of the east and west elongations for the same value of $L_\odot$ is the total angle subtended by the epicycle; we used this fact in the analysis of Figure X.6, for example. This total angle is the angle subtended by the epicycle at the position of the earth, and thus it depends upon the distance from the earth to the center of the epicycle. When the center has a maximum of its distance, which is the circumstance that Ptolemy means by an apogee of Mercury or Venus, the apparent size of the epicycle has a minimum. When the center has a minimum of its distance, which Ptolemy calls perigee, the apparent size of the epicycle has a maximum.

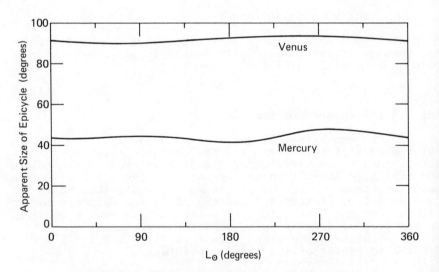

Figure XII.2   The apparent size of the epicycles of Mercury and Venus as functions of $L_\odot$, in Ptolemy's theories. The correct curve for Mercury has only a single maximum and a single minimum, but it has two of each in Ptolemy's theory.

In Figure XII.2 I have plotted the apparent size of the epicycle, derived by adding the east and west elongations from Ptolemy's theories, as a function of the mean longitude $L_\odot$ of the sun, for both Mercury and Venus. The curve for Venus shows a single maximum and a single minimum, as it should. However, the curve for Mercury shows two maxima and two minima.

Figure X.10 also shows the apparent size of the epicycle of Mercury as a function of $L_\odot$, calculated from modern theory rather than from Ptolemy's theory. We see that there should be only a single maximum and a single minimum, rather than the pair of each that Figure XII.2 shows. On the other hand, Ptolemy constructed his model so that it would have a single apogee position and two perigee positions. In between the two perigee positions, however, there must be a place where the distance has a relative maximum, but the distance will be less here than at what Ptolemy calls apogee.

Ptolemy's apogee is at $L_\odot = 190°$, and the apparent size of the epicycle should have a minimum there. It should have another minimum at $10°$, but not as deep a minimum as at $190°$. The curve in Figure XII.2 shows these features. There should be two maxima at Ptolemy's perigee points, and these should be equal and symmetrically located with respect to his apogee at $190°$. We see that the two maxima are equidistant from $190°$, but the values of the apparent size at the two maxima are quite different.

We see from Table X.1 that it is the maximum near $70°$ that is in error.† At $310°$, the apparent size of the epicycle is the sum of the elongations in observations 1 and 4, namely 47;45 degrees. This agrees fairly well with the curve in Figure XII.2. The apparent size when $L_\odot = 70°$ is the sum of the elongations in observations 2 and 3, which is also 47;45 degrees. However, the apparent size shown in Figure XII.2 is less than 45 degrees.

It would be interesting to find the origin of this problem by a careful study of Ptolemy's work in Chapters XII.9 and XII.10, but I shall not lengthen an already long work by taking the space to do so.

3. Conditions at First and Last Visibility of the Planets

I shall pass by the opening chapters of Book XIII of the Syntaxis for the moment and go to the topic with which Ptolemy closes his work. When a planet is in conjunction with the sun, it cannot be seen because the sun is too bright. Thus, as it approaches conjunction, there must be an occasion on which it is last visible. Similarly, as it moves away from conjunction, there must be an occasion on which it first becomes visible again. The times and positions of first and last visibility were a matter of great interest to the Babylonians, and a great deal of Babylonian mathematical astronomy

---

†By an error here, I mean a disagreement with Ptolemy's theory.

is devoted to the subject.† It does not seem to be a topic that interested Greek astronomers greatly, if we can judge by the surviving literature. Ptolemy devotes Chapters XIII.7 to XIII.10 to this topic, but these are short chapters and occupy only a small part of the Syntaxis.

Ptolemy assumes that there is a critical angle that we may designate by η for each planet. If the sun is at the angle η below the horizon when the planet is just on the horizon,‡ the planet is just on the verge of visibility. He takes η to be 10° for Mercury, 5° for Venus, 11°.5 for Mars, 10° for Jupiter, and 11° for Saturn.

If Ptolemy's basic assumption is correct, finding the times of first and last visibility is a matter of calculating from theory when η has the assigned value for each planet. If the planets lay in the plane of the ecliptic, we could find an elongation (that is, a difference in longitude) that would make η take on the required value, and the value of the elongation would depend only upon the time of year, to satisfactory accuracy. However, the planets deviate from the plane of the ecliptic (that is, their latitudes are usually different from zero), and this complicates the situation.

Let us consider a particular time of year, which means that the sun has a particular longitude. If it happens that a planet is in a position of first or last visibility at this time, it must occupy a particular position with respect to the sun. That is, it has a particular heliocentric longitude and the value of this longitude is known when we have assigne the longitude of the sun.‡ Hence the fact that the planets deviate from the plane of the ecliptic does not affect the fact that we can calculate a critical elongation which depends only upon the position of the sun. The latitudes of the planets merely complicate the calculation of the critical elongation.

Because he cannot calculate the critical elongation until he can calculate the latitude of a planet, Ptolemy treats first and last visibility after he has treated the latitudes of the planets.

4.   The Latitudes of the Planets

Ptolemy treats the latitudes of the planets in Chapters XIII.1 to XIII.6. Until this point, he has made the accurate assumption that the latitudes of the planets are too small to affect the theory of their longitudes, to the accuracy needed for naked eye observations.

In a heliocentric theory, the orbit of each planet lies in a plane that passes through the sun, and the theory of

---

†See Neugebauer [1955] for a thorough study of the Babylonia literature on this subject.

‡It is assumed in this that the horizon is perfectly flat.

‡Ptolemy does not word the matter this way, but he reaches the same conclusion, if I understand what he does.

latitudes is relatively simple until we start to worry about
errors of the order of seconds of arc. From this fact, we
can easily see what the Greek theory of latitudes should
have been. In the geocentric theory of the Greeks, a different theory of latitudes is needed for the inner and outer
planets.

For the inner planets, we remember that the deferent is
really the geocentric orbit of the sun and that the epicycle
is the heliocentric orbit of the planet. Hence the deferent
should lie in the plane of the ecliptic and the epicycle
should make a constant angle with the deferent. Specifically,
the plane in which the epicycle lies should have a constant
orientation in space. The angle between this plane and the
plane of the deferent should be about 7° for Mercury and
about 3°24' for Venus. Thus, in the Greek theory, the theory
of latitudes for Mercury and Venus should be quite simple.

Instead, the theory given in the Syntaxis is so complicated that I shall not attempt to describe it here beyond a
cursory summary.† Ptolemy starts by assuming that the plane
of the deferent for an inner planet oscillates about the
plane of the ecliptic instead of coinciding with the ecliptic
as it should. The amplitude of the oscillation is 45' for
Mercury and 10' for Venus. The epicycle undergoes further
oscillations. Dreyer [1905, p. 200] describes a part of the
complex oscillations for Venus and concludes his description
of Ptolemy's theory of latitudes by writing: "Simultaneously
the double rocking of the epicycle is going on, like a ship
pitching and rolling at the same time. For Mercury everything
is reversed, north and south, otherwise the theory is similar."
It does not seem necessary to give more details.

For the outer planets, the deferent is the orbit of the
sun about the planet and the epicycle is the heliocentric
orbit of the earth. Thus two angles are needed in a theory
of latitudes. The angle between the deferent and the plane
of the ecliptic should be the angle that is called the inclination of the planet in modern theory. Since the epicycle lies in the plane of the ecliptic, the angle between
the epicycle and the deferent should be equal and opposite
to the angle between the deferent and the ecliptic.

Ptolemy's theory of latitude for the outer planets seems
to have approximately this property. He puts the plane of
the deferent at a constant angle with the ecliptic. This
angle is 1 degree for Mars, $1\frac{1}{2}$ degrees for Jupiter, and $2\frac{1}{2}$
degrees for Saturn.‡ The angles between the epicycle and the
deferent are not the same as these. They are $2\frac{1}{4}$ degrees,
$2\frac{1}{2}$ degrees, and $4\frac{1}{2}$ degrees, respectively. However, the angles
are arranged so that the epicycle is nearly parallel to the
ecliptic. The angle between the epicycle and the ecliptic is
$1\frac{1}{4}$ degrees for Mars, 1 degree for Jupiter, and 2 degrees for

---

†Dreyer [1905, pp. 198-200] and Pannekoek [1961, pp. 143-
144] give some details, but they do not give an extensive
discussion.

‡The correct values are about 1°51', 1°18', and 2°30',
respectively.

Saturn, if I have understood Ptolemy's treatment correctly.

Ptolemy uses only some very crude statements about latitudes in deriving these angles. For Jupiter, for example, he says only that the greatest latitude is about 2° near opposition and about 1° near conjunction.

If the epicycle makes a constant angle with both the deferent and the ecliptic planes, the angle between the deferent radius and the plane of the epicycle cannot be constant as the epicycle circle moves around the deferent. Thus a special mechanism is required to make the epicycle always remain parallel to itself. Ptolemy drives the plane of the epicycle by a small circle perpendicular to the deferent whose center is in the plane of the deferent. As Dreyer [1905, p. 199] puts it, we should imagine a peg carried on the rim of the small circle which can slide in a slot attache to the epicycle. If we choose the radius and rate of rotatio of the small circle properly, we can make the epicycle always stay parallel to itself.

In Section IV.7, I discussed two considerations that were available to the Greek astronomers which make the heliocentric theory preferable to the geocentric theory as a simpler way to save the phenomena. Here we have still another consideration that favors the heliocentric theory: A satisfactory theory of latitudes is far simpler in the heliocentric theory than in the geocentric theory. In the heliocentric theory, we need only one inclination angle for each planet, and we do not need a complex peg-and-slot mechanism to preserve the orientation of a plane. Simple inertia woul(keep the plane of a heliocentric orbit parallel to itself as the planet moves.

5.   The Double Equant Model

In Chapter X, the most accurate model that we could find for Mercury when we used a circular epicycle had a standard deviation of 1°.70. When we introduced the double equa model of Figure X.14, the standard deviation fell to 0°.32. This accuracy was achieved with the first set of parameters that I tried, and it could almost surely be improved by letting the parameters vary. Now that we have discovered the basic properties of the equant in Sections XI.1 and XI.2, I can describe the reasons that lead to trying the double equant model.

We know that the Greek astronomers were acquainted with the heliocentric theory even if they did not accept it, and thus they had the ability to test it and to examine its consequences. We also know that the epicycle and deferent in a planetary model have natural interpretations in a heliocentric theory. In the model of Venus, the deferent is the orbit of the sun around the earth; if we add 180° to the apogee, we get the orbit of the earth around the sun. For the models of Mars, Jupiter, and Saturn, the deferent is the orbit of the sun around the planet in question; if we add 180° to the apogee, we get the orbit of the planet around the sun.

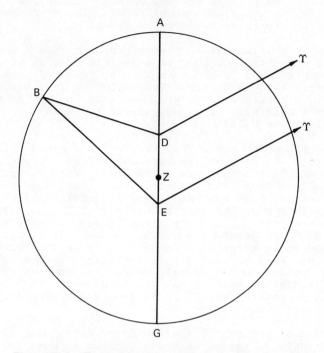

Figure XII.3   The basic equant model. This is the same model
as the one that was called the equant model in Figure IV.4, but
the epicycle is no longer needed. In this model, we concentrate
our attention on representing the motion of point B. In Figure
IV.4, our interest was in point P, while B was merely an auxiliary
construction point.

Thus we have the orbits of four planets, namely the
earth and the three planets outside it, around the sun. The
same model represents all four orbits. This model is shown
in Figure XII.3. It is the same as the equant model of Figure IV.4, except that it lacks the epicycle which is no
longer needed. In the rest of this section, I shall use
"equant" to refer to the model of Figure XII.3. We know
that the two eccentricities in Figure XII.3 should be nearly
equal, but I shall not impose equality as a necessary condition. Ideally, both eccentricities should be found independently from observations.

The most important point in a heliocentric theory is
that there should be no distinction in astronomy between
the earth and any other planet. Hence there should be no
distinction between planets outside the earth and those inside the earth, because the earth is no longer a valid di-
vider between inside and outside. We have seen that the

equant model of Figure XII.3 represents accurately the orbits of the earth, Mars, Jupiter, and Saturn. Hence, if the heliocentric idea is valid, the model should also represent the heliocentric orbits of Mercury and Venus.

Even in a heliocentric theory, we must have a way to represent geocentric motion, because this is what we observe. Hence we still want to use an epicycle and a deferent, but the epicycle and the deferent should be the same kind of curve. In other words, the epicycle that we use for Mercury and Venus should also be an equant, as it is in the double equant of Figure X.14.

When we use an equant for the deferent and a circle for the epicycle, we get the accuracies for Venus, Mars, Jupiter, and Saturn that are shown in Table XI.7. When the parameters are properly chosen, the standard deviations range from $0°.09$ for Jupiter to $0°.36$ for Mars. Now a circular epicycle is the special case of the equant epicycle of Figure XII.3 in which both eccentricities are zero. Hence the models that we have used for Venus, Mars, Jupiter, and Saturn are double equant models in which we have not yet found the best fitting parameters. When we find the best fitting parameters, the standard deviations necessarily decrease. Therefore the double equant can be used for all of the planets, and the accuracy will be better than $0°.36$ in geocentric coordinates for every one of them.

I find it remarkable that no Greek astronomer tried the double equant, so far as we know. I find it even more remarkable that Copernicus did not try the basic idea behind it, namely to treat all of the planets alike. Copernicus was able to defy conventional thinking by trying a heliocentric theory, but he did not manage to rise above convention in other ways. Many people through the centuries since Ptolemy had objected vigorously to the equant because it violated the theory of uniform circular motion in its strict sense. In its strict sense, the motion has to be uniform around the center of the circle. Ptolemy's device of having the rotation be uniform around point D in Figure XII.3, while the center of the circle is at Z, violates this dictum. In contrast, having the earth at the eccentric point E is allowed.

However, the need that was met by the equant model is there whether the model is accepted or not. Hence Copernicus having rejected the equant, had to replace it by a model that satisfies the pure doctrine of uniform circular motion while meeting the observational problem that leads to the equant. He did this in effect by substituting an epicycle† for the second eccentric distance DZ in Figure XII.3. To my eye, at least, the device of Copernicus is more complicated than the equant.

Further, in two important senses, the theory of Copernicus was not a heliocentric theory. In the first place, he did not take the sun to be the fundamental point in his theor

---

†Pannekoek [1961, p. 194] has an excellent discussion of Copernicus's device.

Instead, he took the center of the earth's orbit to be the fundamental point.† In the second place, he did not put the planets, including the earth, on a common basis. He used his replacement for the equant for the orbits of Mars, Jupiter, and Saturn. For the orbit of the earth, he used the single eccentric model of Figure IV.3. For the orbits of Mercury and Venus, he used models that are different from each other and from the models used for the earth and the outer planets. Altogether, he used four different models to represent the motions of six planets. Ptolemy used only three different models to do the equivalent job.

Thus it is not true that Copernicus produced a theory that was much simpler than the Ptolemaic theory and that should therefore have been immediately adopted by all right-thinking persons. His theory, on the contrary, was more complex than the Ptolemaic theory, although it would have been simpler if he had pursued the true idea back of the heliocentric theory with the same rigor that he pursued the idea of uniform circular motion. In view of the complexity of Copernicus's theory, I do not find it surprising that it took a century after Copernicus for the heliocentric idea to gain wide acceptance. In fact, it was Kepler who first accepted the truly heliocentric idea by putting all the planets on a common basis, so far as I know.

There is a limit to the originality that can be shown by a single person, even by a genius like Copernicus. Thus it is not surprising that he accepted the strict doctrine of uniform circular motion and that he developed a substitute for the equant because he followed this strict doctrine. However, I do find it surprising that he did not try the same model for all the planets, particularly for Mercury. If he had done so, he would have developed a truly heliocentric theory, with all the planets on an equivalent basis. This would have been a much simpler theory than the one that he finally presented, and it would probably have been more accurate.

---

†For the earth, he used the simple eccentric model of Figure IV.3. The sun is at the point E in this figure, but Copernicus used point Z as the fundamental point in his theory.

CHAPTER XIII

AN ASSESSMENT OF PTOLEMY

1. Observations That Ptolemy Claims to Have Made

Now that we have reviewed all of the Syntaxis, it is appropriate to undertake an assessment of Ptolemy as an astronomer, and an assessment of his proper place in the development of astronomy. This assessment will be based upon the information that has been developed in this work, and therefore it will be limited to Ptolemy's relation to astronomy. Here we cannot undertake to review or to assess his work in mathematics, geography, or optics, not to mention his work in astrology. Further, with minor exceptions, we deal only with the Syntaxis and not with his lesser works in astronomy.

TABLE XIII.1

OBSERVATIONS THAT PTOLEMY CLAIMS TO
HAVE MADE HIMSELF

Quantity observed	Dates	Error	Conclusions
Autumnal equinoxes	132 Sep 25, 139 Sep 26	28^h	Fabricated
Vernal equinox	140 Mar 22	28^h	Fabricated
Summer solstice	140 Jun 25	36^h	Fabricated
Angle between tropics	Several unspecified dates	21'	Fabricated
Latitude of Alexandria	No details	15'	Fabricated
Triade of lunar eclipses	133 May 6, 134 Oct 20, 136 Mar 6		Fabricated
Longitudes of sun and moon	139 Feb 9	$> 1°$	Fabricated
Longitudes of sun, moon, and Regulus	139 Feb 23	$> 1°$	Fabricated
Inclination of lunar orbit	Several unspecified dates	$\sim 16'$	Fabricated
Meridian altitude of moon	135 Oct 1	41'	Fabricated
Relation between apparent diameters of sun and moon	No details	1'20"	Fabricated
Configuration of α Cnc, β Cnc, α CMi	No details		May be fabricated

TABLE XIII.1 (continued)

Quantity observed	Dates	Error	Conclusion
Other stellar alignments	No details		Not investigated
Longitudes in star catalogue	No details	1°.1	Fabricated
Latitudes in star catalogue	No details	21'	Fabricated
Declinations of 12 stars stated but not used	No details	7'	Genuine
Declinations of 6 stars used to find precession	No details	20'	Fabricated
Conjunction of moon with each planet	139 May 17, 138 Dec 16, 139 May 30, 139 Jul 11, 138 Dec 22	40'	Fabricated
Seven longitudes of Mercury at maximum elongation	132 Feb 2 to 141 Feb 2	1°	6 are fabricated; 1 cannot be tested
Two longitudes of Venus at maximum elongation	140 Jul 30, 136 Dec 25	1°	1 is fabricated, 1 cannot be tested
Longitude of Venus at maximum elongation	136 Nov 18	1°.5	Fabricated
Two longitudes of Venus at maximum elongation	134 Feb 18, 140 Feb 18	0°.5, 1°.5	1 is fabricated, 1 cannot be tested
Three longitudes at opposition for each each outer planet	127 Mar 6 to 139 May 27	~ 1°	1 is fabricated for each planet, the others cannot be tested

We may usefully start by summarizing the observations that Ptolemy claims to have made himself. These observations are listed in Table XIII.1. The first column in the table describes the nature of the observations and the second column lists the dates when Ptolemy gives them; he did not date some of the observations. The third column lists my estimate of the errors in the observations, as determined by comparison with calculations from modern data and theories, and the last column lists the conclusions that I have reached about the authenticity of the observations.

I did not investigate the observations that Ptolemy claims to have made of certain stellar alignments, with the exception of the first one that he describes. I did not investigate them because they have little connection with the rest of the Syntaxis, and they do not tell us much about Ptolemy's procedures. The one that I did investigate showed

-342-

suspicious circumstances, and I have labelled it "may be fabricated" in the table.

I feel rather sure that one set of observations that Ptolemy claims to have made is genuine. That is, I believe that the observations were really made, but whether they were made by Ptolemy and not by someone else is another question. They are the declinations of 12 stars that Ptolemy does not use in finding the precession of the equinoxes. These declinations show errors of a reasonable size, and they lead to accurate values of the precession and of the obliquity. Further, they are not at all compatible with Ptolemy's theories.

The errors in almost all of the observations labelled "fabricated" in Table XIII.1 are unreasonably large, but this is not the main basis for deciding that they were fabricated. Instead, the procedure that I followed in testing the authenticity in most cases is the following: First we locate genuine observations made by other astronomers that are relevant to the theories that Ptolemy wants to establish with the aid of his own observations. Next we find the parameters in Ptolemy's theories by using the observations made by others, and we calculate the circumstances of Ptolemy's alleged observations from his theories, using these parameters to do so. Finally we compare the calculated circumstances with those which Ptolemy claims to have observed.

In many cases, the calculated and "observed" circumstances agree to every digit that Ptolemy writes down. When they do not agree exactly, the discrepancy is one that could easily arise from uncertainty in reading tables, from making numerical approximations, from rounding or truncating numbers, and from other details of computation. In all cases that can be tested in this way, we can estimate the probability that the agreement is consistent with observation. We always find that the probability is small. In most cases, it is ridiculously small, such as 1 chance in 100 000 000, or some such number. In some cases, it is a probability of the order of 1 in 10. We could accept a probability of this sort occasionally, but we cannot accept it at the frequency with which it occurs. If Ptolemy's observations are genuine, we expect to find many probabilities as large as 1 in 2. However, with all of Ptolemy's claimed observations, we never find a probability this large.

In calculating the probabilities, it is necessary to assume standard deviations of the errors. However, the probabilities found do not depend critically upon the sizes assumed for the standard deviation for the following reason: When an observation agrees with a calculated value, the precision used in stating the result defines a pre-assigned range† within which the alleged observation lies. The probability that an observation should lie within the pre-assigned

---

†The word pre-assigned is the key word here. Any value lies in some range. The particular range matters only if our attention is focussed on it before we know what the value is.

range is the product of two factors. If we increase the standard deviation that we use in calculating the probability, we increase one factor while we decrease the other. To the accuracy that we need, the product does not depend upon the standard deviation assumed.

Many students of Ptolemy have looked for observational procedures that could explain the size of Ptolemy's errors, but they have generally been unsuccessful. Even if they were successful, however, this would not change the validity of the argument used in this work. That is, it would still be highly improbable that Ptolemy's observations would agree with his theories in the way that they do, even if the size of his errors were reasonable. No theory of observation can explain such an agreement; it can only be the result of fabrication and not of observation.

In this way, we can prove that all of Ptolemy's own observations connected with the sun and moon are fabricated. All of his observations of the stars that he uses are also fabricated. Some star observations that he does not use are almost surely genuine. We cannot test all of his planetary observations, because we do not have enough observations attributed to others to let us start the testing process. However, we can test more than half of his alleged planetary observations, and every one that we can test is fabricated.

Thus we can say that all of Ptolemy's observations that he uses and that can be tested are fabricated. Further, we can say that all of his theories depend heavily upon fabricated data, and some of them seem to depend completely upon such data.[†]

2.   Observations That Ptolemy Attributes to Other Astronomers

TABLE XIII.2

OBSERVATIONS THAT PTOLEMY ATTRIBUTES TO OTHERS

Observer or Place	Quantity observed	Date	Conclusion
Meton	Summer solstice	-431 Jun 27	Fabricated
Aristarchus	Summer solstice	-279 Jun 26	Genuine
Hipparchus	Summer solstice	-134 Jun 26	Genuine
Hipparchus	Six autumnal equinoxes	-161, -158, -157, -146, -145, -142	Genuine
Hipparchus	Fourteen vernal equinoxes	-145 to -127	Genuine
Alexandria	Vernal equinox	-145 Mar 24	May be genuine

[†]Even when Ptolemy pretends to base his theories completely upon his own observations, he must have actually based them upon observations made by others. However, we cannot reconstruct the fabrication in these cases.

TABLE XIII.2 (continued)

Observer or Place	Quantity observed	Date	Conclusion
Eratosthenes	Obliquity of ecliptic	ca. -225	May be genuine
Babylon	Triad of lunar eclipses	-720 Mar 19, -719 Mar 8, -719 Sep 1	One is certainly fabricated, the other two may be fabricated
Babylon	Triad of lunar eclipses	-382 Dec 23, -381 Jun 18, -381 Dec 12	Fabricated
Alexandria	Triad of lunar eclipses	-200 Sep 22, -199 Mar 19, -199 Sep 12	Fabricated
Babylon	Lunar eclipse	-490 Apr 25	May be genuine
Alexandria	Lunar eclipse	125 Apr 5	May be genuine
Babylon	Lunar eclipse	-501 Nov 19	May be genuine
Hipparchus	Longitudes of sun and moon	-127 Aug 5	Probably fabricated
Hipparchus	Longitudes of sun and moon	-126 May 2, -126 Jul 7	Fabricated
Babylon	Lunar eclipse	-620 Apr 22	Fabricated
Babylon	Lunar eclipse	-522 Jul 16	Fabricated
Alexandria	Lunar eclipse	-173 May 1	Fabricated
Rhodes	Lunar eclipse	-140 Jan 27	Fabricated
Hipparchus	Configuration of stars	No details	Not tested
Timocharis	Longitude of α Vir	ca. -290	May be genuine
Hipparchus	Longitude of α Vir	ca. -130	May be genuine
Hipparchus	Longitude of α Leo	ca. -130	May be genuine
Timocharis or Aristyllus	Declination of 18 stars	ca. -290	Genuine
Hipparchus	Declination of 18 stars	ca. -130	Genuine
Timocharis, Agrippa, or Menelaos	Seven lunar conjunctions or occultations	ca. -290 to ca. +95	Fabricated
Dionysios	Longitude of Mercury at maximum elongation	-261 Feb 12, -261 Apr 25	May be genuine

-345-

TABLE XIII.2 (continued)

Observer or Place	Quantity observed	Date	Conclusion
Dionysios	Longitude of Mercury at maximum elongation	-256 May 28, -261 Aug 23	Fabricated
Alexandria	Longitude of Mercury at maximum elongation	-236 Oct 30, -244 Nov 19	Fabricated
Theon	Longitude of Mercury at maximum elongation	130 Jul 4	May be genuine
Dionysios	Longitude of Mercury	-264 Nov 15 and 19	May be genuine
Theon	Longitude of Venus at maximum elongation	132 Mar 8, 127 Oct 12, 129 May 20	May be genuine
Timocharis	Longitude of Venus	-271 Oct 12 and 16	May be genuine
Alexandria	Conjunction of Mars with β Sco	-271 Jan 18	May be genuine[a]
Alexandria	Jupiter occulted δ Cnc	-240 Sep 4	May be genuine
Alexandria	Conjunction of Saturn with γ Vir	-228 Mar 1	May be genuine

[a]However, the correct date is probably -271 January 16. See Section XI.6.

Table XIII.2 summarizes the observations in the Syntaxis that Ptolemy attributes to other astronomers. The first column in the table gives the observer when Ptolemy names him; otherwise it gives the place where the observation was made. The second column gives the type of observation, the third column gives the dates if they are known, and the last column gives my conclusion about the authenticity of the observation.

In assessing the conclusions shown in Table XIII.2, we must remember that the critical test to use is not always obvious. For example, we can sometimes test a lunar eclipse by calculating the center time of the eclipse from Ptolemy's tables of the sun and moon. If the calculated time is suspiciously close to the "observed" time that Ptolemy states, under circumstances when Ptolemy's theories should not give accurate results, we may be rather sure that the eclipse observation is fabricated.

However, Ptolemy is not always interested in the center time of an eclipse. For the eclipses listed in Table VIII.2, for example, he is interested in the latitude of the moon. He calculates the time when the moon had the desired latitude according to his tables, and he then asserts that this is the

-346-

time of the eclipse, paying no attention to the relation between the sun and moon at this time. If we test the eclipse by calculating the time at the middle of the eclipse, and compare this with the time that Ptolemy states, as I have done in the second and third columns of Table VIII.2, we find that the times do not agree. Hence we might conclude that the eclipse is genuine because the result of "observation" does not agree with Ptolemy's theory.

The conclusion would be wrong, however. The difficulty is that the test was not based upon the appropriate one of Ptolemy's various theories. Thus, in the cases in Table XIII.2 where I have not shown fabrication, the reason may be that I have not found the appropriate test. In other words, failure to prove fabrication is not evidence of validity.

It is clear by now that Ptolemy's assertion about an observation, whether it was one made by himself or by another, is also not a proof of validity. Our guide at this point must be the following: We cannot accept an observation in the Syntaxis as a valid observation unless it has independent verification from ancient sources. The Syntaxis can no longer be considered as a source book of Greek astronomy.

If I have developed strong evidence that an observation in Table XIII.2 is fabricated, I enter "fabricated" for it in the last column. In the triad of lunar eclipses in -720 and -719, there is strong evidence that the middle eclipse is fabricated. There is contradictory evidence about the other eclipses in this triad, and I enter "may be fabricated" for these eclipses. The other observations are marked either "genuine" or "may be genuine".† The guide stated above is the basis for the distinction. If I have found evidence for an observation that is independent of the Syntaxis, I mark it "genuine". Otherwise, I mark it "may be genuine".

In preparing Table XIII.2, I have not had time to consult all the surviving Greek literature on astronomy, or even a representative sample of it. So far, I have had the time to consult only two works, and I have not been able to study them thoroughly. They are Geminus [ca. -100] and Censorinus [ca. 238]. The dates of Geminus are uncertain, but he seems to be earlier than Ptolemy and hence independent of him. Censorinus is later than Ptolemy and can be considered independent of Ptolemy only for information that is not found in the Syntaxis.

Geminus, in his Chapter I, gives some confirmation of the solar observations that Ptolemy attributes to Hipparchus. He says that the lengths of the seasons, beginning with the vernal equinox, are $94\frac{1}{2}$ days, $92\frac{1}{2}$ days, 88 1/8 days, and 90 1/8 days. These are identical with values that Ptolemy attributes to Hipparchus, and they agree with the intervals between equinox measurements that he says Hipparchus made. Thus, although Geminus does not specifically name Hipparchus as the source of the seasons, it seems safe to assume that

---

†Except for the entry that I did not test because Ptolemy does not use it.

he took the values from Hipparchus.†

Censorinus [238, Chapter XVIII] says that Hipparchus
devised a calendar with a cycle of 304 years, of which 112
contained 13 months while the others contained 12 months;
thus the cycle had 3760 months. In order to see the significance of this, we must go back to Section V.2. There we
saw that Meton, presumably in -431, devised a calendar based
upon a cycle of 19 years that contained 235 months. The
length of the year in this calendar was 365 + (1/4) + (1/76)
days. About a century later, Callippus combined 4 Metonic
cycles into a Callippic cycle of 76 years, with 940 months.
By omitting 1 day in 76 years, Callippus reduced the length
of the year to $365\frac{1}{4}$ days. If Hipparchus combined 4 Callippic
cycles into one Hipparchan cycle, and again omitted one day,
the Hipparchan cycle would contain 304 years and 3760 months,
as Censorinus says. The length of the year in the Hipparchan
calendar would then be 365 + (1/4) - (1/304) days, which is
sufficiently close to 365 + (1/4) - (1/300) days. Hipparchus
probably used 304 years rather than 300 years in his calendar
cycle so that he could build upon the earlier cycles.‡

Thus we have confirmation of the length of the year as
well as of the lengths of the seasons that Ptolemy attributes
to Hipparchus.

We have altogether four observations of the summer solstice that are consistent with Hipparchus's length of the
year to the nearest hour. They are the ones attributed to
Meton in -431, to Aristarchus in -279, to Hipparchus in -134,
and to Ptolemy in +140. The fact that we have confirmation
of the length of the year does not prove that any of these
are valid; they could all have been fabricated by using
365 + (1/4) - (1/300) days for the length. However, the
average error in the solstices attributed to Aristarchus and
Hipparchus is close to zero, while the errors in the other
two are more than a day. Hence it seems fairly safe to say
that the solstices of -279 and -134 are genuine. If these
are genuine, the equinox observations attributed to Hipparchus are probably genuine also.

The only other confirmation I have found relates to the
star declinations that Ptolemy gives in Chapter VII.3. Here
he gives the declinations of 18 stars as measured by Timocharis or Aristyllus around -290 and again by Hipparchus
around -130. Hipparchus [ca. -135] gives the declinations
of about 40 stars as measured by himself, and three of these
are included in the list that Ptolemy gives. Two of the

---

†He could have taken the values directly from Hipparchus's
writing or through an intermediate source. Geminus knows
something about Hipparchus, whose name appears three times
in the work of Geminus. See the index to the cited edition.

‡According to a source that I forgot to record, Censorinus
says that the Hipparchan cycle contained 111 035 days. If
Hipparchus constructed his cycle in the way I have just
speculated, the cycle does contain 111 035 days. However,
this is an inference; Censorinus does not say this, at
least not in the edition I have used.

declinations agree exactly in both sources while the third differs in one digit, in a way that could easily be a copying error.† Further, of the declinations that agree exactly, one is in the set of six that Ptolemy uses and the other is in the set of twelve that he does not use. Thus Hipparchus gives confirmation of a star in each set, and thus it is plausible to assume that all of the declinations attributed to Hipparchus are genuine. This argument is weak, however.

The preceding paragraph does not directly tell us anything about the declinations that Ptolemy attributes to Timocharis or Aristyllus. However, Ptolemy does not use these declinations, and the other unused data in the same context seem to be genuine. Hence I have called these data genuine in Table XIII.2, but this conclusion is rather shaky.

All entries in Table XIII.2 that are not marked "fabricated" or "genuine" are marked "may be genuine". These are observations for which I have found no evidence indicating fabrication and for which I have found no confirmation in a highly limited search. A search for confirmation of these observations should be valuable, as well as a search for further tests that might indicate fabrication.

Table XIII.2 contradicts a conclusion that I have stated in earlier work. On page 156 of APO, for example, I concluded that most of the observations which Ptolemy attributes to others were genuine; I wrote this before I had discovered most of the critical tests of authenticity that I have used in this work. The fact that the observations passed the tests I did use was not my principal argument, however. My principal argument was based upon the fact that many Greek works that are now lost still existed in Ptolemy's time and were widely distributed. Thus I did not see how he could fabricate data which would contradict these widely distributed works. While I am now convinced that he did fabricate many observations allegedly made by others, I am still astonished that he dared to do so. I can only assume that the sources of data were not in fact widely distributed and that few people were in a position to test his data by direct comparison with original sources.

Ptolemy makes a few remarks of a general nature that are not included in Table XIII.2. In Chapter V.7 of the Syntaxis, for example, Ptolemy says that Hipparchus has shown that 5° is the inclination of the lunar orbit. Since Ptolemy gives no details, I have not included this among the observations attributed to others. However, I have included the alleged measurement of the obliquity by Eratosthenes, even though Ptolemy is equally sparing of details. I do so in order to call attention to this famous measurement which, so far as I know, has no evidence to support it once we have brought Ptolemy's evidence into question.

---

†Details are given by Newton [1974] and in Section IX.4.

3.   The Role of Observation in Greek Astronomy

    Some people with whom I have discussed Ptolemy's fabri-
cations object to calling them fraudulent. According to
them, using the adjective "fraudulent" rests upon modern
ideas about the relation between theory and observation; they
would not be considered fraudulent from the viewpoint of
Greek astronomers. Other writers, in contexts that have
nothing to do with Ptolemy's fabrications, have also emphasized a relation between theory and observation in Greek
science that stands in strong contrast to the modern relation. As well as I can make out, two somewhat different
ideas are involved in these objections and writings.

    One idea is well expressed by Pannekoek [1961, p. 120],
in a discussion of the work of Aristarchus [Aristarchus,
ca. -280] on the size of the sun and moon, and the fact that
Aristarchus used a severely wrong value for the angles that
the sun and moon subtend at our eye. There Pannekoek points
out that Greek scholars like Aristarchus were primarily geometers who happened to pick out the solar system for the application or illustration of their geometry. "Hence", he
writes, "the astronomical quantities were treated somewhat
superficially, their precise values did not matter; ingenuity
was exhibited in the solution of the geometrical problem."
This idea, I believe, is closely related to an idea that I
have cited several times before. This is the idea that a
number or quantity used in various Greek works on astronomy
or geography is introduced only as a specimen, with no intention that it be taken as correct.†

    I have not found a succinct statement of the second idea,
but I believe that the following summarizes it adequately:
Observation is one way of gaining knowledge, but it is not
the only way. Another way, of equal validity, is through
insight and introspection. The value of the latter way has
received reinforcement in modern times from studies of the
earth's flattening. Isaac Newton, on the basis of his rev-
olutionary theories of physics, predicted among many other
things that the earth is a flattened or oblate spheroid
rather than a sphere. However, careful geodetic traverses
made in France‡ in Newton's time indicated that the earth
has an elongated or prolate shape. To go to extremes, Newton
said that the earth is a pancake while the French surveys
said that it is a cigar.

    To settle the question, the French Academy of Sciences
sent out two famous geodetic expeditions in 1735 and 1736,
one to Lapland and one to Ecuador. The results of the expeditions were that Newton was right and the first French
surveyors were wrong: the earth is more like a pancake than

---

†In this case, however, as we should remember from Section
VIII.2, Aristarchus wanted neither a specimen nor an accurate value. He wanted a value that was certainly too
large, in order to prove certain inequalities. Thus the
example does not support Pannekoek's point. See Appendix
B for more details.

‡See the survey by Fischer [1975].

a cigar. Voltaire thereupon commented about the expeditions:

Vous avez trouvé par de longs ennuis
Ce que Newton trouva sans sortir de chez lui.

There is heavy irony in this situation. The errors in the
measurements made by the expeditions were so large that they
could not really settle the question. As later analyses of
their work have shown, it was purely a matter of luck that
they "confirmed" Newton's pancake rather than the early French
cigar. Nonetheless, their results were accepted as definitive.

In other words, if I understand this idea, we can learn
as much about the universe by thinking carefully about it as
we can learn by observing it, and sometimes more.

A thorough exploration of these ideas is beyond the scope
of this work, but it is possible to make a few comments. There
are situations in modern science in which we cannot rely upon
observation and in which we can rely only upon the product of
thinking about the situation. These situations are present
in most if not all areas of science, but they are especially
prominent in cosmology. Ellis [1975], for example, points
this out in a recent survey article. As he says, many cosmological theories rest upon the following assumptions, among
others, that we have no chance of verifying in the present
state of affairs: (1) whenever normal physical laws† can be
applied, they correctly predict the structure of the universe
and (2) the universe is spatially homogeneous. There is no
way in which we can hope to test these assumptions in the fore-
seeable future yet they, or assumptions like them, are necessary in the development of cosmological theories. Without
them, we cannot begin to construct a cosmology.

In commenting on these ideas, it is useful to continue
the distinction between astronomy and physics that was brought
out in Section I.1. To the Greek scientists, simple description was the goal of astronomy while truth was the goal of
physics. That is, physics must discover the first causes and
the basic laws of the universe. When we undertake fundamental
investigations in physics today, as cosmology shows us, we
must still supplement observation with conclusions that we
reach only by thinking about the nature of things. In other
words, we need both thought and observation in developing our
most fundamental theories.

Greek scientists had the same need, and I am not sure
that their relation between thought, observation, and theory‡
----
†In this context, "normal physical laws" mean the laws of
physics that we have deduced from observations made on the
earth or from artificial space vehicles.

‡To speak of "their relation" (meaning a Greek relation) between thought, observation, and theory without some further
comment could be highly misleading. For one thing, the words
are likely to have different meanings for us and for the Greek
scholars. For another, there was a wide diversity of opinion
on this subject among Greek scholars, and there is no such

was basically different from ours, although the relation in Greek hands led to forms that seem strange or even outlandish to us. Even if their relation of thought to observation were different from ours, however, Greek scientists certainly did not encourage lying about the situation. But this is what Ptolemy did. I have given many examples of this, and I will repeat one of them here.

If Ptolemy had simply asserted that the moon has a particular parallax when it is at the first or last quarter, we might accept the statement as an honest one even if it is seriously in error. But Ptolemy does not simply do this. In Chapter V.12 he describes in considerable detail how he built an instrument for measuring the parallax of the moon, how he put it in place and aligned it correctly, and how he made observations with it. Then in Chapter V.13 he claims that he used this specific instrument at a precisely stated time in order to measure the lunar parallax when the moon is at a quarter, and he even gives the "raw data" from which he deduces the parallax. But we showed in Section VIII.5 that Ptolemy did not make this observation at all. If he had indeed pointed his instrument in the direction he claims, he could not even have seen the moon through it. To put the matter bluntly, Ptolemy lies about what he has done, and his elaborate description of his procedures is false. Presumably he inserts the description of the parallactic instrument only to provide convincing detail that will make us think that he did make the claimed observation. I do not know of any principles of science or philosophy, ancient or modern, that justify such conduct.

Further, Ptolemy was not primarily a mathematician or geometer who used the solar system only to provide examples for his mathematical ingenuity, in which accurate values obtained from observation did not matter. In fact, I do not know of any instance in which this attitude toward the data has been demonstrated on the part of any Greek scholar. As I have remarked several times before, the work of Aristarchus, in which he uses 2° for the apparent diameter of the moon, is often cited as an example of this attitude, but I have shown that using this as an example is based upon a misunderstanding of his work. The reader should consult Section VIII.2 and Appendix B for a discussion of what Aristarchus does. I have not seen any other work cited as an example of this attitude.

Let us return to Ptolemy. When he writes the Syntaxis, he is not writing as a mathematician who happens to have chosen astronomy for a source of examples, although he may write primarily as a mathematician in other works. He is writing the Syntaxis as an astronomer. He tells us the plan of the Syntaxis in the first two chapters: It is to be a complete presentation of astronomy. It will start with the relation of the earth to the heavens, it will then give the theory of the motions of the sun and the moon, of the

thing as a "Greek opinion" on the matter. However, with regard to the severely restricted topic of Ptolemaic astronomy, I believe that the preceding comments are valid.

positions of the stars, and finally of the motions of the
five planets. He also emphasizes that the theories must be
firmly based upon observation. For example, in Chapter III.1,
in a passage that I have already quoted in Section I.1, he
writes: "We think that it is right to explain the phenomena
by the simplest possible hypotheses, provided that they do
not conflict with the observations in any important way."

Many other passages in the Syntaxis show the same attitude, which is that astronomy must be based upon the observations, that it must "save the phenomena". Perhaps the most
convincing evidence about the existence of this attitude comes
from the fact that Ptolemy, when he is about to give a fabricated observation, frequently describes the instruments and
the methods that he claims to use in making the fabricated
observation, and he frequently emphasizes that he made the
observations carefully and accurately. I can think of only
one reason why he should do this. He knows that the readers
for whom he is writing the Syntaxis will expect a system of
astronomy to be based upon observations. Therefore he tries
to convince the readers that the Syntaxis is firmly based
upon the most accurate observations possible.

A different example of Ptolemy's actions may be more
revealing. I doubt if anyone would maintain that the latitudes and longitudes of 1030 stars are quantities that can be
found in any way except observation. The detailed coordinates
of the 1030 stars are not quantities to be considered as demonstration examples, because they are not used as examples,
with only a few exceptions. There is no indication that the
coordinates had mystic, religious, or astrological significance, and hence there is no reason why they should be given
specific values.† Hence we find the latitudes and longitudes
of the stars by observation and by observation alone; theory
and introspection do not enter into the matter.

In Chapter VII.4 of the Syntaxis, Ptolemy says that he
has measured the coordinates of all the stars that it is
possible to observe, down to stars of the sixth magnitude.
He identifies the instrument with which he made the measurements, he describes the procedure that was followed, and he
presents the alleged results in his star catalogue. However,
we proved in Chapter IX above that the coordinates were not
obtained by measurement at all. They were not obtained with
the instrument that Ptolemy claims to have used, they were
not obtained by the method that he claims to have used, and
they were not obtained by any other instrument or procedure
of observation. They were fabricated, and Ptolemy lied about
what he did.

In summary, Ptolemy professed the same general attitude

---

†The positions of the planets, and perhaps the relations of
some of the stars to their rising or setting points, at
specific instants such as the instant of birth of a person,
did have astrological significance. However, this does not
dictate the coordinates of the stars with respect to the
vernal equinox and the ecliptic, which is what we find in
the star catalogue.

toward observations and their role in astronomy that we do. That is, observations form the proper base of astronomy and no theory can be accepted in astronomy unless it is founded upon observations and is consistent with them. However, Ptolemy did not practice what he preached. He developed certain astronomical theories and discovered that they were not consistent with observation. Instead of abandoning the theories, he deliberately fabricated observations from the theories so that he could claim that the observations prove the validity of his theories. In every scientific or scholarly setting known, this practice is called fraud, and it is a crime against science and scholarship.

4. The Accuracy of Ptolemy's Theories and Tables

In earlier parts of this work, I have determined the accuracies of Ptolemy's theories and/or tables for the sun, the moon, the stars, and the planets. It is useful to summarize these results in Table XIII.3. Because the entries in the table are not all on comparable bases, it is necessary to give some explanation of the table.

TABLE XIII.3

THE ACCURACY OF PTOLEMY'S THEORIES

| Body | Standard deviation, degrees | |
	With Ptolemy's parameters	With the best-fitting parameters
Sun		
bias error	1.10	0
periodic error	0.27	0.01
Stars	1.21	0.51
Moon	0.58	0.56
Mercury	2.99	1.70
Venus	1.01	0.14
Mars	0.44	0.36
Jupiter	0.17	0.09
Saturn	0.38	0.10

Because Ptolemy uses a wrong value for the length of the year, there is a bias in his tables of the sun that grows steadily with time. I have entered $1°.10$ for this bias error in Table XIII.3; this is close to the value that applies at the middle of Ptolemy's career. The bias error could have been reduced to a negligible level if Ptolemy had used a reasonable number of observations which had the accuracy expected in his time. Hence I have entered 0 for the bias error in the column headed "with best-fitting parameters"

In addition, Ptolemy's theory of the sun has an error
that varies periodically, with a period of a year. This
error comes mostly from the errors in the eccentricity of
the sun and in the position of its apogee, and it shows up
as the error in the equation of the center for the sun. In
Ptolemy's theory, the equation of the center has a maximum
value of $2°.39$ although the correct value is $2°.01$. The
maximum error is $0°.38$ and the standard deviation of the
error is $0°.27$. Because the eccentric model cannot represent the longitude of the sun exactly, there would still be
a small error, about $0°.01$, even if the parameters had their
best values.

Three kinds of error must be considered for the stars.
There is a bias error in all the longitudes that traces back
to the error in the length of the year; I have taken this to
be $1°.10$ in preparing Table XIII.3. From Table IX.1, we see
that there is also a random error in the longitudes of the
stars with a standard deviation of $22'.3$ and a random error
in the latitudes with a standard deviation of $20'.8$. When
all three errors are present, the standard deviation of a
star position in Ptolemy's catalogue is $1°.21$. When we
eliminate the bias error by using the correct position of
the equinox, the standard deviation is the resultant of the
random errors in latitude and longitude; this is $0°.51$.

The other accuracies in Table XIII.3 refer only to
longitudes. Errors in the latitudes of the moon and the
planets are appreciable, but they are typically less than
the errors in the longitudes. The errors listed have been
discussed earlier in the appropriate sections. In order to
calculate them, I prepared computer programs to compute longitudes from Ptolemy's theory, and used the programs to find
the standard deviation between longitudes from his theories
and those from modern theory. I then allowed all of the
parameters in the models to vary until the standard deviation was a minimum.

We see that Ptolemy comes fairly close to realizing the
potential of his theory for the moon and for Mars. Otherwise,
the errors in his model are much greater than those found
with the best-fitting parameters. This is particularly so
for the inner planets, Mercury and Venus.

The errors in Ptolemy's models are comparable with
errors of observation for the outer planets, but they are
considerably greater than reasonable errors of observation
for all the other bodies in Table XIII.3. There is no basis
for the claim, quoted in Section VII.6, that Ptolemy's theories represent positions as well as the observations of the
time could measure them.

As we move from Mars to Saturn in Table XIII.3, the
ratio of Ptolemy's error to the error with the best-fitting
parameters steadily increases. This is probably the consequence of Ptolemy's treatment of the two eccentricities.
As Table XI.8 shows, the two eccentricities should be nearly
equal for Mars, but the difference between them should steadily increase as we move to Jupiter and then to Saturn, for

-355-

the theoretical reason that was given in Section XI.7. However, Ptolemy forces the two eccentricities to be equal for each planet, instead of letting them be determined separately. This seems to have been a deliberate decision. As we have seen, he let the eccentricities be determined separately for Venus, but he fabricated the data so that they would come out equal. On this basis, he then took them to be equal for the outer planets.

5. Ptolemy's Competence as an Astronomer

In all the relevant literature that I have seen, Ptolemy is considered to be an astronomer of the first rank, and the main question seems to be: Should we consider Hipparchus or Ptolemy to be the greatest astronomer of antiquity? The realization that Ptolemy's work is fraudulent forces us to think about the previously unthinkable. We must now ask: Was Ptolemy a competent astronomer?

Luckily, scholarly fraud is relatively rare, and I have not seen any statistical studies about perpetrators of it. The number of perpetrators may be so small that we cannot use statistical methods in their study. However, the discovery of fraud warrants a tentative judgment. A competent scholar can achieve valuable results by honest methods. Thus, when we find that a pretended scholar is dishonest, we are entitled to suspect that he is not competent. Otherwise, why would he resort to fraud?[†]

In this section, I want to outline a number of places in which Ptolemy's grasp of astronomical theory seems to be defective.

(1) In his Chapter I.10, Ptolemy explains how he has constructed a table of chords for every angle that is a multiple of $\frac{1}{2}$ degree. The most important part of his method is the conclusion that

$$\text{chord } 1° = 0;1,2,50,$$

in sexagesimal notation. As we saw in Section II.3, this value is indeed correct when it is rounded to the third sexagesimal position. Unfortunately, as we saw, Ptolemy's proof is incorrect. However, it is possible that this was not really a theoretical error on Ptolemy's part. He may have originally solved the problem correctly but, in attempting to find a succinct way of explaining the solution for his book, he may simply have overlooked the fact that his explanation was deficient. Many scholars have experienced this kind of difficulty, and it does not redound to their discredit.

There is another error in Ptolemy's discussion of this problem. He says that it is not possible to find chord $\frac{1}{2}$

[†]I want to emphasize that the argument of this paragraph has only heuristic value. Leaving Ptolemy aside for further study, I do not know whether any proven scholarly frauds are also persons of outstanding ability.

degree (or chord 1 degree) directly from chord $1\frac{1}{2}$ degrees, which was already known. However, this is not correct. We can find chord $\frac{1}{2}$ degree from chord $1\frac{1}{2}$ degrees by solving the simple equation given at the end of Section II.3. We know that Ptolemy knew the basis of this equation, because it is founded upon a method that he describes.

(2) As we saw in Chapter VI, Ptolemy starts his study of the moon by studying the longitude of the moon at the syzygies. From this study, he finds how the mean longitude of the moon varies with time and he finds the form that the equation of the center of the moon has when it is restricted to the syzygies.† At the beginning of his study, in Chapter IV.1 of the <u>Syntaxis</u>, Ptolemy says that it is necessary to use only observations made during lunar eclipses, because all other observations are rendered invalid by parallax. This is not true, for several reasons.

First, the effect of parallax can be calculated and removed from the measurements. Doing so requires knowledge of the distance to the moon, but this can be found without knowing the theory of its longitude.

Second, even if the parallax could not be found in general, it is straightforward to find the conditions when the parallax in longitude is zero. Doing this requires only that we know the relation between the equator and the ecliptic, and this was already known from the theory of the sun. It would have been straightforward to base the theory upon measurements of the lunar longitude made when the moon was full and when it was in the position of no parallax. These conditions are met with sufficient accuracy almost every month, whereas a lunar eclipse visible at a particular place happens only once in about 16 months.

Third, Ptolemy does not avoid the use of calculated corrections to observation by using lunar eclipses. The reason is that he must know the time when the eclipse is a maximum, but this cannot be measured accurately. If the times of both beginning and end of the eclipses were measured accurately, he could take the maximum to be the midpoint. However, in almost all the eclipses that he claims to use, only the time of the beginning or end was measured, and the time between the middle and the beginning or end had to be calculated from the magnitude. Ptolemy's average error in this calculation is about 10 minutes of time, and the moon moves about 5' during 10 minutes. In contrast, the average error in calculating the parallax should be practically zero.

Fourth, the edge of the earth's shadow is diffuse because of refraction in the atmosphere. Hence the time when an eclipse begins or ends is not well defined, as I point out in Section VI.5. It would be more precise to line up

---

†As I pointed out in Chapter VI, Ptolemy really studies only the full moon in this part of his work, and he assumes, with no evidence that I noticed, that the results also apply to the new moon.

the moon with the sights in an astrolabe or, still better, to observe when an occultation begins and ends.

Thus Ptolemy chooses to use what is probably the poorest rather than the best kind of observation for his purposes, although he has the basic information necessary to decide upon the best kind.

(3) We should note the order in which Ptolemy establishes the coordinates of the various celestial objects. First he finds the positions of the equinoxes and solstices, and then he finds the mean longitude and the equation of the center of the sun at any time. Next he constructs the theory of the moon. In the basic observations, he measures the longitude of the moon by using the sun as the basic standard when he uses eclipses, and by using the equinox as the basic standard when he uses the astrolabe.† Then he uses the moon as the standard to measure the longitudes of the stars; that is, he measures the separation of a star from the moon and adds this to the calculated longitude of the moon. Finally, he measures the longitudes of the planets by referring them to the stars.

This is clearly the wrong sequence to use. It is certainly necessary to use the sun in establishing the position of the equinoxes. However, we should invert the order of using the moon and the stars. The best procedure is essentially the one that he claims to have followed in measuring the longitude of Regulus (α Leonis). In this, he establishes the equinox by sighting on the sun; this gives the absolute longitude of the sun, and he does not need to know the longitude of the sun from tables to do this. He then measures the position of the moon just before sunset and then relates the star to the moon by an observation just after sunset. This relates the star to the equinox, and we do not need to calculate anything except the small amount the moon moves between the observations. When the measurement is done in exactly this way, the position of the equinox is established by an observation made near the horizon, so that refraction can introduce a significant error. However, there are ways to avoid this that I shall not take the space to describe, and thus there are ways to relate the stars directly to the equinox with no systematic errors.

When the stars are located, they then serve as the references for locating the moon and the planets, preferably by observing occultations or conjunctions.

Actually we know that Ptolemy did not observe the stellar coordinates by the method he describes, nor by any method at

---

†It is written in many discussions of the subject that Ptolemy measures the elongation of the moon from the sun when he uses the astrolabe. However, as we saw in Section VII.1, Ptolemy first uses the astrolabe to establish the position of the equinox from the declination of the sun. He then reads the longitudes of both the sun and moon independently. I should say that this is what he claims to do; he actually fabricates the observations.

all. Instead he took an older star catalogue and simply added a fixed quantity to the longitudes. We saw in Section IX.2 that the original star coordinates were more accurate than the theory of the moon. This means that the coordinates were not measured with reference to the moon. They must have been measured either by reference to the sun or directly to the equinox, but we have lost the method that was followed.

It is clear that Ptolemy does not understand how the co-ordinates of stars should be determined, since the method that he claims to have used is an inferior one.

(4) He claims that he measured the minimum zenith distance of the moon an unspecified number of times, and that he always found it to be 2 1/8 degrees. He does not seem to realize that the minimum recurs only at intervals of 19 years. As we saw in Section VIII.4, there is no plausible set of dates when the measurement could have been made often enough to warrant the word "always".

(5) We saw in Section VII.4 that Ptolemy makes a serious theoretical error in analyzing the observation of -127 August 5. He states that the moon was in the position where there is no parallax in longitude. Actually the parallax was 9', and including the parallax in his analysis of the observation would destroy the agreement that he claims to find with this observation.

He makes the same mistake in connection with the alleged observation of 139 February 9. Again the correct parallax is 9', but in the opposite direction to the parallax on -127 August 5. If Ptolemy had included the parallax correctly on both occasions, the observations would not have confirmed each other. They would have disagreed by 18', a quite serious discrepancy.

Ptolemy makes a further serious theoretical error in analyzing the alleged observation of 139 February 9. He finds that the anomaly of the moon was 87;19 degrees, and he says that this is the anomaly that makes the equation of the center a maximum. Actually, the anomaly when this happens is 97;40 degrees. The resulting error in the longitude of the moon is about 7'.

(6) In finding the parallax of the sun, which is equivalent to finding its distance, Ptolemy uses Equation VIII.2 in Section VIII.2. This equation is

$$\rho_\odot + \rho_U = \Pi_\odot + \Pi_\mathbb{C}.$$

Other astronomers apparently used this equation before him.†
As others do, he finds $\rho_\odot$ and ρ_U from various observations, so that he knows the sum of the solar and lunar parallaxes, $\Pi_\odot + \Pi_\mathbb{C}$.

---

†More accurately, Ptolemy and others solved the geometric relations that lead to this equation without using the equation itself. See Section VIII.7 and Appendix B.

From this point on, Ptolemy's use of the equation is incompetent. Since the sum of the two parallaxes is known, we need one other relation between them in order to find both. Ptolemy supplies the extra relation by allegedly measuring the parallax $\Pi_{\mathbb{C}}$ of the moon by direct observation on 135 October 1. He then subtracts $\Pi_{\mathbb{C}}$ from the sum to find $\Pi_{\odot}$. The trouble with this method is that it involves finding the small quantity $\Pi_{\odot}$ by subtracting two nearly equal quantities, namely $\Pi_{\mathbb{C}}$ and $\Pi_{\odot} + \Pi_{\mathbb{C}}$. It does not take much insight to see that this process magnifies the inevitable errors in measurement. In fact, Ptolemy's method can and does lead to a negative value for the solar parallax and hence for the solar distance, as I showed in Section VIII.8. A negative distance is nonsense, of course, but Ptolemy deliberately adopts a method that is likely to give nonsense for an answer, even though he claims to have studied the merits of the various methods available.

Aristarchus's method is far superior. He measured the ratio of the parallaxes. Once we know the ratio, we can combine it with the sum to find both. Although Aristarchus's value of the ratio was inaccurate, even an inaccurate value leads to the sensible conclusion that the sun is much larger than the earth and much farther away than the moon. The method does not lead to physical nonsense, as we saw Ptolemy's method do in Section VIII.8.

(7) Ptolemy chooses the worst way rather than the best way to get the apparent diameters of the sun and moon, as we saw in Section VIII.6. The best way is to construct a sighting instrument of some sort that allows direct comparison of the moon with a circle of known size and distance. In Chapter V.14 of the Syntaxis, Ptolemy says that Hipparchus used this method. He goes on to say that the method is not reliable, but what he says on this point does not make sense, at least to me.

The method that Ptolemy recommends involves measuring the magnitudes of partial lunar eclipses which occur when the lunar anomaly has the same value.† From two such eclipses, it is possible to deduce both the apparent radius $\rho_{\mathbb{C}}$ of the moon and the radius ρ_U of the earth's shadow for the specified value of the anomaly. This is a quite poor method, for two reasons. For one reason, the magnitude is hard to measure accurately because the edge of the shadow is not sharp. For the other, $\rho_{\mathbb{C}}$ must be found by subtracting two nearly equal numbers, a process that magnifies observational error.

However, Ptolemy may have been led to making this error from a reason other than ignorance. He could not recommend an accurate method of measuring $\rho_{\mathbb{C}}$ because an accurate method would immediately show up the fatal error in his lunar theory. He had to recommend a method that would not disprove his theory, and this had to be a method that could be used only at the syzygies.

---

†More accurately, either at the same value or at symmetric values that correspond to the same lunar distance.

(8) Ptolemy also chooses a poor way to find ρ_U. It is
necessary to find ρ_U from lunar eclipses, but one should not
use Ptolemy's method, for reasons that were explained in
Section VIII.9. He has to use two eclipses in order to find
one estimate of ρ_U. It is better to start by measuring $\rho_{\mathbb{C}}$
without using lunar eclipses. When $\rho_{\mathbb{C}}$ is known, each partial
eclipse yields an independent estimate of ρ_U, and the result
is to improve the accuracy considerably. However, Ptolemy
may have been constrained in his freedom on this point, just
as he may have been constrained about the preceding one.

(9) In several places, Ptolemy measures or pretends to
measure the same quantity more than once. In spite of this,
he does not seem to understand the significance of measurement error. We see this because Ptolemy's repeated "measurements" always agree with almost impossible accuracy. If he
had understood the situation, he would have known that his
"confirmations" did not confirm; they contradicted the possibilities of measurement.

Appreciating the significance of measurement error was
within the scientific capacity of the time. Ptolemy reveals
this in his discussion of some of Hipparchus's results, as
we saw in Sections VI.6 and VI.7. Hipparchus had found two
different values of the lunar eccentricity from two different
sets of data, and the difference between the values is consistent with measurement error. Ptolemy does not understand
this. He says that the difference arose because Hipparchus
made mistakes in his calculations, and he fabricates a revision of the data that leads to identical values of the eccentricity.

(10) In Chapter IV.2 of the Syntaxis, Ptolemy gives a
value of ψ' which he says Hipparchus had found; ψ' means the
average amount by which the argument of the latitude of the
moon changes in a day. In Chapter IV.9, Ptolemy determines
the argument of the latitude at the middle of the eclipses
of -490 April 25 and 125 April 5. The difference is 9'
greater than the change that we calculate from Hipparchus's
value of ψ', and Ptolemy changes the value of ψ' accordingly.
He does not realize that the change is illusory, because the
error in measuring the change (Section VI.9 above) is about
20'.

Further, although Ptolemy has only one significant figure (9') in the change to be added to the argument of the
latitude, he adds an amount with five significant figures to
the value of ψ'. It is probably useful to keep some figures
that are beyond significance in order not to lose any of the
accuracy available from the observations, but keeping four
extra figures seems unreasonable. Before we make a judgment
on this point, however, we should like to know whether Ptolemy does something unusual for his time or whether he is
following a custom.

Ptolemy, in Chapter IV.7 of the Syntaxis, does a similar
thing for the rate at which the anomaly changes. This alter-
ation of Hipparchus's value is totally illusory, because
Ptolemy has fabricated all the data involved.

(11) In Chapter V.2 of the Syntaxis, Ptolemy discusses the fact that E, the maximum equation of the center for the moon, is greater at the quarters than it is at the syzygies. Now sin E = r/R, in which r is the radius of the epicycle and R is the radius of the deferent curve. As Ptolemy points out explicitly at the end of Chapter V.2, this means that the ratio r/R must be bigger at the quarters than at the syzygies. He then tries a model which makes R smaller at the quarters, so that the ratio will be bigger there, and he soon finds that this model does not work.

At this point, it seems to me, a well-qualified astronomer would have tried the effect of varying r; if Ptolemy had done so, he would have found that the resulting model succeeds well. However, Ptolemy does not do this, as we have seen. Instead, he fabricates data in an attempt to make his defective theory seem correct.

(12) In Chapter VII.3 of the Syntaxis, Ptolemy quotes seven occultations or conjunctions involving stars and the moon, which I have discussed in Section IX.5. All these observations are attributed to astronomers other than Ptolemy. As I said earlier, carelessness or even slovenliness is the main feature in Ptolemy's treatment of these observations. It would take too much space even to summarize the situation here, and I shall cite only one example.

At a certain hour on 98 January 11, the star α Virginis was hidden by the moon, so that the uncertainty in its position was equal to the apparent diameter of the moon. A short time later, the star was again visible, and the observer estimated the separation between the star and the center of the moon. It is obvious that the second observation is more precise than the first and that it is the one to use. However, Ptolemy ignores it and uses only the imprecise first observation. Even after we realize that the observations are fabricated, we see that this is not sensible. If a fabrication is to be plausible, it must correspond to a sensible observational procedure.

(13) Ptolemy gives a fair amount of detail about some of his observing instruments and procedures, but he omits many important details. For example, in Chapter I.12 of the Syntaxis, he describes two instruments that he uses to measure the meridian elevation of the sun. He tells us enough about the instruments to let us draw pictures of them, but he does not tell us their dimensions nor does he tell us the size of the divisions on their graduated circles. Yet, as he should realize, this is probably the most important information that we need in order to judge the accuracy of the observations.

(14) Only for Mars and for the moon do the parameters that Ptolemy finds come close to providing the potential accuracy of his models. I find this inexplicable. We know that Ptolemy exaggerates the accuracy of his models by using fraudulent data. Still, within the limits allowed by his models, we expect that he would try to find the parameters as accurately as possible with the genuine data that were

available. However, he has not done so.

(15) Ptolemy fails to understand almost all aspects of
the theory of Mercury. He does not even use the idea of
symmetry correctly. I cannot attempt to summarize the situation here, and I can only refer the reader back to Chapter X.

(16) Ptolemy has two or more small distances in his
models for the moon and each of the planets. He fabricates
the data so that the small distances in each model will turn
out equal. As we have seen in the relevant tables (Tables
VII.2, X.2, XI.2, XI.4, XI.5, and XI.6), they should be nearly
equal for Mars, but they should be appreciably different for
the other bodies, and greatly different for Mercury. In one
sense, this error of Ptolemy's is a special case of item (14)
above. It also illustrates another propensity that Ptolemy
has, which is to force parameters to take on simple values,
regardless of the data. For example, he fabricates his data
so that the precession of the equinoxes will come out 1° per
century, even though his genuine data show that this is not
correct, and even though it is not the value found by Hipparchus. Hipparchus merely sets this as a lower limit, apparently because of doubts about the observational accuracy of his
data. Because of the short time span of the data available
to Hipparchus, he was correct in allowing the possibility of
a value this low. Three centuries later, the data left no
doubt that the value was much larger. However, Ptolemy does
not seem to understand the limitations imposed by observational
error, and he takes Hipparchus's limit as the actual value,
probably because of its simplistic nature. In doing so, he
ignores the real data and relies upon fabrication to support
a false hypothesis.

(17) In both items (9) and (16), Ptolemy does not seem
to understand the significance of observational error. In
many places, he also does not seem to understand the significance of rounding, of approximations, and of other types of
arithmetical error. This shows up perhaps most strongly in
his treatment of the model which yields the radius of the
universe, which was described in Section IV.8. As we saw
there and in Section VIII.7, the lunar, solar, and planetary
parameters that Ptolemy finds in the Syntaxis agree with the
model rather closely, and we concluded that he may have fabricated some of the data in order to produce this agreement.
However, when he came to the detailed description of his
model, he introduced approximations and made other arithmetical errors that destroyed the agreement. Further, he does
not seem to understand that the disagreement is a result of
his arithmetic, and he does not understand that the arithmetic must be done carefully if the model is to have meaning.
He seems instead to think that a fundamentally new analysis
of the lunar parallax will be needed in order to produce
agreement.

Every scholar, even a genius, is entitled to a few errors, but Ptolemy's errors seem excessive to me. Since
there is no accepted scale for defining competence, I hesi-
tate to say that he was incompetent. However, I think that

-363-

it is safe to say that he was not an astronomer of the first rank. The best we can say for him, it seems to me, is that he was mediocre. In view of the summary above, I believe that most readers will have serious reservations about Ptolemy's capacity as an astronomer.

Some readers will probably raise objections to this conclusion. I suspect that the objection raised most frequently will be the following: Ptolemy shows in the Syntaxis that he is capable of solving quite complex theoretical problems, and he also understands thoroughly just what types of observation are needed in order to find the theoretical parameters with the minimum of labor. This shows that he is an astronomer of the first rank, and the errors that I have enumerated are no more than he is entitled to in a work of the magnitude of the Syntaxis.

The most complicated problem that comes up in the Syntaxis, both from the standpoint of choosing observations and of solving the mathematics involved, is probably one connected with the solar theory. This problem was described in Section V.1. There Ptolemy simultaneously finds apogee, the eccentricity, and the anomaly at some epoch, from three observations of the sun. The same problem comes up for each outer planet, as we saw in Section XI.6. Further, the same problem comes up for the moon in Section VI.5, except that the problem there is worded in terms of the epicycle model rather than the eccentric. Since the eccentric and epicycle give identical results, the lunar problem essentially is the same as the others.

Many of us have known people who could learn to solve certain problems by rote, but without understanding the true significance of the procedures that they were following. As long as they confront only problems that can be solved by the learned procedures, they seem to be quite successful. Their limitations are seen only when they confront a problem that cannot be handled by these procedures, or when they confront aspects of the situation that are not covered by the routines they have learned.

There is no obvious reason to doubt that this is the situation with Ptolemy. The problem that was just described had already been solved by Hipparchus, and it may have been solved by others before him. Thus it was not new, and Ptolemy needed no originality in order either to solve it or to choose the necessary observations.

6.   Did Ptolemy Contribute Anything to Astronomy?

We have seen that Ptolemy is not a first-rate astronomer that all of his own observations have been fabricated, and that many observations which he attributes to others are also fabricated. We must now ask: In spite of all this, is there any aspect of the Syntaxis that can be considered as a positive contribution to astronomy?

One contribution that has been claimed for Ptolemy is the following: The Syntaxis is the only place where many observations and other important parts of Greek astronomy have

been preserved. Even if Ptolemy had made no direct contributions, he made an important indirect contribution to astronomy by preserving this material, which would otherwise have
been lost.

This reminds me of a political cartoon that I saw some
years ago; I do not remember the cartoonist, the source, or
the basic problems, persons, or entities that were commented
on. The cartoon showed a pedestrian who had been knocked
down by a passing car and who was now lying beside the road.
The car had stopped a short distance beyond, and the driver
was running back toward the victim while calling out: "I'm
a doctor. You're lucky I happened to be passing by!"

The idea back of this cartoon can be applied in many
areas. When it is applied to astronomy, the characterization
is clear: Ptolemy is the doctor-driver and the victim is the
science of astronomy. Because he based all his astronomical
work upon fraud, Ptolemy could pretend to a universality that
was denied to honest astronomers. As a result, his "universal"
work displaced almost all of the earlier and valid Greek astronomy. If the Syntaxis had not been written, we can be sure
that much valid Greek astronomy now lost would have been preserved directly.

In other words, we do not owe Ptolemy our thanks for the
small amount of earlier astronomy that he has preserved. Instead, we owe him our condemnation for the large amount of
genuine astronomy that he has caused us to lose.

The people who have claimed this contribution for Ptolemy tacitly assumed that the material he "preserved" is genuine. We have seen that this is not so. All his own observations that are used are found to be fraudulent when we can
test them. Further, a large fraction of the observations
that he attributes to others turn out to be not material
that he has preserved but material that he has forged.

In fact, the "information" that Ptolemy has preserved
has interfered with both astronomy and history instead of
helping them. We saw an example of the damage he has done
to history in Section V.5. He says that the summer solstice
in the year -431 came at 06 hours, Athens time, on June 27.
Whether Ptolemy is responsible for fabricating the hour or
not, he is the only source that has preserved it, and it has
caused many scholars to construct erroneous theories of the
Athenian calendar. We can no longer accept as evidence anything Ptolemy says unless we have independent confirmation,
and historians must now confront the task of identifying all
historical material that rests upon the unsupported word of
Ptolemy. At a guess, the realization of Ptolemy's fraud
destroys half of what we have been accepting as Greek astronomy.

There are many examples of the damage that Ptolemy has
done to astronomy by his fabricated data. Because he accepted
the observations that Ptolemy used, and because he thus had
to reconcile these data with genuine data, Copernicus [1543]
had to make his heliocentric theory much more complicated

than it needed to be; see for example his theory of the oscillatory motions of the obliquity and the equinox in his Chapters II.2 and III.2. It is quite possible that this unnecessary complexity interfered with the acceptance of Copernicus's ideas; certainly it did not help. Since the time of Copernicus, many writers including myself have used Ptolemy's fabricated material in studying the accelerations of the sun, the moon, and the planets. All this work must now be redone.

Several writers say, explicitly or implicitly, that Ptolemy made an important contribution by accepting and propagating the precession of the equinoxes, which had been discovered by Hipparchus.† For example, <u>Dreyer</u> [1905, pp. 203-204] writes: "It is very remarkable that so important a discovery should not have become universally known; and yet we find that precession is never alluded to by Geminus, Kleomedes, Theon of Smyrna, Manilius, Pliny, Censorinus, Achilles, Chalcidius, Macrobius, Martianus Capella!‡ The only writers except Ptolemy who allude to it are Proklus, who flatly denies its existence, and Theon of Alexandria, who accepts the Ptolemaic value of one degree in a hundred years, ..."‡

It seems to me that the implication does not follow from the evidence that Dreyer cites. I have not read the relevant part of Proklus myself, but I understand that he did accept the steady motion of the equinox between Hipparchus's time and his own.* However, he advocated, for reasons that are now unknown, an early form of the trepidation of the equinoxes, in which the motion which had been going on for many centuries would shortly reverse and go in the opposite direction at the same rate and by the same total amount. Thus both Proklus and Theon of Alexandria did accept Hipparchus's observation of the equinoctial motion, but Proklus did not accept his interpretation of it.

I have not read most of the other writers mentioned by Dreyer, but the subject of precession is irrelevant to the purposes of those whom I have read.* Thus it is not surprising that they <u>do</u> not mention it. Further, we can hardly

†Do we know that Hipparchus discovered the precession? Is there any testimony independent of Ptolemy's on this point?

‡The exclamation point is Dreyer's.

‡Theon of Alexandria is 4th century and Proklus is 5th century.

*See <u>Dreyer</u> [1905, p. 204].

*The writing of Censorinus suggests to me that he did not have a significant knowledge of astronomy. For example, in his Chapter XIX, he gives the lengths of the year as determined by various early astronomers, without mentioning either Hipparchus or Ptolemy, and apparently without realizing that later values should be preferred because they are based upon a longer time span. The lengths he cites range from $364\frac{1}{2}$ to 366 days. In his Chapter XXII, he concludes that the year has 365 days plus a fraction that he does not know, because the astronomers have not yet found it. Thus we do not expect him to understand the significance of

conclude that precession was passed over by serious students of the subject just because it is absent from many works. We have lost almost all astronomical writing between Hipparchus and Ptolemy, and writing of a character likely to mention precession is also of the character most likely to be displaced by the Syntaxis. That is, specifically astronomical writing is the most likely to be lost because of the influence of the Syntaxis.

Now we can turn to direct contributions to astronomy that Ptolemy might have made. Most writers take these to consist of his three models of motion, namely the crank models of the moon and of Mercury and the equant model for the other planets.

We studied the crank model of the moon thoroughly in Chapters VII and VIII. It gives the geocentric longitude of the moon more accurately than does a simple eccentric or epicycle, but at the same time it introduces large errors in the parallax, both in latitude and longitude. The parallax enters into the topocentric latitude and longitude, which are the quantities that are observed, and thus the effect of parallax must be considered in judging the true accuracy of Ptolemy's lunar model. Since the maximum error in the parallax is close to 1° in each coordinate, the overall accuracy of Ptolemy's model in the observed angular coordinates differs little from that of the simple epicycle or eccentric. Further, any possible gain in angular accuracy is achieved only by introducing a gross error into the lunar distance and hence into the apparent size of the moon. As we concluded earlier, Ptolemy's model of the moon cannot be considered as a success in a meaningful way and hence it is not a contribution to astronomy.

Ptolemy's model of Mercury is actually inferior in accuracy to that achieved by simpler models and thus it is certainly not a contribution to the subject. Further, Ptolemy's model is based upon a serious misunderstanding of the subject, and Ptolemy's misunderstanding has caused many later writers to misunderstand the subject also.

Finally, we come to the equant model that Ptolemy uses for Venus and the outer planets. Here the accuracy achieved by the equant is indeed greater than that which we get from a simple epicycle plus an eccentric,† so the equant is indeed a contribution to astronomy. However, we must ask whether this contribution was made by Ptolemy. There is no evidence that it was made by anyone else, but there is also no evidence that it was made by him.

---

precession, and it would be accidental if he mentioned it. He mentions Hipparchus only once, saying about him only that he devised a calendar with a cycle of 304 years, and he does not mention Ptolemy at all. We cannot infer from this that Hipparchus and Ptolemy were almost unknown to later Greek astronomers, nor can we infer that precession was unknown to them.

†An epicycle moving on an eccentric deferent is the minimum model needed for a planet.

First let us look at what Ptolemy says. He says in Chapter IX.2 of the Syntaxis that Hipparchus put the observations of the planets into order and that he showed that the observations did not agree with the planetary models which had been proposed by earlier astronomers. Ptolemy then says nothing more about the origin of the theories of the planets,† and this is the only basis for attributing the equant model to him. This is not a satisfactory basis, for two reasons.

First, as we have seen, we cannot accept the unsupported testimony of Ptolemy, whether it is about his own activities or those of others. We have seen too many examples in which he has lied to us.

Second, even if we could accept the testimony of Ptolemy, we must recognize that he has not really given any on this matter. All we have is silence. His silence about the origin of an idea does not suggest that the idea originated with him, and we can give a demonstrable example. In Chapter I.13 of the Syntaxis, Ptolemy states an important theorem in spherical trigonometry. He does not attribute the theorem to anyone else, and, if anything, he implies that he is the originator. He does this by using the first person singular when he states the proposition that is to be proved.‡ However we should not read much into this fact; it may have been a custom or even an accident of writing with no intent behind it.

For a long time this theorem was attributed to Ptolemy, but we know now that it occurs in Menelaos [ca. 100], who is earlier than Ptolemy.‡ Hence Menelaos either discovered the theorem or he took it from a still earlier source. Thus Ptolemy's silence tells us nothing about the origin of a theorem or of an idea.

Even if Ptolemy did originate the equant, he does not seem to understand it well. Although the original form of

---

†At least, he says nothing on the subject that I find clear. There are some vague statements that one might interpret as imputing a role to himself in the development of the planetary theories.

‡That is, he states the hypotheses, then he writes "I state" (λεγω) followed by the statement of the proposition.

‡I have seen it stated that the discovery of this fact was made by Mersenne (1588-1648), but I have not seen a statement about where the discovery occurs among Mersenne's works. It is interesting that the lettering of the figure that accompanies the theorem is the same in both the Syntaxis and the cited edition of Menelaos, except that two minor construction points have been interchanged. However, this does not prove anything about priority, since the text of the cited edition was prepared by a medieval astronomer who was well acquainted with the Syntaxis. That is, in preparing his text, the medieval astronomer might have borrowed the lettering from Ptolemy, instead of Ptolemy borrowing it from Menelaos.

the model allows the two eccentricities to be different,
Ptolemy forces them to be equal when they should not be.
As we have seen, this causes Ptolemy to lose much of the
potential accuracy of the model.

To summarize Ptolemy's possible contributions to astronomy, he has almost surely caused us to lose an important
part of Greek astronomy. There is no reason to believe that
he has preserved anything important that we should otherwise
have lost. Three ideas occur in his work that do not occur
in any earlier work that has survived. Two of the ideas,
namely the models of the moon and of Mercury, violently contradict elementary observation and thus they are not contributions. The third is an improvement over known Greek competitors, but Ptolemy makes it lose much of its value by his
inept treatment of it. So far as I know, there is no evidence that Ptolemy originated this idea.

7. Ptolemy's Missed Opportunities

While it is doubtful that Ptolemy actually made any
contribution to astronomy, there is no doubt that he had the
opportunity to make many important ones. To do so, he needed
only to use honest data instead of fabricated data.

His opportunity arises to a great extent from the time
in which he lived. Hipparchus was limited in what he could
accomplish by the relatively short time span of accurate
data available to him. Ptolemy, three centuries later, had
access to all the data available to Hipparchus and to all the
data gathered in the meantime. The additional time span covered by the data is probably more important than the additional volume of data.

For example, Hipparchus estimated the length of the
year by using observations made 145 years apart. The error
in the interval covered by these years was about 15 hours,
so that the error in his length of the year was about 6 minutes. Ptolemy should have been able to make a more accurate
measurement, but let us suppose that he could not. The total
time span available to him was about 420 years rather than
145 years. Simply making a measurement with the same accuracy as the old ones would have given him the year with an
error of only about 2 minutes.

Similarly, he should have been able to find the motion
of the solar apogee, or at least to prove that it moves. It
is possible that he could have discovered that the apogees
of the planets also move with respect to the stars, instead
of simply sharing in the precession of the equinoxes, as he
"proved" with false data. He should have been able to measure the obliquity of the ecliptic accurately and to show
that it is decreasing slowly. Finally, he actually gives
us data that yield an accurate value for the precession of
the equinoxes, but he refuses to use them. Instead, he
"proves" an erroneous value by the use of fabricated data.

He also had the opportunity to make two important theoretical advances, one for the moon and one for Mercury. For

both, he states that either the epicycle radius r or the
deferent radius R must vary systematically with longitude.
When he found that varying R led to results that contradict
observation, he should have tried the consequences of varying r, but he did not. If he had done so, he could have
discovered accurate theories of both bodies.

8. Missed Opportunities in Greek Astronomy

One peculiarity that I have noted several times is that
the accuracy of the observations, so far as we can judge
from those which have survived, did not improve during almost the entire course of Greek astronomy. The known course
of this astronomy, aside from a period whose history rests
mostly upon conjecture, runs from Meton and Euctemon in
Athens around -430 to Heliodorus around 510. Although Greek
observations seem to be more accurate than Babylonian ones,†
failure to improve them over a span of nine centuries seems
like a missed opportunity.

Another missed opportunity comes from maintaining the
basic idea back of the epicycle or eccentric. As I showed
in Sections IV.2 and VIII.9, the epicycle (and the eccentric)
give a variation of distance that is twice what it should be
for the correct variation of longitude. Further, we saw that
the Greeks had a considerable body of evidence which showed
that this was the case, both for the moon and the planets,
and we saw that the equant was one answer to this problem.
While the Greeks, or at least Ptolemy, did adopt the equant
for some of the planets,‡ they never applied it to the moon,
so far as we know.

However, the most important opportunity that the Greeks
missed is undoubtedly the opportunity to adopt the heliocentric hypothesis. As we have seen in earlier sections, many
important lines of evidence known to the Greek astronomers
lead directly to the fact that the heliocentric hypothesis
saves the phenomena more easily than the geocentric one:
(1) The sun is much bigger than the earth. (2) The mean
longitude of the sun (Section IV.7) is related to the motions
of the planets in a way that has no explanation in a geocentric theory but that is obvious in a heliocentric theory.
(3) The configuration of the mean sun, the earth, and an
outer planet at opposition (Section XI.6) must be proved as
a theorem under the geocentric hypothesis, but it is a tautology under the heliocentric hypothesis. (4) A theory of
latitudes is complicated in a geocentric system, but the most
obvious theory in a heliocentric system is also a highly
accurate one. See Section 3 of Appendix A for what the theory of latitudes should be in a geocentric system.

This failure of the Greeks has been the major subject
of much historical writing, and the question of why they
failed is probably no more susceptible of answer than the
question of why the Roman Empire fell. There is undoubtedly
no simple answer. However, the present work does suggest a
facet of the problem that has not been noticed before, so

---

†This is shown by the results of <u>APO</u>, in which a large body
of Babylonian observations is analyzed.

‡But in an unduly restricted form.

far as I know. We see now that Greek astronomy was rather
restricted in both time and participants. I wrote a while
back as if Greek astronomy endured for nine centuries, from
Meton to Heliodorus, and this is correct if we base the duration only upon the known observations.

However, from the standpoint of known ideas, it really
endured only from Meton to Hipparchus. The only innovation
that entered astronomy after Hipparchus, so far as we know,
is the equant, and this statement is made insecure by Ptolemy's
fraudulent writing. Thus, so far as concepts, depth of understanding, and inquisitiveness are concerned, Greek astronomy
probably had a lifetime of little more than three centuries.
Further, during its lifetime, it had few innovators. If we
leave aside those who contributed only observations, the participants during its three centuries can probably be counted
on the fingers, or at most on the fingers and toes.

This is not intended to mean that more participants
necessarily mean more progress; the quality of the participants is more important than their number. However, if there
are no more than, say, a score of participants, there are
drastic limits to the ideas that they can conceive and explore,
even if all the participants are geniuses. In view of all
else they did, the innovative Greek astronomers may not have
had the time to explore the implications of the heliocentric
hypothesis and hence to realize that it was needed to save
the phenomena.

Whether this explanation is correct or not, it is unfortunate that the Greek astronomers did not explore the
heliocentric hypothesis further.† As we saw in Section XII.5,
a careful study of its implications would lead almost immediately to the adoption of the equant, which is close to an
ellipse, for the heliocentric orbit of each planet. One consequence of this would have been improved accuracy in representing the planetary motions. Another consequence, which is
probably more important in the history of ideas, would have
been greater freedom in looking at the universe around us,
and a realization that the earth does not have a privileged
position in the universe.

9.   The *Syntaxis* and Chronology

The Syntaxis has been used extensively in two areas of
chronology. We saw in Section V.5 that the summer solstice
attributed to Meton in the summer of -431 was fabricated.
The fabrication may have been done in two stages. The day
of the solstice was already fabricated by about the year
-108 for an innocent reason. The hour may have been fabricated at that time also, or it may have been fabricated
later by Ptolemy for his own purposes. This fabricated solstice has played an important role in the study of the
Athenian calendar, with unfortunate consequences. This
would probably have happened even without the Syntaxis,
since there is an earlier document which gives the dates

---

†We must recognize that deeply seated prejudices probably
  stood in the way of this exploration.

involved in the calendar study.

Ptolemy also gives the dates of seven other observations (if I have not overlooked any) in both the Athenian and Egyptian calendars. These have also been used extensively in studies of the Athenian calendar, and the interested reader may consult Meritt [1961] and van der Waerden [1960] for additional information and references. Three of these observations are the lunar eclipses in -382 and -381. The others are the four earliest occultations or conjunctions that Ptolemy uses in establishing the rate of precession; they are studied in Section IX.5.

Unfortunately, all seven observations are fabricated. This fact does not in itself mean that the equivalent Athenian and Egyptian dates are incorrect, but it does not give us confidence in the situation. Even though Ptolemy fabricated the observations, he may well have started with the dates of genuine observations. Further, even if he fabricated the dates themselves, he would first fabricate the date in the Egyptian calendar and then convert it to an Athenian date for added verisimilitude. He could have used correct tables of conversion in doing this.

However, he could also have used incorrect tables of conversion. Our basic rule in dealing with the chronological aspects of the Syntaxis must be the one that we adopted in Section XIII.2 for the astronomical observations: We cannot accept any statement in the Syntaxis as evidence. We can accept only those statements that have confirmation from independent sources, and this means that we are not using the Syntaxis itself; we are using only the independent sources. We must now extend this guide to all aspects of the Syntaxis, whether astronomical or chronological or other.

This means that all studies of the Athenian calendar which have been based in whole or in part upon the Syntaxis must be redone, so that their dependence upon it can be removed.

The Syntaxis has also been used extensively in Babylonian chronology. Ptolemy says that he has a copious collection of astronomical observations made in Babylon. The Babylonian calendar is discussed in some detail in Appendix C. The feature of the calendar that concerns us here is the method of stating the year. The year is always given as being the year Y of a certain king. In order to use the Babylonian observations, Ptolemy prefaces the Syntaxis with a list of Babylonian kings and the number of years assigned to each.† Since the list is brought down to Alexander, and since we can date Alexander, we can then assign years to each Babylonian king in Ptolemy's list. If his list is accurate, it takes us back to -746.

Ptolemy states the dates of seven lunar eclipses with the aid of the Babylonian kings. However, as we point out

---
†Strictly speaking, the last several kings in the list are the Persian kings who ruled Babylonia and Assyria.

in Appendix C, he never gives any more of the Babylonian
date than the year. This contrasts strongly with his treatment of other calendars. In dealing with any other calendar,
Ptolemy gives the full date in that calendar, and he then
gives the equivalent in the Egyptian calendar. The exceptions are so few that they can easily be accidental.

His practice with regard to the Babylonian calendar
does not arise from defects in the Babylonian records. In
all Babylonian astronomical records that I have examined, the
year, month, and day are all stated. Thus there is no osten-
sible reason why Ptolemy should treat Babylonian dates differently from other dates.

However, as I show in Appendix C, there is a peculiarity about the Babylonian calendar which is not shared by
other calendars. With the Athenian astronomical (Callippic)
calendar, for example, one can convert dates in either direction with relative ease.† This is not so for the Babylonian
calendar. If one has the Babylonian date of a lunar eclipse
in the Babylonian calendar, it is easy to find the Egyptian
(or Julian) date if one has the list of kings. The converse
is not true. If one has the Egyptian date of a lunar eclipse,
one can determine the Babylonian year from the king list.
That is as far as one can go unless one has certain extensive
records, which it is doubtful that Ptolemy had. Even if he
had them, it would be exceedingly laborious for him to find
the Babylonian month and day. The nature of the needed records is described in Appendix C.

TABLE XIII.4

LUNAR ECLIPSES FOR WHICH PTOLEMY GIVES THE BABYLONIAN YEAR

Date	King	Year	Authenticity
-720 Mar 19	Mardokempad	1	May be fabricated
-719 Mar 8	Mardokempad	2	Fabricated
-719 Sep 1	Mardokempad	2	May be fabricated
-620 Apr 22	Nabopolassar	5	Fabricated
-522 Jul 16	Kambyses	7	Fabricated
-501 Nov 19	Darius	20	May be genuine
-490 Apr 25	Darius	31	May be genuine

The seven eclipses in question are listed in Table
XIII.4. The table gives the Julian date of the eclipse,
the Babylonian king as stated by Ptolemy, the year of that
king as stated by Ptolemy, and my assessment of the validity

---

†At least the Callippic calendar followed specific numerical
rules which were probably still known in Ptolemy's time.
In view of the confusion that Ptolemy has introduced into
the subject, it is possible that we are not sure of the
rules today.

of the record. Three of the records are almost certainly
fabricated, two may well be, and for two I have found no
strong evidence; the latter pair are marked "may be genuine".
Nothing seriously contradicts the assumption that all seven
records are fabricated.

Let us see how Ptolemy would go about fabricating a
Babylonian record of a lunar eclipse. He would start by
determining the Egyptian date of an eclipse that he wants
to use, and he would then fabricate the exact circumstances
(magnitude and hour) as he wants them. It is important to
realize that this process gives the date of an actual eclipse,
and that the fabricated circumstances are fairly close to the
truth. He then wants to give the date in the Babylonian calendar, but he cannot for the reasons that have been outlined.
All he can give is the Babylonian year.

It is also important to realize that Ptolemy does not
need an authentic king list in order to give a year in the
Babylonian fashion. Even if his king list is fabricated, he
can still use it in order to assign a specific year of a
specific king to his fabricated eclipse record.

Now let us see what happens to a modern historian or
chronologist who studies Ptolemy's eclipse records. He sees
that there is a list of kings and their reigns. He also sees
that Ptolemy dates a lunar eclipse in the first year of Mardokempad, for example, on a certain month and day in the
Egyptian calendar,† at a certain hour on that day, and he
states the fraction of the moon that was shadowed during the
eclipse. The historian uses Ptolemy's king list to find the
year in our calendar and he uses the Egyptian month and day
to find the complete date in our calendar. He then finds by
astronomical calculations that there was an eclipse on that
date, that it came close to the hour that Ptolemy states,
and that the stated amount of shadowing is also close to correct. This agreement between Ptolemy and modern astronomy
happens not just once but seven times.

The historian or chronologist naturally concludes that
there is overwhelming evidence confirming the accuracy of
Ptolemy's king list, and he proceeds to use it as the basis
for Babylonian chronology. Yet there is no evidence at all.
The key point is that there may have been no Babylonian record at all. Ptolemy certainly fabricated many of the aspects
of the lunar eclipses, and he may have fabricated all of them.
When he fabricated them, it did not matter whether he used a
correct king list or not. Any king list he used, regardless
of its accuracy, would seem to be verified by eclipses.

For example, according to Ptolemy's king list, Ilulaeus
reigned for 5 years and his successor Mardokempad reigned
for 12. Suppose that Ptolemy's list had omitted Mardokempad
but assigned 17 years to Ilulaeus. Instead of putting an
eclipse in the 1st year of Mardokempad, Ptolemy would put
the same eclipse in the 6th year of Ilulaeus. From the altered list, we would still establish that the eclipse was on

---

†This is the eclipse of -720 March 19; see Table XIII.4.

-374-

-720 March 19, and we would still have the same apparent verification of the king list.

It follows that Ptolemy's king list is useless in the study of chronology, and that it must be ignored. What is worse, much Babylonian chronology is based upon Ptolemy's king list. All relevant chronology must now be reviewed and all dependence upon Ptolemy's list must be removed.

Luckily, the later part of his king list has independent verification. I mentioned in Section VIII.8 that there is a Babylonian record of the lunar eclipse of -522 July 16, which is one of the eclipses that Ptolemy fabricated. More accurately, I should have said that there is a Babylonian record of a lunar eclipse in the 7th year of Kambyses, which is the same year that Ptolemy states. The document was published by Kugler [1907, pp. 70-71] and the astronomical observations in it are analyzed in APO, Chapters IV, X, and XIV. The document gives the times and magnitudes of two lunar eclipses,† a conjunction of Mercury with the moon, 5 statements of the dates when Venus had its first or last visibility after or before passing the sun, and 4 such statements for Mars. If we assume that the 7th year of Kambyses began in the spring of -522, the times and magnitudes of the lunar eclipses agree fairly well with the stated values, 3 statements about Venus are accurate while 2 are impossible, and 3 statements about Mars are accurate while 1 is impossible. The most likely situation is that the year is -522/-521 and that there are some scribal errors in the record. Nonetheless, the confirmation of the year is not as strong as we would like.

However, there is another document from the 37th year of Nebuchadrezzar‡ [Neugebauer and Weidner, 1915]. According to Ptolemy's list, this year began in the spring of -567. The document records 9 measurements of the times of moonrise or moonset, 5 times of conjunctions of the moon with specified stars, plus 1 conjunction of Mercury, 2 of Venus, and 3 of Mars, all with specified stars. When I analyze these on the assumption that the year is -567/-566, I find that the times of moonrise or moonset agree with calculated values within about 10 minutes. The longitudes of the moon and planets inferred from the conjunctions agree with calculated values within 1° or less for most observations, although there is a discrepancy of about 3° for one lunar conjunction.

Thus we have quite strong confirmation that Ptolemy's list is correct for Nebuchadrezzar, and reasonable confirmation for Kambyses. Since the beginning of Nebuchadrezzar's reign takes us back to -603 if -567 is correct for his 37th year, it seems likely that any error in Ptolemy's list is no more than a few years for dates after -603. So far as I know, there is no astronomical confirmation for earlier dates. I have not attempted to study the evidence available from sources other than Ptolemy for earlier years.

---

†The other eclipse is -521 January 10.

‡I believe that English-speaking Assyriologists now prefer this spelling to the traditional Nebuchadnezzar.

In summary, studies of Babylonian chronology need to be reviewed in order to remove any dependence upon Ptolemy's king list. It is unlikely that there is any serious error in his list after -603, but errors before that year can have any size, so far as information from astronomy can tell us.

10. The Reception of the *Syntaxis*

Several colleagues with whom I have discussed this work have asked what could be the motive for Ptolemy's fraud. We shall never know the actual motive, but it is not hard to think of possible ones. For example, it is possible that Ptolemy was a devotee of some religious fanatic who thought that he had divined the true system of the world, and that all other truth was unimportant. In this case, Ptolemy's actions, while justified as viewed by an adherent of the fanatic, are still fraudulent as viewed by serious astronomers.

Another answer is probably the most likely: Ptolemy wanted to be known as a great astronomer, perhaps as the greatest of all time. He may have found, early in his career, that he did not have the qualifications, and so he turned to the only remaining way of satisfying his ambition, which was to replace ability by fraud.

We can even make a highly tentative reconstruction of a few of the events in his career. As we saw in Section IX.4, Ptolemy made genuine measurements of the declinations of 12 stars, although he does not use them since they conflict with the rate of precession that he chooses to adopt. It is possible that these measurements come from an early period when he was an honest apprentice astronomer. As an apprentice, he could have become aware of the problem of the lunar evection, and could have learned that variations of r or R were the onl possible approaches within the framework of Greek astronomy. He devised the "crank" model as a way of varying R. By the time he discovered that it did not work, he was so enamoured of it that he could not give it up. Instead of giving it up and trying the alternative, he decided to "prove" its validit by generating fraudulent data. From here to the rest of his fraud is a small step. I emphasize that this is highly tenta tive, and I do not press this reconstruction.

The hard problem, as it seems to me, is to account for the favorable reception that the Syntaxis has almost universally received. We must distinguish three historical stages

The first stage lasts from the writing of the Syntaxis until the close of Greek astronomy and philosophy four centuries later. In this stage, we find some criticism of the accuracy of the tables. Heliodorus, for example, observed [APO, p. 212] a conjunction of Mars and Jupiter on 509 June 13. He comments that the conjunction should not have come until the 17th, according to the tables, but by that time the planets were far apart. However, this was only a minor point, and most Greek reception of the Syntaxis was favorabl so far as we can judge from the surviving literature.

By studying the accuracy of Ptolemy's theories, we have seen that observations of even moderate competence made in or shortly after his own time would have revealed that they are seriously defective. Why were these measurements not made and his fraud immediately revealed? The only answer that I can see is that there were no astronomers left who were able to make competent measurements in the critical period, say in the century following Ptolemy. Even if an astronomer like Heliodorus showed a defect by means of observation, he apparently could not deduce the consequences.

The actors on the second stage are the medieval astronomers of Islam and Europe. According to tradition, it was the early Islamic astronomers who applied the name Almagest (which means The Greatest) to the Syntaxis.† Many Islamic astronomers were skilled observers, and I find it hard to understand why they did not realize the situation. Perhaps they did not question the quality of the Syntaxis because it was already hallowed by tradition when they came to know it.

Medieval astronomers can be divided broadly into two classes. One class, of whom the great Islamic astronomer ibn Yunis is an example, realized that Hipparchus's solar data were superior to Ptolemy's In his work, ibn Yunis [1008, p. 142] concludes that Ptolemy's data are not consistent with Islamic observations while Hipparchus's data are. Thus he rejects Ptolemy's solar data (and also his rate for the precession of the equinoxes), but he continues to use Ptolemy's models. He does not realize the gross defects in Ptolemy's model of Mercury, and he continues to use it, along with Ptolemy's values of the parameters [ibn Yunis, 1008, pp. 216ff]. He does not even realize the enormous error in Ptolemy's value of apogee.

The second class of medieval astronomers accepted both Ptolemy's data and his methods, and they were forced into tortuous maneuvers in trying to do so. Copernicus [1543], who rejected the geocentric hypothesis in favor of the heliocentric one, still continued to follow Ptolemy closely in his general methods, and he still accepted Ptolemy's data.

The third stage brings us to the modern period. Delambre [1817], along with several other writers of about the same time, argued that Ptolemy derived some of his observations from his tables instead of the other way around, but his arguments were based upon the size of Ptolemy's errors. Arguments of this kind can be countered by explaining how such large errors of observations are possible, and I did not notice any compelling arguments in this work. In a later work, however, Delambre [1819, p. lxviii] did find the overwhelming argument. He calculated the times of some of Ptolemy's alleged equinox

---

†It is possible that Almagest was originally intended to refer to the size rather than to the quality of the book in question. However, the term has by now acquired a distinct implication of high quality. Since Ptolemy's book is fraudulent rather than great, it does not deserve to be called Almagest. This is the reason why I use the neutral term Syntaxis.

and solstice observations as I did in Section V.4, and he
found that the calculations give exactly the times that
Ptolemy claims to have measured. This argument cannot be
answered by demonstrating the possibility of large errors
of observation, or in any other way. Such agreement can
never be the result of observations, no matter how bad they
may be. It can only come from fabrication.

Since Delambre's argument is unanswerable, his work
should have led to a systematic study of Ptolemy's work from
this new viewpoint. However, so far as I have found, his
argument has been ignored. I cannot pretend to have read
all of the literature on Ptolemy written in the century and
a half since Delambre, but I have read a good sample of it,
and I have never seen a published citation or even an allu-
sion to Delambre's devastating argument except in my own
writing. I do not know why this should be so. To be sure,
several writers refer to suspicions that have been raised,
but only as part of an attempt to allay them.† They do not
cite the strong evidence on the matter; they mention only
weak arguments that are easy to refute.

This, then, remains the most difficult problem connected
with the <u>Syntaxis</u>: Why has it been accepted as a great work
throughout the eighteen centuries since it was written?

11. A Final Summary

All of his own observations that Ptolemy uses in the
<u>Syntaxis</u> are fraudulent, so far as we can test them. Many
of the observations that he attributes to other astronomers
are also frauds that he has committed. His work is riddled
with theoretical errors and with failures of comprehension,
as we saw in Section XIII.5. His models of the moon and
Mercury conflict violently with elementary observation and
must thus be counted as failures. His writing of the <u>Syntaxi</u>

---

†This course is still followed in the most recent book-length
work on the subject of Ptolemy [<u>Pedersen</u>, 1974]. The author
of this work does not mention the evidence given by Delambre
nor does he include the relevant work of Delambre in his bib
liography. Further, he does not mention any of the evidence
that I have published. O. Neugebauer follows the same cours
in <u>A History of Ancient Mathematical Astronomy</u> (Springer-
Verlag, New York, 1975). This work deals with mathematical
astronomy in Babylonia, Egypt, and Greece, but about a thir
of it is devoted to the work of Ptolemy.
    This is a good place to point out that many writers hav
tested the calculations given in the <u>Syntaxis</u>, in whole or
in part, and have studied the manner in which Ptolemy's
theories are derived from observations. None of them, so
far as I know, has pointed out the conclusion to be drawn
from the fact that Ptolemy always gets exactly the answer
he needs in spite of his computational or theoretical error
I have not attempted to cite these works. <u>Pedersen</u> [1974]
gives an extensive bibliography of them; the number of such
works is probably too large for anyone to attempt an exhaus
tive bibliography.

has caused us to lose much of the genuine work in Greek astronomy. Against this we can set only one possible asset, and it is questionable whether this is a contribution that Ptolemy himself made. This possible asset is the equant model used for Venus and the outer planets, and Ptolemy lessens its value considerably by his inaccurate use of it.

It is clear that no statement made by Ptolemy can be accepted unless it is confirmed by writers who are totally independent of Ptolemy on the matters in question. All research in either history or astronomy that has been based upon the Syntaxis must now be done again.

I do not know what others may think, but to me there is only one final assessment: The Syntaxis has done more damage to astronomy than any other work ever written, and astronomy would be better off if it had never existed.

Thus Ptolemy is not the greatest astronomer of antiquity, but he is something still more unusual: He is the most successful fraud in the history of science.

APPENDIX A

SOME TECHNICAL MATTERS

1.   Units of Time and Angle

Two basic units for measuring angles are used in this
work. One is the degree and the other is the radian. If we
divide the circle into 360 parts, the result is the degree.
If we divide the circle into 2π parts, the result is the
radian. The radian is more useful than the degree in certain
types of theoretical discussion, such as finding the equation
of the center in Section IV.2. The main reason for this is
related to the fact than $\tan \alpha$ and $\sin \alpha$ are nearly equal to
α when α is small, if α is expressed in radians.

The degree may be divided by means of elementary fractions, by means of the decimal system, or by means of the
sexagesimal system. When it is divided by means of the
decimal system, the symbol that denotes a degree is usually
placed before the decimal point; thus $5°.2$ denotes 5 degrees
plus a fifth of a degree.

When the degree is divided by means of the sexagesimal
system, there are two ways of writing. One is to write a
sexagesimal number followed by the name of the unit. Thus
37;29,17 degrees denotes 37 degrees plus 29/60 of a degree
plus 17/3600 of a degree. The other way is to use symbols
to denote the sexagesimal positions. Thus the same angle
can be written $37° \ 29' \ 17''$.

Angles must be written in a variety of contexts in this
work, and the form that is most convenient is not always the
same. This applies both to the form of writing the numerical
value and to the designation of the unit. I make no apology
for failing to be consistent in my manner of writing angles.

In some situations, the minute, which is 1/60 of a degree, becomes the basic unit. All the remarks just made
about the degree also apply to the minute when the minute is
the unit.

The basic unit for measuring time is usually the hour.
It too can be divided in all the ways that the degree can be
divided, and the same methods of writing also apply. When a
symbol is used to denote an hour, that symbol is "h" placed
in the same position as the degree mark. Thus $5^h.2$ is the
same as 5.2 hours. When the hour is divided into minutes
and seconds, the symbols used for the divisions are "m" and
"s". Thus $5^h \ 12^m$ is the same as $5^h.2$. One point should be
emphasized: The symbols m and s always refer to time while
$'$ and $''$ always refer to angles in this work.

There are three kinds of hour. In one kind, we divide
a full alternation of daylight and darkness into 24 hours.
When "hour" is used without a modifier, this kind of hour is
meant. In the second kind, we divide the interval from sunrise to sunset into 12 parts. This kind is called an "hour

of the day". Its length obviously varies with the seasons, and it can be used only for designating a time that comes during daylight. In the third kind, we divide the interval from sunset to sunrise into 12 parts. This is called an "hour of the night"; its length is variable and it can be used only at night.

I assume that the reader is familiar with time zones and with the idea that the hour, at the same instant of time, varies in a regular fashion as we go around the earth.

Finally, we need to distinguish between "mean" and "apparent" time. If we say that the time is always noon (12^h) when the real sun is in the meridian, we are using apparent time. Because the sun does not move uniformly in a circle around the equator, the interval between successive apparent noons is not constant. When these variations are averaged out, the result is called mean time.

In current usage, we usually refer to mean time unless we specifically say "apparent time". In Greek usage, which is the general usage of this work, the opposite is true. Within the accuracy of Greek astronomy, the difference is less than the accuracy of observation except when we are deal ing with the motion of the moon.

2. Coordinate Systems; Precession

I assume that the reader is familiar with the system of latitude and longitude for locating points on the surface of the earth. This system is based upon the plane called the equatorial plane. Latitude measures position in a direction perpendicular to the equator and longitude measures position in a direction parallel to the equator. The equatorial plane itself furnishes a natural reference that is called 0° latitude. However, we must pick arbitrarily the reference that we call 0° longitude. The reference line where the longitude is 0° is the line where the meridian of Greenwich crosse the Equator.

When we want to locate a point on the celestial sphere, we do so in a similar way. We choose a plane and we choose a reference line in that plane. We then build a coordinate system like that of latitude and longitude, starting with this plane and reference line. Two coordinate systems, based upon different choices of the plane, are used in locating points on the celestial sphere.

In the first system, we again use the equatorial plane, which we imagine to be extended until it crosses the celesti sphere. However, we must now pick a reference line; we canne use the line that defines 0° longitude on the earth because is rotating with respect to the celestial sphere. The reference line that we pick is the line that points toward the sun at the time of the vernal equinox, and we call this line the vernal equinox. Thus we use "vernal equinox" to denote either a certain time or a certain direction. The context always makes it clear which usage is meant.

The coordinate that is analogous to latitude on earth is called declination in the heavens and the coordinate that is analogous to longitude is called right ascension.† There is only one matter that needs particular attention. In drawing a map of the earth, if north is at the top, east is to the right. In drawing a map of the heavens, we look at the sphere from the inside rather than the outside. Hence, if north is at the top, east is to the left.

In the second system, the reference plane is the plane in which the sun appears to move around the earth. This is the plane called the ecliptic. The reference line is again the vernal equinox. This line is now seen to be the line where the equatorial and ecliptic planes cross, and thus it lies in both planes. The coordinate that is analogous to latitude on the earth is called celestial latitude and the coordinate that is analogous to longitude on the earth is called celestial longitude. Usually there is no confusion about whether we are locating a point on the earth or in the heavens. When there is no danger of confusion, we usually omit the adjective "celestial" in front of "latitude" or "longitude". That is, in a purely celestial context, latitude and longitude denote celestial latitude and celestial longitude.

The phenomenon of parallax can affect the measured values of any angular coordinate. I analyzed parallax in considerable detail in Section VI.1, and I shall not repeat any of the analysis here.

The vernal equinox, considered as a direction in the heavens, is slowly moving with respect to the stars. With high but not complete accuracy, this motion is along the plane of the ecliptic, and the motion is called the precession of the equinox (the plural is often used, yielding the precession of the equinoxes). The motion is westward, and the rate of motion at the present time is about $50''.3$ per year. The rate of precession is increasing slowly with time, and about 2000 years ago it was about $49''.8$ per year. The change in the rate is probably oscillatory, but with a very long period of oscillation. Over a period of only a few thousand years, the change is in the same direction.

3. Orbital Planes; Inclination and Node

The plane in which the sun revolves around the earth (or the earth revolves around the sun, if the reader prefers this usage) is called the ecliptic, as we just saw. The moon and each of the planets moves in a plane which is different from the ecliptic plane. The first task in describing an orbit is to locate the plane in which the orbital motion occurs. We do this by means of two angles.

---

†At one time, astronomers used a general quantity called ascension, particularly in solving problems related to the point where the sun or some other object rises or ascends. There were oblique ascensions and right ascensions. Only right ascensions have survived in common use.

The first angle is called the inclination. It is simply the angle between the plane of the orbit and the plane of the ecliptic.

To define the second angle, we first locate the line in which the two planes intersect. The orbit crosses this line twice, once when the body is passing from the south to the north side of the ecliptic and once when it is passing from the north to the south side. We call the first point the ascending node and the second point the descending node. The ascending node, like any other point in the heavens, has a celestial longitude, and this longitude of the ascending node is the second angle that we need. It and the inclination, taken together, locate the orbital plane completely.

When we are dealing with the moon, the orbital plane passes through the center of the earth. When we are dealing with the planets, the orbital plane passes through the sun. This tells us what a correct theory of planetary latitudes should be in the geocentric theory used by the Greeks. For the inner planets, the deferent is the orbit of the sun and it therefore lies in the plane of the ecliptic. The epicycle should then make an angle with the deferent equal to the inclination of the planetary orbit. For an outer planet, the deferent is the orbit of the sun around the planet, so it should make an angle with the ecliptic equal to the planetary inclination. The epicycle is the orbit of the earth around the sun, so it should be parallel to the ecliptic.

The nodes of the planetary orbits move slowly with respect to the stars. This means that the longitudes of the ascending nodes increase at a rate nearly equal to the precession of the equinox. The node of the lunar orbit, in contrast, moves rapidly westward and makes a full circle with respect to the equinox in about 18.6 years.

4. Motion in the Orbital Plane; Mean Motion

With an error that does not exceed about 1', and that is usually much less, the heliocentric orbit of every planet is an ellipse that lies in the orbital plane. Specifying this ellipse requires four parameters.

The first parameter locates the position of the major axis of the ellipse (Figure IV.2). Since the axis has two ends, we must first choose which end to identify, and we choose the end that lies closer to the sun; this end of the axis is the point called the perihelion of the planet. If the ellipse lies in the plane of the ecliptic, as the orbit of the earth does, we let the line S♈ in Figure IV.2 denote the line from the sun to the vernal equinox, and we locate the perihelion point G by giving the angle GS♈. This angle is called the longitude of perihelion.

If the orbit does not lie in the plane of the ecliptic, we let the line S♈ denote the line from the sun to the ascending node of the orbit, and we again locate perihelion by giving the angle GS♈. However, in this case, we say that the angle GS♈ is the argument of perihelion.

Many planetary tables give a quantity called the longitude of perihelion even for planets other than the earth. For the other planets, this quantity is the sum of the longitude of the ascending node and the argument of perihelion. Although it is not an angle that can be drawn in a figure like Figure IV.2, it is a useful parameter.

The second parameter of the ellipse is half of the length of the long axis of the ellipse. This is the distance CG in Figure IV.2. It is usually called the semi-major axis, meaning half of the major axis. It would probably be less ambiguous to call it the major semi-axis, but this usage is rare.

The third parameter of the ellipse is called the eccentricity. It is the ratio of the distance CS in Figure IV.2 to the semi-major axis CG. The significance of the point S in Figure IV.2 is that it is the point occupied by the sun; it is also called a focus of the ellipse. The term "eccentricity" has analogous but not identical meanings in ancient and modern astronomy.

The fourth parameter can be and is often specified in various ways. Perhaps the most common way is to give the value of some important coordinate at a convenient epoch. In modern astronomy the epoch most commonly used for this purpose is noon, Greenwich mean time, on 1899 December 31. Astronomers usually write this date as 1900 January 0. Ptolemy usually chose his epoch to be noon, Alexandria apparent time, on -746 February 26.

An ellipse does not give as good a fit to the lunar orbit as it does to the orbits of the planets. The maximum deviation of the moon from the best-fitting ellipse is about $1°.5$. However, the best-fitting ellipse still furnishes a convenient starting point for describing the motion of the moon. All the preceding discussion applies to this best-fitting ellipse, with one exception. Since the earth, rather than the sun, is at the focus of the lunar ellipse, the point G is called perigee rather than perihelion.

The positions of the perihelia of the planets move slowly with respect to the fixed stars. Hence their longitudes increase at a rate that is close to the precession of the vernal equinox. The position of the lunar perigee moves rapidly, however, and makes a complete circuit of the heavens in about 8.85 years.

With each celestial body we associate a moving point called the mean body. Thus we have a mean moon, a mean Mercury, and so on. The mean body moves in the same plane as the real body, but with a constant angular velocity. Obviously we take the constant angular velocity of the mean body to be the average angular velocity of the real body. We position the mean body so that it is sometimes ahead of the real body and sometimes behind; on the average, the angle between the mean body and the real is zero.†

---

†Average here refers to the algebraic average.

Two forms of speech are used in referring to the mean
body and its motion. We may refer to the motion of the mean
body or to the mean motion of the body, and we may apply
similar usage to any important coordinate. Thus we may
speak of the longitude of the mean moon or of the mean longitude of the moon. Both usages denote the same thing and I
have not tried to be consistent in choosing between them.

In referring to the moon, perigee and apogee mean the
points where the moon is at its least distance from the earth
or at its greatest distance, respectively. In referring to
a planet, the terms should strictly have the same meaning.
However, Ptolemy generally uses perigee and apogee for a
planet to mean the points where the deferent, rather than
the planet itself, has its least or greatest distance from
the earth. I have tried to distinguish carefully between
these two usages.

5. Standard Deviation

I have often used the term standard deviation as a measure of the error in a theory or in a set of measurements,
but without giving an accurate definition of the term. This
is a good place to repair the omission.

Suppose that we have a set of measurements that we can
designate by x_1, x_2, x_3, and so on. Suppose further that we
have a way of assigning an error to each measurement. For
example, if all the x's come from attempts to measure a particular quantity, we may assume that the average value of
the x's is the "correct" value, X say. Then the errors are
x_1 - X, x_2 - X, x_3 - X, and so on. Let us denote the errors
by r_1, r_2, r_3, and so on.

To give another example, suppose that the x's are values
of the longitude of Mercury calculated at a set of different
times from Ptolemy's theory of Mercury. The modern theory of
Mercury is so accurate that we can neglect the errors in it
in comparison with those in Ptolemy's theory. Then the errors, which we can again call r_1, r_2, r_3, and so on, are the
differences between the values calculated from the two theories.

In order to calculate the standard deviation, we first
find the squares of all the errors. The squares are r_1^2,
r_2^2, r_3^2, and so on. We then find the average of the quantities r_1^2, r_2^2, r_3^2, and so on. Finally we take the square
root of this average value. The square root is the standard
deviation, which we denote by σ.

In most circumstances, it turns out that about a third
of the errors r_1, r_2, r_3, and so on are greater than σ and
about two thirds of the errors are less. This is the basis
for an approximate definition of σ that I have used several
times in the main text.

In many places in this work, I have calculated an observ
quantity with the use of modern theory and I have then called
this calculated value the "true" or "correct" value of the
quantity in question. If we include a reasonable estimate o

the effects of the secular accelerations of the moon in its orbit and of the earth's spin about its axis, we have strong reasons to believe that the errors in the calculated values are negligible compared with errors in ancient observations, whether genuine or fabricated. Thus it is reasonable to use these calculated values as if they were "true" values for the purposes of this work.

In many other places, when I have calculated the probability that a particular result could have been the result of chance errors of observation, I have assumed that there are no sources of error in the observations except chance errors. This is not strictly correct, but it is adequate for our purposes except perhaps in a few places that I have noted explicitly. For example, when an observer measures a declination, he has a systematic error in his latitude and he probably has errors in the calibration of his measuring scales. These errors should amount to only a few minutes of arc. Thus they can be neglected in comparison with, for example, the error in Ptolemy's measurement of the moon on 135 October 1 (Section VIII.5), which was 41 minutes of arc.

I have also assumed that the chance errors of observation obey what is called the normal law of error. This is also not strictly correct, but the errors in the assumption clearly do not matter for our purposes, since we need the probabilities only to the order of magnitude.

6. Notation

In this section, I shall summarize the principal notation used in this work. A symbol that is used only a few times within a short space may not be included. First I give the symbols that are not letters of an alphabet, then the symbols that are Latin letters, in alphabetical order, and finally the symbols that are Greek letters, again in alphabetical order.

In many cases, a symbol is used with another symbol as a subscript. When this happens, the main symbol and its subscript must be looked up separately.

Two symbols need separate comment. The symbol "o", used only as a subscript, denotes the value that a variable quantity has at some specified instant. The symbol $'$ has two uses. One is to denote the sixtieth part of a degree, as we have discussed in Section A.1; in this usage it always follows a number. If it follows a symbol, it means a rate. Thus, for example, γ' means the rate at which the variable quantity γ changes; in this work, it is usually the change that occurs in 1 day.

φ: the vernal equinox, except in discussing Figure IV.2

$\u263f$: Mercury

$\u2640$: Venus

$\u2642$: Mars

♃:	Jupiter
♄:	Saturn
☉:	the sun
☽:	the moon
a:	position of apogee
D:	elongation, that is, the angle between the moon or a planet and the sun
E:	maximum value of e_c
e:	eccentricity, in either the ancient or modern sense
e_c:	equation of the center
L:	a mean longitude
M:	anomaly in the modern sense
n:	rate of change of a mean longitude
R:	radius of a deferent curve
r:	radius of an epicycle
T:	time measured in centuries from 1900
Z:	angle between a celestial object and the zenith
γ:	anomaly in the Greek sense
ε:	obliquity of the ecliptic
λ:	a celestial longitude
Π:	horizontal parallax
ρ:	apparent radius of an object seen against the celestial sphere in most contexts. Sometimes it denotes a parameter in one of Ptolemy's models.
σ:	a standard deviation
ϕ:	a terrestrial latitude
ψ:	argument of the latitude of the moon
ω:	argument of perihelion for a planet or of perigee for the moon.

ARISTARCHUS'S METHOD OF FINDING
THE SIZE OF THE SUN

Figure B.1 shows the geometrical relations needed to understand the method that Aristarchus [ca. -280] uses to find the sizes of the sun and moon. The horizontal line that ends at point V on the right joins the centers of the earth and sun. The center of the sun is at the left end of this line. At the distance S from the center of the sun is the center of the earth. The distance L is the (average) distance from the earth to the moon, and the distance U is the remaining distance to the vertex V of the shadow. The line that meets the line of centers at point V is tangent to the spheres of the sun and earth. The radius of the earth is taken as 1, and the radius of the sun is s. The distance u is the radius of the umbra at the average distance of the moon.

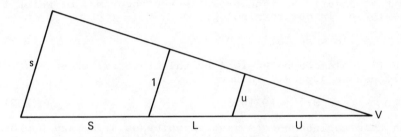

Figure B.1   Similar triangles in the cone of the earth's shadow. The horizontal line passes through the centers of the sun and earth and the line joining it at point V is tangent to the earth and sun. The slanting line at the extreme left of the figure is the sun's radius, the next slanting line is the earth's radius, and the shortest slanting line is the radius of the umbra at the distance of the moon. The earth's radius is taken as 1. S is the distance from the earth to the sun, L is the distance from the earth to the moon, and U is the remaining distance to the vertex. The radii of the sun and of the umbra are s and u, respectively. The angle formed at V is greatly exaggerated; its correct value is less than 1°.

I shall let ℓ denote the radius of the moon, which is not shown in Figure B.1.

In solving the geometrical relations of Figure B.1, neither Ptolemy (see Section VIII.7) nor Aristarchus four

centuries before him used the simple Equation VIII.2 involving angles. Instead, both started from the fact that the
triangles in the figure are similar, and they proceeded to
solve for the dimensions labelled in the figure. The main
difference between their procedures, aside from the fact
that they started from different data, is that Ptolemy had
fairly accurate values for the trigonometric ratios needed
while Aristarchus could only set limits. Heath [1913,
pp. 333-336], following an article by Paul Tannery that I
have not consulted, shows how Aristarchus might have found
the limits that he uses.

Aristarchus starts by stating six hypotheses which I
quoted in Section VIII.2. The first three justify certain
geometrical constructions that Aristarchus uses, and I shall
not repeat them. The fourth says that the angle D shown in
Figure VIII.1 is 87°, the fifth one says that the distance u
subtends, at the center of the earth, an angle twice that
subtended by the radius ℓ of the moon, and the sixth one says
that ℓ subtends an angle of 1°.

Aristarchus next states the three main conclusions that
he will draw from his hypotheses. I shall pass over these
for the moment. He then proves seven theorems, of which the
seventh is, in our terminology,

$$18 < \sec 87° < 20 . \tag{B.1}$$

I shall replace hypothesis 4 by this theorem, except that I
shall use an accurate value:

$$\sec 87° = 19.1073 . \tag{B.2}$$

When this replacement is made, hypotheses 4, 5, and 6 provide
three equations, which I shall write in the form: †

$$\ell = \lambda L, \qquad u = \psi L, \qquad S = \Sigma L . \tag{B.3}$$

The numerical values of λ, ψ, and Σ are obtained from the
hypotheses, and I shall insert them later in the discussion.

There are three similar triangles in Figure B.1, and
the standard theorems about similar triangles give us two
more independent equations:

$$(U + L)/1 = U/u ,$$
$$(U + L + S)/s = (U + L)/1 . \tag{B.4}$$

Now ℓ appears only in the first of Equations B.3, and we can
leave it aside for the present. When we do so, we have left
4 equations connecting the 5 unknown quantities s, u, S, L,
and U. It is impossible to find 5 unknowns from only 4
equations.

This problem is not resolved until we come to Aristarchus's 8th theorem. This theorem says that the apparent

†The notation of this appendix differs from the table of
notation in Section 6 of Appendix A.

size (subtended angle) of the sun is the same as that of the
moon. In proving this theorem, Aristarchus tacitly introduces two new hypotheses, which he says "are manifest from
observation." These hypotheses, which I take the liberty
of numbering 7 and 8, are:

Hypothesis 7: The moon is able to eclipse the sun
totally.

Hypothesis 8: A total eclipse has no duration in time.

Since the moon is able to eclipse the sun totally, its apparent diameter cannot be less than the apparent diameter of the
sun. Since a total eclipse has no duration, the moon's diameter cannot be greater than the diameter of the sun. Hence
the apparent diameters are equal and we have a further relation:

$$s = {_\sigma}S. \tag{B.5}$$

Aristarchus's 8th theorem says that σ is numerically equal to
λ in Equations B.3, but I shall not use this equality for the
present.

At this point, I think we may conclude that Aristarchus
is deliberately dealing in approximations. He assumes that
there are unique values of ℓ, u, L, and U; that is, he neglects the variation in the distance of the moon. I suppose
it is possible that astronomers had not yet discovered this
variation. Hypothesis 7 amounts to a denial of annular
eclipses. Since the first verifiable reference to an annular
eclipse (see Section VIII.6) is nearly contemporary with Ptolemy but probably a bit later, it is possible that astronomers
in the time of Aristarchus had not realized the existence of
annular eclipses. However, I find it hard to believe that
astronomers in Aristarchus's time thought that a total eclipse
is over in an instant, with no measurable duration. I doubt
that observers seeing a total eclipse, but who do not have
accurate time pieces, tend to underestimate the duration of
an eclipse. On the contrary, I think that they would tend to
exaggerate it. Hence I believe that Aristarchus introduced
hypothesis 8 in order to simplify a calculation that was
formidable with the tools available in his time.

His error in hypothesis 6 is also interesting, but it
is convenient to complete the solution of Figure B.1 before
we discuss it. Because Aristarchus did not have access to
much of the mathematical apparatus that we take for granted,
his method of solution seems cumbersome to us. Instead of
presenting his method of solution, I shall outline a simple
method. We first eliminate u and s by using $u = \psi L$ and
$s = {_\sigma}S$, and we then eliminate S by using $S = \Sigma L$. This leaves
us with two equations in the unknowns L and U:

$$U + L = U/\psi L,$$
$$U + L = (U/{_\sigma}\Sigma L) + (1 + \Sigma)/{_\sigma}\Sigma. \tag{B.6}$$

U and L now appear only in the combinations U + L and U/L, so we should regard these as the unknowns for the moment. It is trivial to find

$$U/_\psi L = U + L = (1 + \Sigma)/(\sigma\Sigma - \psi).$$

Since only the combination U + L appears in the second of Equations B.4, we may immediately solve that equation for S, and thence evaluate s, L, and ℓ. The results are

$$S = (1 + \Sigma)/(\sigma + \psi), \qquad L = (1 + \Sigma)/\Sigma(\sigma + \psi),$$
$$\text{(B.7)}$$
$$s = (1 + \Sigma)/[1 + (\psi/\sigma)], \qquad \ell = [(1 + \Sigma)/\Sigma][\lambda/(\sigma + \psi)].$$

Both radii and both distances depend upon Σ. The distances further depend upon σ and ψ, but the radii do not. The radius s depends upon the ratio ψ/σ, but not upon either value alone. The radius ℓ depends upon the ratio of λ to $\sigma + \psi$, but it also does not depend upon the actual value of any one of these quantities. Aristarchus assumes that λ and σ are equal and that ψ is twice either of these. If we use these ratios in Equations B.7, we finally get

$$S = (1 + \Sigma)/3\sigma, \qquad L = (1 + \Sigma)/3\Sigma\lambda,$$
$$\text{(B.8)}$$
$$s = (1 + \Sigma)/3, \qquad \ell = (1 + \Sigma)/3\Sigma.$$

If we take Σ to be sec 87° and use its value from Equation B.2, the value of s is 6.702 times the earth radius. The cube of this gives the ratio of the volumes, which is more than 300. The value of ℓ is 0.3508 earth radii and the ratio of the volumes is 0.0432. These values do not depend upon the value of σ or λ, as I have said. The distances S and L come out to be 384.040 and 20.099, respectively, if we use 2° for the apparent diameter of the sun or moon. The value of L is obviously much too small. However, if we use 30' for the apparent diameters, which is the value that Archimedes [ca. -225] says that Aristarchus actually found, the values of S and L are almost exactly four times as large. The new value of L is now reasonable, although somewhat large.

Aristarchus is mostly interested in the sizes (volumes) of the sun and moon, which do not depend upon the apparent diameter used for the moon. His main conclusions, which he states immediately after his hypotheses, relate only to the sizes. Further, his erroneous assumption about the apparent diameter of the moon does not affect the validity of any intermediate theorem that he states. He uses the apparent diameter only to prove certain propositions that we would probably take as obvious but which posed a problem in his time.

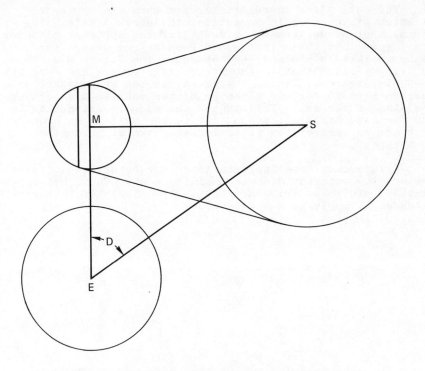

Figure B.2   The configuration of the sun, earth, and moon
when their centers form a right triangle. Because the sun is larger
than the moon, more than half of the moon is illuminated,
as the figure shows.

In order to see his problem, let us look at the configuration of the sun, earth, and moon that is shown in
Figure B.2. In this configuration, the lines from the center M of the moon to the center S of the sun and to the
center E of the earth meet at a right angle. Aristarchus
assumes that the angle D is 87°. The plane through M perpendicular to the line MS passes through E.

Since the sun is larger than the moon, the plane that
divides the light and dark portions of the moon is not the
plane that passes through M. Instead it is the plane that
lies to the left of M in the figure. Aristarchus wants to
prove that he can neglect this fact and assume that the
dividing plane passes through M. To do this, he needs to
set upper limits to two quantities, which are the distance
between the two planes and the difference between the radii
of the circles that they cut from the sphere of the moon.
He then shows that he can neglect the upper limits and,
a fortiori, he can neglect the quantities in the actual
situation.

The only place where Aristarchus uses the apparent diameter of the moon is in setting the upper limits. The limits that he derives by using 2° for the apparent diameter are larger than those that would result from using, say, 30', although this is probably not obvious from Figure B.2. Thus, by using a value that is too large, Aristarchus has strengthened his argument. This, I suggest, is the reason that Aristarchus uses 2° for the diameter of the moon in his work on the size of the sun. This is not the same as using a value for the sake of illustration. It is the well-known procedure of using an exaggerated value for the sake of strengthening an argument.

Aristarchus uses approximations for several quantities besides the apparent diameters of the moon. The other approximations are reasonably accurate, particularly if they are intended to apply to average values.

APPENDIX C

PTOLEMY'S USE OF THE BABYLONIAN CALENDAR

Ptolemy uses a variety of calendars in presenting the observations which he attributes to other astronomers. These include but are not limited to the following:

1. The Egyptian calendar. This calendar was based upon a year of exactly 365 days. The year had 12 months of 30 days each, followed by 5 "epagomenal" (added on) days. In the course of 1461 Egyptian years, which equal 1460 Julian years, the first day of the Egyptian year made a complete circuit of the Julian year.

2. The Callippic calendar. This calendar was based upon a cycle of 76 years, and the average length of a year was $365\frac{1}{4}$ days. The months alternated between 29 and 30 days, with an occasional correction intended to make the average length of a month equal the astronomical month of 29.530 589 days. Some years had 12 months while some had 13 months. The assignments of months and days followed a rigid scheme, so that it was possible to convert unambiguously from the Egyptian calendar to this one. It is possible that this calendar was used only in astronomy and that it was not used in civil life. Meritt [1961] surveys the problems connected with this calendar and other calendars used in Athens.

3. The Babylonian calendar. In the period that interests us in connection with the Syntaxis, the Babylonian calendar did not follow a rigid numerical scheme, as the Egyptian and Callippic calendars did. Instead, it was based upon observations which were probably made by designated officials.

The day began at sunset. The first day of a month began at the sunset when the new crescent moon could be seen for the first time in the west at sunset. In very early periods, it may be that this rule was rigidly followed regardless of clouds which might prevent seeing the new moon. By the times that concern us, the rule was probably modified to the following: Watch for the new moon at the sunset that closes the 29th day of the month. If the new moon is seen, the next month begins at this sunset. If the new moon is not seen, for any reason whatever, this sunset begins the 30th day. The next month then begins at the following sunset.

Most years had 12 such months, whose average length necessarily equals the astronomical month. However, 12 of these months contain about 354 1/3 days, so that a given month would move rapidly through all the seasons if all years contained 12 months. To prevent this, slightly more than 1 year in 3 contained 13 months. The extra month was called the intercalary month. The need for an extra month was determined by some kind of observation, but we are not sure what kind. In the Jewish calendar, which is clearly descended from the Babylonian, the rule is: The vernal equinox cannot

come after the 14th day of the first month. The Babylonian
rule has this general effect, but two other possibilities
have been suggested for the exact rule. Section II.2 of
APO surveys the Babylonian calendar and gives further references.

Since the need for the extra month was probably determined by observation made by officials who had political
interests and political power, we may be sure that the assignment of the extra month was somewhat erratic. It has
been estimated that the first day of the year might have
come any time from the middle of March to the middle of June
in middle Babylonian times. By the times that concern us,
it is possible that intercalation was fairly regular, but it
did not become subject to a rigorous system until after the
times that concern us.

Thus we cannot tell the number of months in a year nor
the number of days in a month without detailed information.
Likewise we cannot convert dates from one calendar to the
other unless we have an event recorded in the Babylonian calendar that we can date in ours. Even when we can establish
the conversion from the Babylonian calendar to ours for a
particular month, we cannot extend the conversion even by a
month without further detailed information.

There is still another problem with the Babylonian calendar. In most and probably all Babylonian documents, the
year is identified only as, for example, the 12th year of the
reign of a particular king. At least there was a regular
rule for dealing with the fractions of a year that arose from
the fact that a king did not usually die on the last day of
a year. If he died (or was deposed) at any time during the
12th year of his reign, for example, that entire year was
considered as part of his reign for chronological purposes.
The first year of his successor began on the "New Year's
Day" that followed his accession.

Now let us suppose that Ptolemy had a Babylonian record
of an astronomical observation, and let us summarize the
information that he would need in order to convert the Babylonian date into an Egyptian date; an Egyptian date is what
he needed in order to use the observation for astronomical
purposes. He would need a complete list of the kings and
the lengths of their reigns, he would need a complete list
of the number of months in each year, and he would need a
complete list of the number of days in each month. Even with
this information, the conversion of a date would be a formidable task.

As it happens, Ptolemy does not use any Babylonian observations, either real or fabricated, except lunar eclipses.
Since the first day of a Babylonian month was always close
to the new moon, a full moon and hence a possible lunar
eclipse could come only near the middle of the month. On
the average, one came on the 14th day of a month, and it
would be unusual if one departed from the 14th by more than
a day.

Further, lunar eclipses can occur only during the so-called eclipse seasons, which are about 6 months apart, and there can never be two lunar eclipses in the same season. Thus, if we know the month and year in which a lunar eclipse occurs, we can identify the season and hence we can identif· the eclipse, even if we do not know how many months were ir that year. In order to identify a Babylonian lunar eclips then, Ptolemy would need the list of kings, and he would have to know the name of the month. However, he would not need the number of months in that year, nor the number of days in that month.

Ptolemy handles Babylonian dates and other dates in remarkably different ways. As an example of the way he handles a date that is not Babylonian, let us consider the occultation of α Virginis by the moon that was observed in Alexandria on -293 March 9. I analyzed this observation in Section IX.5 and showed that it is fabricated. Ptolemy introduces the record, in Chapter VII.3 of the Syntaxis, by saying: "Timocharis reports that he saw in Alexandria, in the 36th year of the first Callippic cycle,† on the 15th day of the month Elaphebolion . . ." He then states the equivalent of this date in the Egyptian calendar. He does the same for all non-Babylonian observations, with exceptions that are so few that they may easily be accidental.

All the Babylonian astronomical records that I have seen, except those that are obviously damaged, give the year, month, and day, and there is reason to believe that the damaged ones originally gave the full information. For example, the Babylonian record of the lunar eclipse of -522 July 16, which I mentioned in Section VIII.8, gives the date as the 7th year of Kambyses, in the night of the 14th of the month Duzu. Ptolemy uses a "record" of this same eclipse in Chapter V.14 of the Syntaxis, and I showed in Section VIII.8 that his record is fabricated. Ptolemy begins this record by saying: "In the 7th year of Kambyses, ..." He then gives the year, month, and day of this eclipse in the Egyptian calendar, without giving any more of the Babylonian date than the year.

If this happened for only one of the Babylonian dates, we could dismiss it as an accident. However, it happens for all of the Babylonian dates that Ptolemy uses, and it can hardly be an accident. When we consider how Ptolemy treats dates in other calendars, this is not what we expect if the Babylonian records are genuine. However, it is exactly what we expect if the records are fabricated.

The reason comes from the assignment of months to the year. The first month of the Babylonian year always starts fairly close to the vernal equinox. Duzu is the fourth month, so it starts fairly close to the summer solstice. The only lunar eclipses in the 7th year of Kambyses are those of -522 July 16 and -521 January 10, and both are

---

†The first year of the first cycle began in the summer of -329.

recorded in existing Babylonian records [APO, Sections IV.6 and X.2]. Hence the eclipse that came in the month Duzu must be the one of -522 July 16, and we do not need to know the exact day when the year began or how many months it contained in order to know this.

Suppose, however, that we do not have the Babylonian record and that we do not have complete records of the months in a year and the days in a month. We do know, however, that there was an eclipse on -522 July 16. What can we say about the Babylonian date of this eclipse?

If we have an accurate king list, we can say that this was in the 7th year of Kambyses, but this is all we can say. We can be fairly sure, if we take the trouble to calculate when the new moons came that year, that the date was in either the 4th or the 5th month, if the intercalation was being done with care. But, unless we know the assignment of months to the year, we cannot tell in which month the date July 16 came in that year.

Thus, if Ptolemy fabricated the lunar eclipses which were allegedly observed in Babylon, he would be unable to give any more of the Babylonian date than the year, unlike his custom when giving dates in other calendars.

In Section VI.7, we found evidence that the triad of eclipses which occurred in -720 and -719, when considered as a unit, definitely contains fabricated elements. We saw that the eclipse record of -719 March 8 was certainly fabricated, but the way in which Ptolemy uses the record leaves the possibility that the other two records in the triad are genuine. We saw in fact some evidence which indicates that the other two records, those of -720 March 19 and -719 September 1, may well be genuine. The evidence given in this appendix, on the other hand, indicates that the records are fabricated. Since the records of these eclipses definitely occur in a context of fabrication, I shall let my assessment of them tip toward the side of fabrication, and I shall place them in a category to be called "may be fabricated".

The records of the eclipses of -501 November 19 and -490 April 25 were studied in Section VI.9, where we found no strong evidence either for or against their validity. The way that Ptolemy states their dates suggests that they have been fabricated. However, I have not found any other evidence of fabrication, perhaps because I have not thought of the correct tests to apply. Since I have found no evidence of fabrication except the way in which the dates are written, I shall leave the scales tilted toward validity for these eclipses, and I shall place them in a category to be called "may be genuine".

REFERENCES

al-Biruni, Abu al-Raihan Muhammad bin Ahmad, Kitab Tahdid
    Nihayat al-Amakin Litashih Masafat al-Masakin, 1025.
    There is a translation into English by Jamil Ali, with
    the title translated as The Determination of the Coor-
    dinates of Positions for the Correction of Distances
    between Cities, published by the American University
    of Beirut, Beirut, Lebanon, 1967.

American Ephemeris and Nautical Almanac, U. S. Government
    Printing Office, Washington, D.C., published annually.

APO = Newton [1976].

Archimedes, The Sand Reckoner, ca. -225. There is a trans-
    lation into English in The World of Mathematics, James
    R. Newman, editor, volume 1, pp. 420-429, Simon and
    Schuster, New York, 1956. The translator is not iden-
    tified.

Aristarchus (of Samos), On the Sizes and Distances of the
    Sun and Moon, ca. -280. There is an edition with a
    parallel English translation in Heath [1913].

Aristotle, De Caelo, ca. -350. There is a translation into
    English by Richard McKeon, The Basic Works of Aristotle,
    Random House, New York, 1941.

Becvar, A., Atlas of the Heavens, Catalogus 1950.0, Czecho-
    slovak Academy of Sciences, Prague, or Sky Publishing
    Corp., Cambridge, Mass., 1964.

Boll, Franz, Die Sternkataloge des Hipparch und des Ptole-
    maios, Bibliotheca Mathematica, Series 3, volume 2,
    pp. 185-195, 1901.

Britton, J.P., On the quality of solar and lunar observations
    and parameters in Ptolemy's Almagest, a dissertation
    submitted to Yale University, 1967.

Britton, J.P., Ptolemy's determination of the obliquity of
    the ecliptic, Centaurus, 14, pp. 29-41, 1969.

Brown, E.W., with the assistance of H.B. Hedrick, Chief
    Computer, Tables of the Motion of the Moon, in 3 volumes,
    Yale University Press, New Haven, Connecticut, 1919.

Bunbury, E.H. and Beazley, C.R., article on Ptolemy, Encyclo-
    paedia Britannica, Eleventh Edition, v. 22, Encyclopaedia
    Britannica, Inc., New York, 1911.

Censorinus, Liber de Die Natali, 238. There is an edition,
    with a parallel translation into French, by Desiré
    Nisard in Celse, Vitruve, Censorin, et Frontin, Dubochet
    et Cie., Paris, 1846.

Copernicus, Nicolaus, De Revolutionibus Orbium Caelestium Libri Sex, 1543. There is an edition, including a facsimile of Copernicus's holograph, by Fritz Kubach, Verlag von R. Oldenbourg, Munich, in 2 volumes: volume 1, facsimile of the holograph, 1944; volume 2, a critical edition of the text, 1949.

Cornford, F.M., Plato's Cosmology, Humanities Press, New York, 1937.

Delambre, J.B.J., Histoire de l'Astronomie Ancienne, in 2 volumes, Chez Mme. Veuve Courcier, Paris, 1817.

Delambre, J.B.J., Histoire de l'Astronomie du Moyen Âge, Chez Mme. Veuve Courcier, Paris, 1819.

Diels, H. and Rehm, A., Parapegmenfragmente aus Milet, Sitzungsberichte der Berliner Akademie der Wissenschaft, Gesammtsitzung vom 14 Januar 1904, pp. 92-111, 1904.

Dreyer, J.L.E., History of the Planetary Systems from Thales to Kepler, 1905. This has been republished under the title A History of Astronomy from Thales to Kepler, Dover Publications, New York, 1953.

Eckert, W.J., Jones, Rebecca, and Clark, H.K., Construction of the lunar ephemeris, in An Improved Lunar Ephemeris, 1952-1959, issued as a Joint Supplement to The American Ephemeris and Nautical Almanac and The Astronomical Ephemeris, U. S. Government Printing Office, Washington, D.C., 1954.

Ellis, G.F.R., Cosmology and verifiability, Quarterly Journal of the Royal Astronomical Society, 16, pp. 245-264, 1975.

Explanatory Supplement to The Astronomical Ephemeris and The American Ephemeris and Nautical Almanac, H. M. Stationery Office, London, 1961.

Fischer, Irene, The figure of the earth - changes in concepts, Geophysical Surveys, 2, pp. 3-54, 1975.

Fotheringham, J.K., assisted by Gertrude Longbottom, The secular acceleration of the moon's mean motion as determined from the occultations in the Almagest, Monthly Notices of the Royal Astronomical Society, 75, pp. 377-394, 1915.

Fotheringham, J.K., The secular acceleration of the sun as determined from Hipparchus' equinox observations; with a note on Ptolemy's false equinox, Monthly Notices of the Royal Astronomical Society, 78, pp. 406-423, 1918.

Geminus, Eisagoge Eis ta Phainomena, ca. -100. There is an edition under the title Elementa Astronomiae, with a parallel translation into German, by K. Manitius, B.G. Teubner, Leipzig, 1898. The date -100 is my guess. Geminus could not have been much earlier than this, but he could have been as much as two centuries later.

Goldstein, B.R., The Arabic version of Ptolemy's planetary hypotheses, <u>Transactions of the American Philosophical Society</u>, <u>57</u>, <u>part 4</u>, 1967.

Heath, Sir Thomas, <u>Aristarchus of Samos</u>, Oxford University Press, Oxford, 1913.

Hipparchus, <u>In Arata et Eudoxi Phaenomena Commentariorum Libri Tres</u>, ca. -135. There is an edition, with a parallel German translation, by K. Manitius, B.G. Teubner, Leipzig, 1894. Page citations refer to this edition.

ibn Yunis, Ali ibn Ahmad, <u>Az-Zij al-Kabir al-Hakimi</u>, 1008. There is an edition of part of this work, with a parallel translation into French, by J.J.A. Caussin de Perceval, under the title <u>Le Livre de la Grande Table Hakémite</u>, with the author's name spelled as Ebn Iounis, Imprimerie de la République, Paris, 1804. Page citations refer to this edition.

Keightley, D.N., <u>The Sources of Shang History</u>, University of California Press, Berkeley, California, in press.

Kugler, F.X., <u>Sternkunde und Sterndienst in Babel</u>, <u>volume 1</u>, Aschendorffsche Verlagsbuchhandlung, Münster, Westphalia, 1907.

Menelaos, <u>Sphaerica</u>, ca. 100. Abu Nasr Mansur bin Ali bin Iraq prepared an Arabic version around the year 1000. There is a German translation of this version by Max Krause under the title <u>Die Sphärik von Menelaos aus Alexandrien</u>, Weidmannsche Buchhandlung, Berlin, 1936.

Meritt, B.D., <u>The Athenian Year</u>, University of California Press, Berkeley, California, 1961.

Morison, S.E., <u>Admiral of the Ocean Sea</u>, in 2 volumes, Little, Brown, and Co., Boston, 1942.

Moulton, F.R., <u>Introduction to Celestial Mechanics</u>, <u>Second Edition</u>, The MacMillan Co., New York, 1914. All editions probably contain the information we need for this work.

Muller, P.M., An analysis of the ancient astronomical observations with the implications for geophysics and cosmology, a dissertation presented to the University of Newcastle upon Tyne, 1975.

Needham, J., <u>Science and Civilization in China</u>, <u>volume 3</u>, Cambridge University Press, Cambridge, 1959.

Neugebauer, O., <u>Astronomical Cuneiform Texts</u>, in 3 volumes, Lund Humphries, London, 1955.

Neugebauer, O., <u>The Exact Sciences in Antiquity</u>, <u>Second Edition</u>, Brown University Press, Providence, Rhode Island, 1957.

Neugebauer, P.V. and Weidner, E.F., Ein astronomischer Beobachtungstext aus dem 37. Jahre Nebukadnezars II. (-567/66), Berichte über die Verhandlungen der Königlichen Sachsischen Akademie der Wissenschaften zu Leipzig, Philologie-Historie Klasse, Bd. 67, Heft 2, pp. 29-89, 1915.

Newcomb, S., Researches on the motion of the moon, Washington Observations, U. S. Naval Observatory, Washington, D.C., 1875.

Newcomb, S., Tables of the motion of the earth on its axis and around the sun, Astronomical Papers Prepared for the Use of the American Ephemeris and Nautical Almanac, VI, Part I, U. S. Government Printing Office, Washington, D. C., 1895.

Newcomb, S., Tables of Mercury, Astronomical Papers Prepared for the Use of the American Ephemeris and Nautical Almanac, VI, Part 2, U. S. Government Printing Office, Washington, D. C., 1895a.

Newton, R.R., Secular accelerations of the earth and moon, Science, 166, pp. 825-831, 1969.

Newton, R.R., Ancient Astronomical Observations and the Accelerations of the Earth and Moon, Johns Hopkins Press, Baltimore, 1970.

Newton, R.R., The earth's acceleration as deduced from al-Biruni's solar data, Memoirs of the Royal Astronomical Society, 76, pp. 99-128, 1972.

Newton, R.R., Astronomical evidence concerning non-gravitational forces in the earth-moon system, Astrophysics and Space Science, 16, pp. 179-200, 1972a.

Newton, R.R., The authenticity of Ptolemy's parallax data - Part I, Quarterly Journal of the Royal Astronomical Society, 14, pp. 367-388, 1973. This is frequently cited as Part I.

Newton, R.R., The obliquity of the ecliptic two millenia ago, Monthly Notices of the Royal Astronomical Society, 169, pp. 331-342, 1974.

Newton, R.R., The authenticity of Ptolemy's parallax data-Part II, Quarterly Journal of the Royal Astronomical Society, 15, pp. 7-27, 1974a. This is frequently cited as Part II.

Newton, R.R., The authenticity of Ptolemy's eclipse and star data, Quarterly Journal of the Royal Astronomical Society, 15, pp. 107-121, 1974b. This is frequently cited as Part III.

Newton, R.R., Ancient Planetary Observations and the Validity of Ephemeris Time, Johns Hopkins Press, Baltimore, 1976. This is frequently cited as APO.

Oppolzer, T.R. von, Canon der Finsternisse, Kaiserlich-
      Königlichen Hof- und Staatsdruckerei, Wien, 1887.
      There is a reprint, with the explanation of the tables
      translated into English by O. Gingerich, by Dover Pub-
      lishing Co., New York, 1962.

Pannekoek, A., Ptolemy's precession, Vistas in Astronomy, 1,
      pp. 60-66, 1955.

Pannekoek, A., A History of Astronomy, Interscience Publishers,
      New York, 1961.

Part I = Newton [1973].

Part II = Newton [1974a].

Part III = Newton [1974b].

Pedersen, Olaf, A Survey of the Almagest, Odense University
      Press, Odense, Denmark, 1974.

Peters, C.H.F. and Knobel, E.B., Ptolemy's Catalogue of Stars,
      Carnegie Institution of Washington, Washington, D.C.,
      1915.

Plato, Timaeus, ca. -380. There is an edition with commentary
      and a parallel English translation by R.D. Archer-Hind,
      MacMillan and Co., London, 1888.

Plutarch, De Facie Quae in Orbe Lunae Apparet, ca. 90. There
      is an English translation by Harold Cherniss in Plutarch's
      Moralia, volume 12, H. Cherniss and W.C. Helmbold, editors,
      Harvard University Press, Cambridge, Mass., 1957.

Ptolemy, C., 'E Mathematike Syntaxis, ca. 142. There is an
      edition by J.L. Heiberg in C. Ptolemaei Opera Quae Ex-
      stant Omnia, B.G. Teubner, Leipzig, 1898. There is also
      an edition with a parallel French translation by N.B.
      Halma, Henri Grand Libraire, Paris, 1813. There is a
      translation of Heiberg's edition into German by K. Mani-
      tius, B.G. Teubner, Leipzig, 1913. The chapter numbering
      used here is that of Heiberg and Manitius, which differs
      in places from that used by Halma. This reference is
      often cited as the Syntaxis.

Ptolemy, C., Geographikes Yphegeseos, year unknown. I mark
      this "year unknown" because I do not wish to imply any
      relation in time between this and the Syntaxis, which we
      can date fairly closely. There is an edition with a
      parallel Latin translation by Carolus Mulleros, Claudii
      Ptolemaei Geographia, in 3 volumes, A.F. Didot, Paris,
      1883.

Seidelmann, P.K., Doggett, L.E., and DeLuccia, M.R., Mean
      elements of the principal planets, The Astronomical
      Journal, 79, pp. 57-60, 1974.

Smith, D.E., and Ginsburg, J., From numbers to numerals and from numerals to computation, Part III, Chapter 3 in The World of Mathematics, J.R. Newman, editor, Simon and Schuster, New York, 1956.

Syntaxis = Ptolemy [ca. 142].

Tannery, Paul, Recherches sur l'Histoire de l'Astronomie Ancienne, Gauthier-Villars et Fils, Paris, 1893.

Thompson, J.E.S., Maya astronomy, in The Place of Astronomy in the Ancient World, Philosophical Transactions of the Royal Society, 276A, pp. 83-98, 1974.

Times Atlas of the World, Mid-Century Edition, in 5 volumes, The Times Office, London, 1955.

van der Waerden, B.L., Greek astronomical calendars and their relation to the Athenian civil calendar, Journal of Hellenic Studies, 80, pp. 168-180, 1960.

Vogt, H., Versuch einer Wiederherstellung von Hipparchs Fixsternverzeichnis, Astronomische Nachrichten, 224, columns 17-54, 1925.

INDEX

A

Abu Qir, 98

Abu Sahl al-kuhi, 97

Agrippa, 226

Alexandria, Egypt, 34, 38, 42ff, 76, 92, 98-99, 115, 136, 137, 184, 194, 202, 217, 220, 228 320

al-Biruni, 48, 96, 97

Almagest, 377
  = Syntaxis

alphabet, Greek, 17

American Ephemeris and Nautical Almanac, vi

angle, units of, 381ff

Antoninus Pius, 213

Apollonius of Perga, theorem of, 329ff

apparent time, 382

Aratus, 211

Archimedes, 177, 178, 181, 392

Aristarchus of Samos, 27, 54, 65ff, 83, 86, 88, 133, 172ff, 198ff, 207, 348, 350, 360, 389, 392ff

Aristotle, 34, 40, 41, 73, 193, 211

Aristyllus, 217, 220ff, 349

arithmetic
  Greek, 17ff
  Egyptian, 20

armillary sphere, 145

Asia, 48

astrolabe, 144-145, 149

astronomy
  Chinese, 139
  elements of, v, 1ff
  Greek, 1, 6, 51, 54, 59, 65, 67, 76, 80, 109ff, 143, 147, 170-171, 193, 207ff, 299, 330, 335, 336, 347, 350-354, 364-365, 369-371
  Islamic, 48, 65, 86, 139, 140
  theory and goal of, 1ff
  medieval, 93

Athens, Greece, 34, 38, 40, 76, 80
  Academy of, 193

autumnal equinox, 6, 75ff, 85

B

Babylon or Babylonia, 48, 194
  kings of, 372ff
  literature on astronomy, 334
  observations in, 110, 111, 136

Babylonian astronomy, 4, 53, 110, 115, 204, 270, 330, 333-334, 370
  see: calendar, Babylonian

Baghdad, 97

Bayer, Johann, 4, 5

Beazley, C.R.: see Bunbury, E.H. and Beazley, C.R.

Becvar, A., 5

Bede, 35

Bithynia, 226, 227

Boll, Franz, 239ff

Boötes, 221

Brahe, 73

Britton, J.P., 93, 100, 232

Brown, E.W., 161

Bunbury, E.H. and Beazley, C.R., 48

Bush, George, 145

C

calendar
  Athenian, 80, 95-96, 365, 372
  Babylonian, 128, 140, 372-373, 395-398

The entries do not distinguish between a writer and the works that are cited under his name.

Heiberg, J.L., 145

heliocentric system, 54, 65ff

Heliodorus, 370, 376ff

Hellespont, 178

Hipparchus, 3, 18, 31, 45, 76ff,
91-92, 95-96, 107, 110, 113,
115, 117, 119ff, 125, 128-129,
131ff, 136, 137, 144ff, 154,
170-171, 178-179, 184, 190,
202, 208, 211, 213ff, 217,
218, 220ff, 237ff, 285, 326,
348-349, 356, 361, 363, 366ff
measurements analyzed, 85

Hispalensis: see Isidorus
Hispalensis

Holland, B.B., vi

hour of the day, defined, 146

hour of the night, defined, 146

Howe, J.W., vi

I

ibn ash-Shatir, 167

ibn Yunis, 377

inclination, 383-384

inner planets, 11

instruments, 38-39, 79, 85, 88,
92, 96, 99-102, 144-145, 182,
192, 194, 217

Isidorus Hispalensis, 35

Islamic astronomy: see
astronomy

J

Jenkins, R.E., vi

Jones, Rebecca: see Eckert,
W.J., Jones, Rebecca and
Clark, H.K.

Julian year: see year

Jupiter, 11, 68, 263, 304, 316ff,
334ff

Justinian, 193

K

Keightly, D.N., 42

Kepler, J., 206

kinematics, 49

king list (Ptolemy's), 372ff

Knobel, E.B.: see Peters,
C.H.F. and Knobel, E.B.

Kugler, F.X., 204, 375

L

Lapland, geodetic expedition, 350

latitude
defined, 382
measurements, 36, 49
theory of, 335

Leo, 218

Libra, 214

line of nodes, 8

longitude, 126ff
defined, 382
measurement, 47ff

lunar: see moon

Lysimachia, Thrace, 46

M

Madeira, 47

Manitius, 145, 239

Mars, 11ff, 68, 263, 304, 316ff,
334ff, 362, 375

Marseilles, 48

mean motion, 384-386

mean time, 382

Menelaos, 30, 229, 230, 330, 368

Mercury, 11ff, 333ff, 375
double equant model of,
295-299, 336-339
model for orbit of, 257ff
motion of, 257-299
Ptolemy's study of, 265-277,
303-304

meridian circle, 88

Meritt, B.D., 80, 94, 372, 395

Meton, 76, 80, 84, 91, 94-96,
131, 370

Meton's solstice, 94

LIBRARY OF CONGRESS
CATALOGING IN PUBLICATION DATA

Newton, Robert R.
  The crime of Claudius Ptolemy.

  Bibliography: p. 399 .
  Includes index.
  1. Ptolemaeus, Claudius. I. Title.
QB36.P83N47   520'.92'4   77-4211
ISBN 0-8018-1900-3

ERRATA

Page 19, line 45: for ουδεν   read ονδεν

Page 177, 2nd footnote: for ευρηκοτος   read ευρηκοτος

Page 325, lines 8, 18, and 21: for six observations
                                                           read five observations

   "   line 9: for five read four

   "   line 22: for observations give
                             read observation gives

Page 327, line 22: for five observations
                             read four observations